BABY BIRD IDENTIFICATION

BABY BIRD IDENTIFICATION

A North American Guide

Linda Tuttle-Adams
Foreword by Rebecca Duerr

Comstock Publishing Associates
an imprint of
Cornell University Press
Ithaca and London

First published 2022 by Cornell University Press

Printed in China

Design and composition by Chris Crochetière, BW&A Books, Inc.

Library of Congress Cataloging-in-Publication Data
Names: Tuttle-Adams, Linda, 1955– author.
  | Duerr, Rebecca S., writer of foreword.
Title: Baby bird identification: a North American guide /
  Linda Tuttle-Adams, foreword by Rebecca Duerr.
Description: Ithaca [New York]: Comstock Publishing Associates, an
  imprint of Cornell University Press, 2022. | Includes bibliographical
  references and index.
Identifiers: LCCN 2021054749 | ISBN 9781501762857 (paperback)
Subjects: LCSH: Birds—North America—Identification.
  | Birds—Infancy—North America.
Classification: LCC QL673.T88 2022
  | DDC 598.13/92—dc23/eng/20211123
LC record available at https://lccn.loc.gov/2021054749

CONTENTS

FOREWORD

Countless people are engaged in birding and observing bird behavior. Avian natural history aficionados might delve into identifying birds by their vocalizations or dig into the finer details of complicated molt cycles and subtle clues for aging or sexing individual birds. Yet even an extremely experienced avian enthusiast might be stumped when presented with a chick of a species that they can instantly identify in its adult form. It is difficult to see the adult in the features of the infant. The young of most species of birds are hidden away for their own protection for much or all of their childhood. Though wildlife rehabilitators who specialize in the care of these delicate young birds are in the privileged position of being able to observe them through the entirety of this period of their lives, even the most observant and experienced rehabilitator may struggle to identity the chicks of uncommon or nonlocal chicks.

I first learned to differentiate the hatchlings, nestlings, and fledglings of San Francisco Peninsula's wild bird species while working at a small wildlife center in Palo Alto, California. I was absolutely smitten by these fascinating creatures and the way they would mature from day to day, especially my favorites, American Robins. These charming chicks have an endearing enthusiasm for eating and growing that fuels their transformation from tiny hatchlings to adult-size fledglings that top out more than ten times heavier in less than two weeks. We had a chart on the wall showing various developmental characteristics of local species of baby songbirds that was tremendously helpful in resolving confusion between common birds, but it covered only a limited number of species.

In 2007, the first edition of *Hand-Rearing Birds*, a book that I cocreated and coedited with Laurie Gage, was published. One chapter stemmed from an effort to expand the earlier helpful chart to include more species. Even so, space limitations meant we were able to include only a small selection of species from each taxonomic order, and only in text form, with no space for images. We knew that to include images of even a small fraction of each species at even a single point along their journey from hatchling to independence would be a massive undertaking. No doubt that was why such a resource did not already exist.

I first met Linda Tuttle-Adams more than a decade ago at a wildlife rehabilitation conference and enjoyed the sensible advice on chick identification she dispensed to other rehabilitators in online forums. She had a gloriously nerdy approach to digging into the ornithology literature for clues and was a natural choice to author the chick identification chapter in the second edition of *Hand-Rearing Birds* (2020). Space limitations again favored text rather than imagery, and the chapter refers the reader to this new identification guide for more detailed information.

This book vastly expands the scope of baby bird identification guides. When Linda asked me to write this foreword, I was honored and excited to get a first look at such a comprehensive and needed work. I never dreamed she was skillfully painting hundreds of images of chicks at various ages; I certainly had no idea what a talented artist she is! Her paintings capture plumage details, typical body postures, facial expressions, subtle skin tone, and the very real charm and beauty of the birds. The accompanying text fills in information not conveyable in images alone and consolidates what is known about the young of each wild bird species that breeds in the continental United States, Canada, and Alaska. Linda presents information on chick anatomy, estimating age, and normal developmental stages, as well as crucial instructions for how to look at these young so as to pick out the characteristics that enable species identification. Those of us who work with baby birds have been in need of a truly comprehensive identification guide for a very long time; Linda has delivered the guide we need!

This book will be invaluable for all wildlife rehabilitators, biologists, veterinarians, and others who directly work with these species. For everyone else who loves birds, it will provide a sweet glimpse into the private lives of wild birds.

Rebecca Duerr
Director of Research and Veterinary Science,
International Bird Rescue

PREFACE

The evolution of this project began with an experience I had while studying birds in the Mojave Desert in southeastern California. While observing a Verdin (a tiny desert songbird) building her nest, I heard something "plop" into the pond nearby. I plucked the drenched creature off of a stick it had climbed onto and as the sun dried it, I could see a fuzzy body like a bat but with long, pointed, feathered wings. Although I'd never seen a White-throated Swift, one clue in particular stood out: all four toes were pointing forward. Tiny feet gripped my finger, and enormous dark eyes stared into mine. Soon the swift began to flap its wings, and with a gentle boost of my hand it flew into the sky as if nothing ever happened. Years later I realized how that connection affected my life.

As long as I can remember, I have always felt a deep sense of empathy for animals, especially those most misunderstood and most affected by anthropocentrism (human-centered views). As a kid, my show-and-tell days included frogs and snakes that were easy to identify, like the now rare "horny toads" (Horned Lizards, *Phrynosoma*). But baby birds were puzzling

White-throated Swift nestling

Baby Horned Lizard

and, unfortunately, were no more revered than the slimy snails stepped on by little kids. Yet I found my heart drawn to them so much so that I have dedicated much of my life to revealing their magnificence.

Many people and experiences paved the way for this guide. During my time at the University of California, Santa Cruz, I concentrated on the behavior of birds and mammals. During my employment at the California Academy of Sciences, San Francisco, I had the opportunity to meet and work with many ornithologists and ethologists. In the ancient realm of the Suriname, South America rain forests, I witnessed some of the most extraordinary bird behaviors. My favorite, the Musician Wren (*Cyphorhinus arada*), was like a tiny old man playing a piccolo while dancing a Scottish jig through the forest, stopping to give a concert in the rain on a mossy log.

Later in life, after raising four children, I became involved in the profession of wildlife rehabilitation. As a home rehabber, I recall desperately thumbing through photos in guides and online to find any resemblance of the nestling I received so that I could feed it accordingly. That feeling of helplessness inspired me to create an identification chart (arranged by mouth colors) of the baby birds in my area. That first chart in 2009 evolved into this guide. However, the chart was like putting a cart before the horse. A process to enable one to identify a bird starting with the family group it belongs to was needed.

Another reason for the guide arose from direct experiences with the public as a rehabilitator. As a biologist, I was initially conflicted about saving birds that people found. We all learned about natural selection, how "only the strong survive." However, I accepted the fact that many injured or orphaned wild animals found by humans are the result of anthropogenic (human-caused) events. Without a place to take wild animals, their fate would be left in the hands of people with little knowledge or experience. In spite of the laws protecting wildlife, there are still "well-intentioned" people who take matters into their own hands and consult the internet instead of a professional. Often, the animal has been misidentified and fed an improper diet. One unforgettable case was a nestling Cliff Swallow that arrived in a large pink bakery box resting on top of a bologna sandwich generously sprinkled with bird seed. Since the boy who saved the bird's life had no idea what it was, he did not know that swallows eat only insects.

I have painted a world of birds that most have never seen. Baby birds are hidden from view, deep in tree cavities, in clumps of tall grass, in domes or pendulous sacs, in dense thickets and shrubs, on cliff ledges, on floating rafts, and high in tree canopies. Some nests

Cliff Swallow on sandwich

are merely a scrape in gravel where eggs perfectly match surrounding pebbles. Some are meticulously woven, finely sewn, packed with mud and feathers, or dug into the earth. They are decorated with moss, leaves, bits of lichen, spiders' webs, string, animal hair, and feathers. All this expertise is required to provide warmth and safety for young birds. We may be fortunate to see wild baby birds in their nest on our porch or if we happen upon a nest elsewhere. Fledgling birds are most noticeable shortly after leaving their nest when they are not adept at flying.

Writing and illustrating this guide has been a work of astronomical proportion. As I painted, I discovered each bird as a unique design of perfection that in a very short period of time would learn to master survival, in defiance of any human ability. My experience with the swift is a good example of the steps I now teach in identifying a baby bird: gather clues, determine a family to which it might belong, eliminate other possibilities, and then put all clues together. While I have spent thousands of hours writing this guide, developing charts, photographing, studying, and illustrating baby birds, I admit that with all the knowledge and information available to us, it will not always be possible to identify every nestling with absolute certainty; it may require waiting a few days for more features to develop.

As I have been finishing this book, the future of many animals is uncertain, and the laws protecting them have been under attack. Birds are barometers of the health of the world. The devastation from climate change that has already occurred, and is predicted to occur in the next thirty to fifty years, has validated the importance of continuing to preserve wildlife habitats. It is my wish that this work inspires others to "listen" to birds and speak on their behalf.

ACKNOWLEDGMENTS

This book grew largely from an obsession, beginning as a child, to save a helpless animal. My parents, Barbara and Richard Tuttle, instilled and nurtured my inclinations for art, song, science, and nature. Kyle, my husband, deserves my eternal gratitude for sharing a fascination of wild things, for tolerating my mental absences while writing and painting, and for believing in my project, which provided me the strength to conquer many obstacles along the way. We are blessed to have four children, each unique and talented in their own way. Kyle, Jeremy, Noelle, and Kendall, I thank you for sharing your living space with a compendium of wild creatures coming and going and for tolerating the spare freezer dedicated to carcasses under study.

The challenges I faced in painting undeveloped birds were beyond extraordinary. As a science illustrator, I draw what I see, not what I *think* I see. It was much easier to do this with pencil. But as I evolved to watercolor, I had to dig deep into literature and into the recesses of natural history museums to uncover undescribed details. This project would not have been possible without information from millions of researchers, only a fraction of whom can be mentioned in this book. I am indebted to the Berkeley Museum of Vertebrate Zoology (MVZ), University of California; the California Academy of Sciences (CAS), Department of Ornithology and Mammalogy (San Francisco, CA); the Cornell Lab of Ornithology's Birds of the World website; and the University of California Davis Museum of Wildlife and Fish Biology's Putah Creek Nestbox Highway (Evelien De Greef).

I am especially grateful to Dr. Steven Bailey, Dr. Luis Baptista, Dr. Burney Le Boeuf, Rebecca Duerr, DVM, Dr. Jane Goodall, and Dr. Pepper Trail. A special thanks to Peter Pyle, who edited this work early on and clarified the muddy sections on molt in young birds. Connie Black provided insights from her perspective as a wildlife rehabilitator of nineteen years. Her contributions of data and photographs of Ontario, Canada, birds were invaluable. Tedious specimen studies at museums could not have been completed without the help of my sister-in-law Carrie Adams-Arai, who also helped with proofreading and editing. We will always share special memories of stinky jars of 100-year-old specimens and our silly amusement from the bizarre expressions of tiny "clown-lipped" (pickled) Boreal Chickadee hatchlings.

Without the expertise of the following, there would be no book. Kitty Liu and Allegra Martschenko helped bring this book to publication. Kitty, editorial director for Comstock Publishing at Cornell University Press, advocated for my early draft as an "unhatched egg," which, under her patient guidance, is now a fully fledged bird. Allegra, acquisitions assistant at Cornell University Press, addressed technical aspects required for submission. I am grateful to readers Susan Elbin (emeritus at NYC Audubon) and an anonymous contributor. I also extend my sincerest thanks to Susan Specter (production editor), Nancy Raynor (copyeditor), and the entire production team for their work in the final stages of the guide.

The photos in the photo section and those used as references for many of the paintings would not have been possible without generous contributions from the following wildlife rehabilitators and professional photographers: Alice Abela, Andrea Aiuto, Connie Anderson, Vicki Anderson, Janice Barbary, Thomas A. Benson, Fred Bentler, Simon Boivin, Jenni Boonjakuakul, Sandra Boyd, Jim Brighton, Cara Brown, Patricia Clark, Georgia Conti, Travis Cooper, Peggy Costa-Smith, Evelien De Greef, Denise Finnegan Dewire, Maureen Eiger, Jen Osburn Eliot, Michelle Whitfield English, Chris Gates, Lisa Gibson, Elizabeth Gow, Bobbie Hefner, Lee Hiller, Tim Jasinski, Deb and John Kirkpatrick, Ashton Kluttz, Randy Koeber, Lisa Landrie, Garrett Lau, Debbie Lefebre, Judy Lehner, Tony LePrieur, Kristin Marini, Paul McDonald, Brendan McGarry, Jamie Davis Meyer, Maranda Mink, Karen Irvan Montalvo, SeEtta Moss, Karen O'Connor, Ruth Plenty, Annette Purther, Samuel G. Roberts, Steve and Diane Rose, Mary Rumple, Cameron Rutt, Leckie Seabrooke, Patricia Velte, Alan Vernon, Martin Wall, Kelly Ward, Dustin Welch, Heather Wilkerson, Kathleen Willis, Elise Wolf, Kaycee Wood, Hilary Yu, and Jessica Zorge.

The following contributors and organizations provided feedback, data, or photos of birds they have special expertise with: Nancy Barbachano (Bushtit, woodpeckers, and others); the Bird Rescue Center, Santa Rosa, CA (Mario Balibit); Connie Black (eastern species, Destined to Fly Ontario); East Valley Wildlife; Kim Franza (raptors); Carrie Laxon (Cooper's Hawk); Monterey County SPCA; LouAnn Partington (Common Nighthawk and others); the Place for Wild Birds (Kathleen Frisbie, president, and Walter Bezaniuk, photographer); Salmon Creek Tree Swallow Project; Sialis.org (Elizabeth Zimmerman); Songbird Care and Education Center (Vicki Anderson); TriState Bird Rescue and Research; USFWL (Brette Soucie); USGS (Rebecca Lazarus).

A NOTE TO READERS: ETHICAL CONSIDERATIONS REGARDING WILDLIFE

DO NO HARM

Do not disturb wild nesting birds unless you are legally permitted to do so. Wild birds are protected by state/province and federal regulations; rehabilitating them requires appropriate permits. If you find a wild animal that appears to be injured, sick, or orphaned, you must contact a wildlife rehabilitation agency immediately for advice and instructions. Veterinarians may provide emergency care but must transfer the animal to a permitted wildlife rehabilitator within 24 to 48 hours after stabilization (the amount of time varies by state/province). Birders, wildlife rehabilitators, and those conducting field studies have codes of ethics in addition to these rules. First, we are to do no harm. Unless you are positive that the bird really does need help and that your intended action will cause no harm, do not proceed (for example, do not force-feed food or water). Wildlife rehabilitation techniques are based on a sound scientific understanding of how animals live in the wild. Regardless of how you have come into contact with a wild animal, you should always seek the advice of someone who is experienced with the animal of concern.

HANDLING GUIDELINES

If you have found a baby bird and have followed the guidelines above, place the chick in a secure container, provide warmth, and call a licensed rehabilitator. Line a box no larger than twice the size of the bird with tissue or a soft cloth and set it in a warm, quiet, darkened room.

When birds have come out of their nest and have been exposed to cold or heat too long, they are already stressed. Regardless of age, birds experience some level of stress in captivity. Injured birds may be in shock, which can be fatal. Shock may be a factor if they've been caught by a cat and then "caught" by another predator (human), taken for a ride in a car, further handled by another human, and then put in a strange "house" with unfamiliar sounds. Some species, such as turkey, quail, some rails, and many ground-nesting birds such as fledgling California Towhees, exhibit high stress behaviors when they feel vulnerable. They may not eat and can injure themselves attempting to escape.

When handling a nestling whose eyes have not yet opened, although it may seem unaffected, it is a highly sensitive being. The innate drive for survival causes nestlings to sense a change in their environment, especially if they are without their parents and siblings. Baby birds are very different from mammals. Bird parents do not nurture their young by cuddling. Overhandling or treating any bird like a pet causes stress. Fledgling birds will first try to escape. Even if they are unable to fly well, some are capable of running very quickly. Never allow a young bird to sit on an open hand as it can jump off and become injured.

For wildlife rehabilitators, identification can easily take place during the initial exam and treatment. It should not be an *extra* step in the process of initial care resulting in excessive handling.

GUIDELINES FOR PHOTOGRAPHING NESTS AND YOUNG

If you have found a young bird and have located a nest that you suspect it came from, before taking a photo of the nest contents, be mindful of your disturbance. If there are fully feathered nestlings in the nest, they may be frightened into premature fledging. Again, there are laws prohibiting certain disturbances of the nest, its contents, and the surrounding area. Any time surroundings near a nest are disturbed there is a possibility of exposing the nest, eggs, or young to predators. Many corvids, especially crows and jays, are opportunistic foragers and are very attuned to their environment. Alarm calls made by parents or young attract the other's attention as well as that of predators, which may rob the nest after you leave. Do not spend too much time getting that perfect photo. Parent birds may be reluctant to tend to their young if there is a predator lurking about. Some baby birds depend on feedings every 15 minutes! If your photograph serves no significant purpose, then it's better to enjoy the memory.

PHOTOGRAPHING WILD BABIES IN CAPTIVE SETTINGS

Many wildlife rehabilitation centers do not allow volunteers to take photographs or video of birds in care because it can involve excessive handling and unnecessary exposure, which adds to the bird's stress level. A few good reasons for taking photos include for identification purposes, to document an injury or surgical technique, or to video an abnormal behavior. If photo documentation is absolutely necessary, there are a few things to keep in mind. If the bird becomes startled or distressed or attempts to escape its enclosure, then stop what you are doing. The single lens of a digital camera may appear as a large eye of a predator, especially to birds that are prey to other species, so try to take the photo at a reasonable distance. For identification purposes, three views are optimum: dorsal (entire top of bird including tail), ventral (underside), and side (showing leg and toes). Provide a size reference (metric ruler or coin), and photograph under natural lighting against a light-colored background. A description of parts not visible in the photo should be provided, such as weight of the bird, mouth interior color, ventral feather color, and the like. Include locality, habitat type, and other clues, such as nest and eggs. Natural lighting is preferable because a camera flash can startle birds, and opinions vary as to whether a flash can cause harm to a young bird's eyes. Myelinization (forming a sheath around) the optic nerve in altricial young is not completed until several days after hatching (Suzuki and Miller 2004). Baby birds are "blind" only in the sense that they cannot see because their vision is undeveloped. In general, in cavity-nesting species where the lighting is dark to very dim, the eyes take longer to open than in open-cup nesters. Best to err on the side of caution: one or two photos may not be harmful whereas continual or rapid flash exposure may be.

Photography guidelines for precocial chicks (those that are mobile shortly after hatching) and other highly sensitive species such as wrens should be restrictive. Most precocial chicks, such as quail and Killdeer, exhibit a high stress level when captured, especially when something large is grasping them or staring down at them in the container. They are easily startled, scramble to the corners of the container, attempt to jump out, peep incessantly, and exhibit rapid breathing.

In my opinion, photos or video for publicity or fund-raising purposes should be taken conservatively and discreetly. Consider the necessity and purpose of the photo and put the bird first before any other goals. Check with colleagues or consider copyright-free or open-access photos on the internet. Taking a photo for no reason than to snap a cute picture or "selfie" is not ethical or appropriate. Rehabilitation centers and individual rehabbers should set clear guidelines for photography.

ILLUSTRATED GLOSSARY

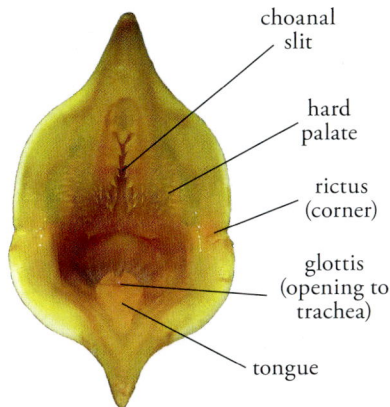

choanal slit

hard palate

rictus (corner)

glottis (opening to trachea)

tongue

Fig. G1. Parts of the mouth (Northern Mockingbird nestling)

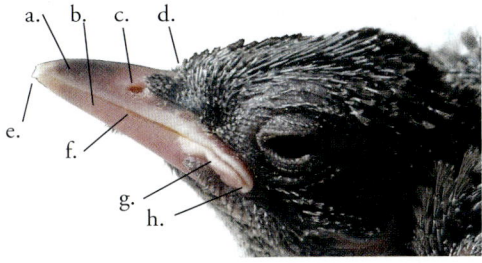

a. upper mandible (maxilla)
b. lower mandible
c. nare (nostril)
d. base of beak (meets beginning of feathers)
e. beak tip (on maxilla), with egg tooth
f. commissure line (maxilla meets lower mandible) = e. to h.
g. gape (oral) flanges
h. rictus (corner of mouth)
Beak measurement method 1: the culmen (the ridge on top of the beak): from tip of beak e. to base of beak d.
Beak measurement method 2: from tip of beak e. to nare at c.

Fig. G2. Parts of the beak (Steller's Jay nestling)

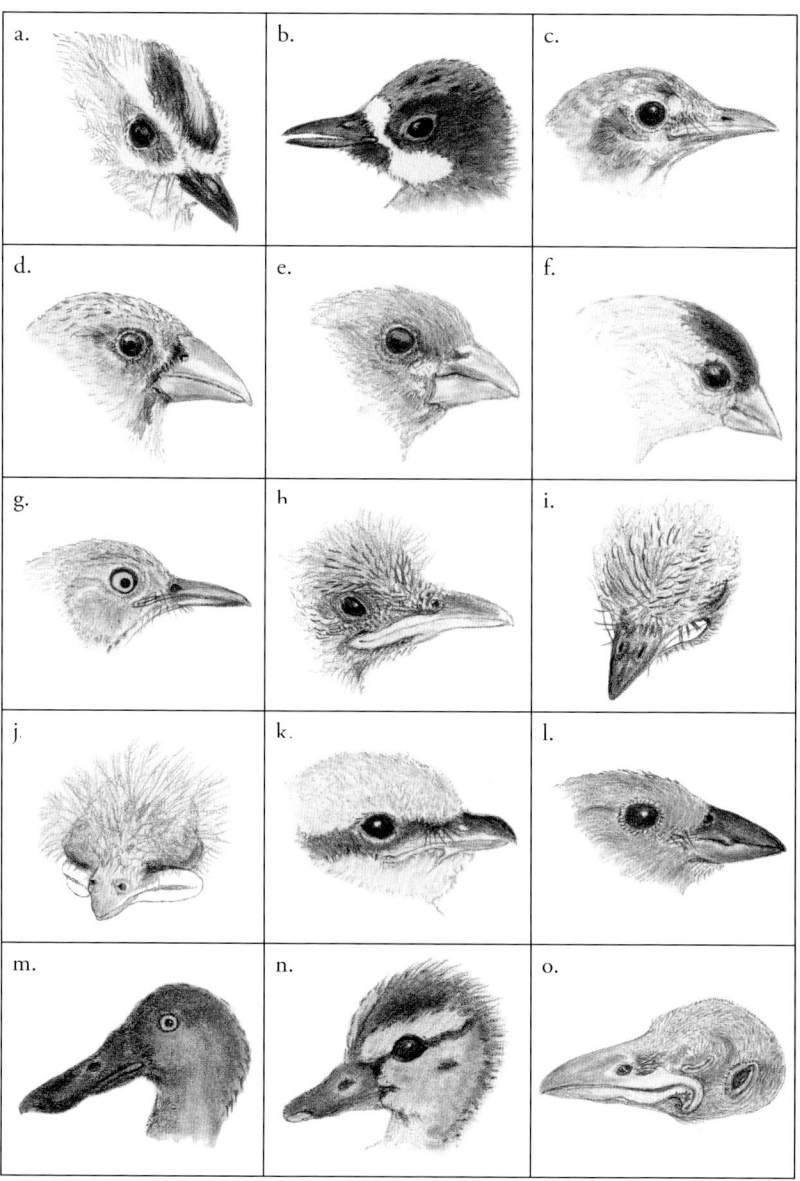

Fig. G3. Beak shapes: a. acute (kinglet adult); b. chisel-like (woodpecker fledgling); c. compressed (robin fledgling); d. conical large (grosbeak adult); e. conical medium (finch adult); f. conical small (goldfinch adult); g. decurved (thrasher juvenile); h. decurved (thrasher nestling); i. depressed (flycatcher nestling); j. depressed (flycatcher hatchling); k. hooked (shrike fledgling); l. notched (tanager juvenile); m. spatulate (shoveler adult); n. spatulate (shoveler duckling); o. stout (raven nestling)

Bones of the Leg
- ☐ femur
- ☐ tibiotarsus & fibula
- ☐ tarsometatarsus (tarsus)

Bones of the Wing
- ☐ humerus
- ☐ radius & ulna
- ☐ carpometacarpus

Foot: ☐ tarsus + ☐ toes

Fig. G4. Bones of the legs, feet, and wings (House Sparrow nestling)

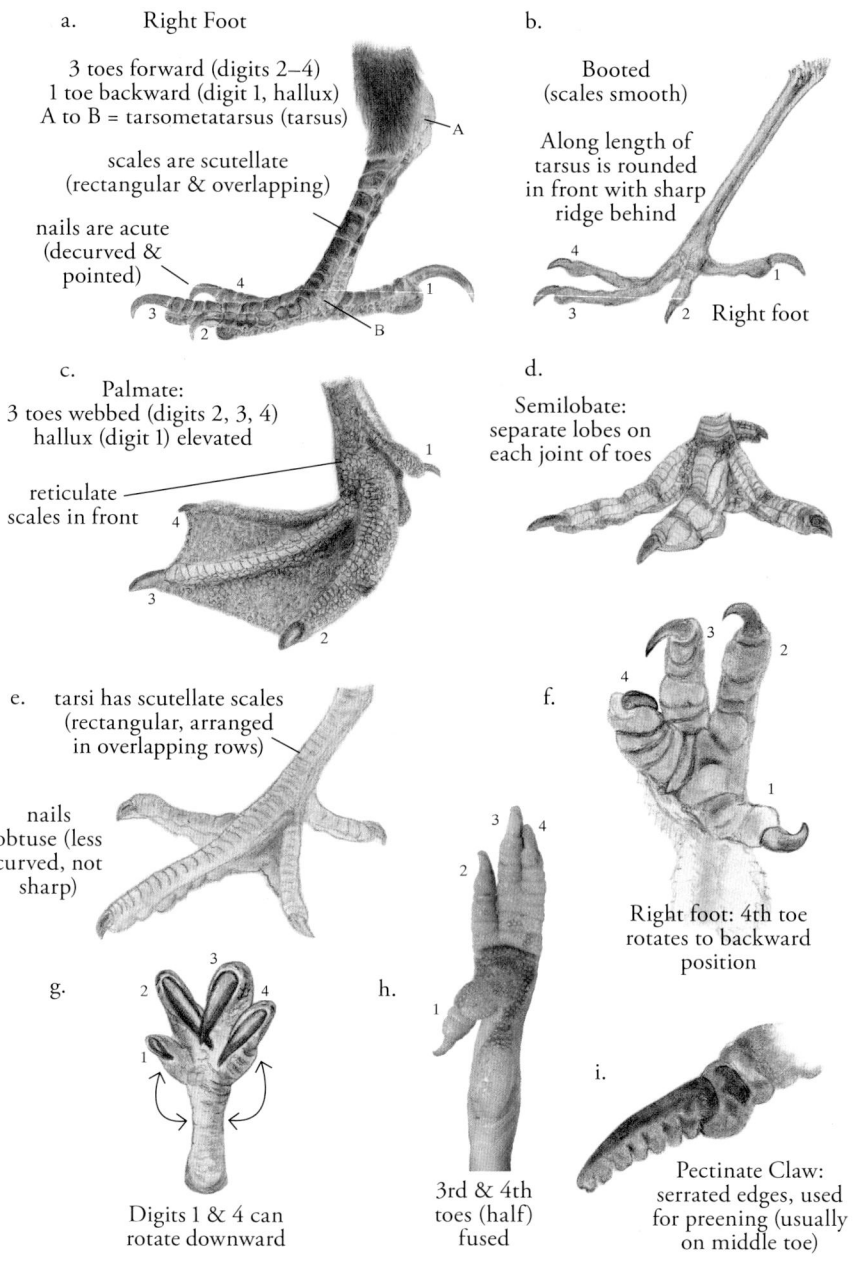

a. Right Foot

3 toes forward (digits 2–4)
1 toe backward (digit 1, hallux)
A to B = tarsometatarsus (tarsus)

scales are scutellate
(rectangular & overlapping)

nails are acute
(decurved &
pointed)

4

3

2

1

A

B

b.

Booted
(scales smooth)

Along length of
tarsus is rounded
in front with sharp
ridge behind

4

3

2

1

Right foot

c.
Palmate:
3 toes webbed (digits 2, 3, 4)
hallux (digit 1) elevated

reticulate
scales in front

4

3

2

1

d.
Semilobate:
separate lobes on
each joint of toes

e. tarsi has scutellate scales
(rectangular, arranged
in overlapping rows)

nails
obtuse (less
curved, not
sharp)

f.

4

3

2

1

Right foot: 4th toe
rotates to backward
position

g.

2

3

1

4

Digits 1 & 4 can
rotate downward

h.

2

3

4

1

3rd & 4th
toes (half)
fused

i.

Pectinate Claw:
serrated edges, used
for preening (usually
on middle toe)

Fig. G5. Toe arrangement, nails, and scales: a. anisodactyl, scutellate (raven);
b. anisodactyl, booted (thrush); c. webbed, palmate (goose); d. webbed,
semilobate, hallux elevated, tarsi scutellate, posterior edges serrate (coot);
e. semipalmate (turkey chick); f. zygodactyl (Barred Owl nestling);
g. pamprodactyl (swift); h. syndactyl (kingfisher); i. pectinate claw
(whip-poor-will)

a.

Dorsal Side

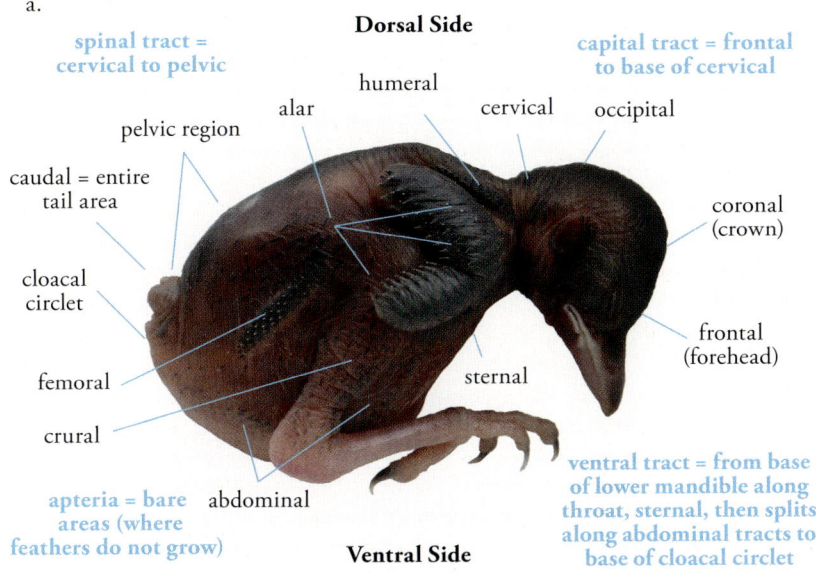

spinal tract =
cervical to pelvic

capital tract = frontal
to base of cervical

humeral

alar cervical occipital

pelvic region

caudal = entire
tail area

coronal
(crown)

cloacal
circlet

frontal
(forehead)

femoral

sternal

crural

ventral tract = from base
of lower mandible along
throat, sternal, then splits
along abdominal tracts to
base of cloacal circlet

apteria = bare   abdominal
areas (where
feathers do not grow)

Ventral Side

b.

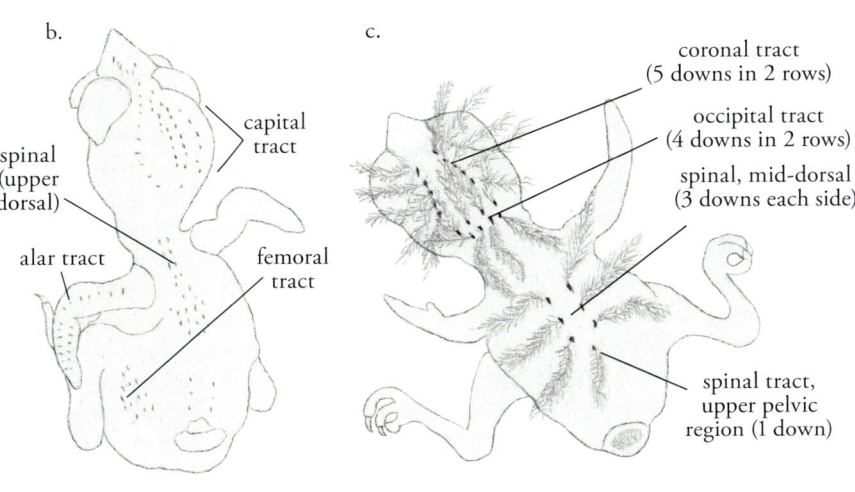

capital
tract

spinal
(upper
dorsal)

alar tract

femoral
tract

feather tracts of a hatchling
(lacking down at hatch) where
first (juvenile) feathers will grow

c.

coronal tract
(5 downs in 2 rows)

occipital tract
(4 downs in 2 rows)

spinal, mid-dorsal
(3 downs each side)

spinal tract,
upper pelvic
region (1 down)

Fig. G6. Feather tracts (pterylae) (see Table 2.8):
a. basic feather tracts of altricial nestling Steller's Jay (see also Table 2.8);
b. pterylography (description of pterylae) of Bushtit hatchling (no down);
c. pterylae, areas of neossoptiles (natal down) of House Wren hatchling

xxv

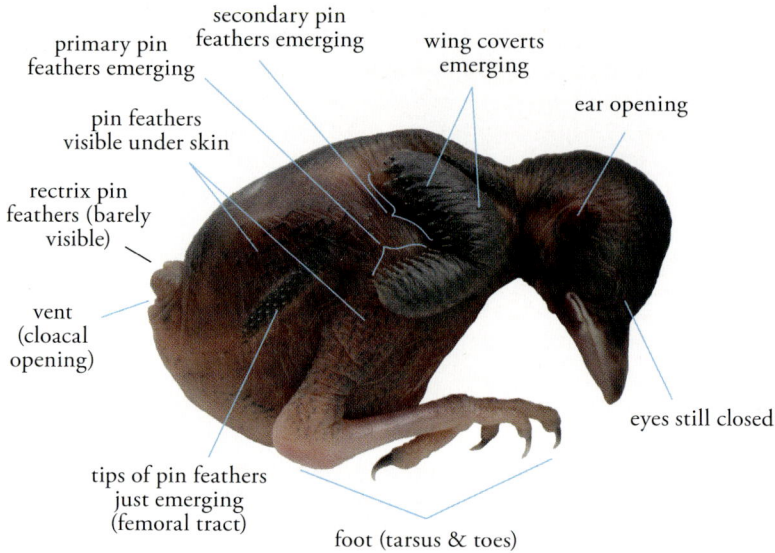

secondary pin
feathers emerging

primary pin
feathers emerging

wing coverts
emerging

ear opening

pin feathers
visible under skin

rectrix pin
feathers (barely
visible)

vent
(cloacal
opening)

eyes still closed

tips of pin feathers
just emerging
(femoral tract)

foot (tarsus & toes)

Fig. G7. Feather development (Steller's Jay, 5 days old)

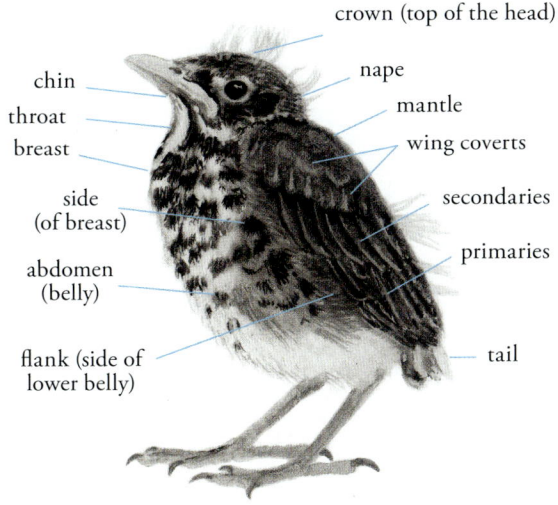

crown (top of the head)

chin

throat

breast

nape

mantle

wing coverts

side
(of breast)

secondaries

primaries

abdomen
(belly)

flank (side of
lower belly)

tail

Fig. G8. Feather topography (American Robin fledgling)

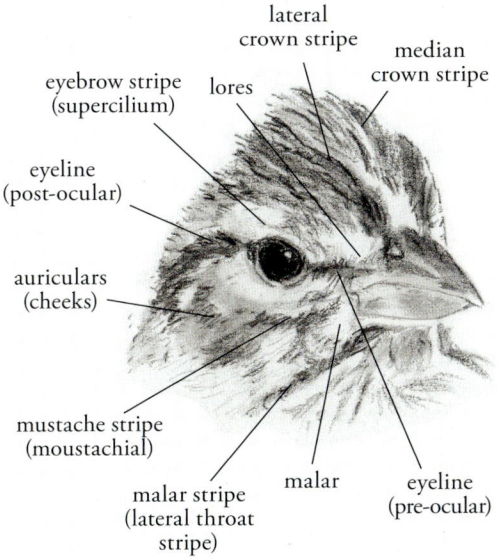

lateral
crown stripe

median
crown stripe

eyebrow stripe
(supercilium)

lores

eyeline
(post-ocular)

auriculars
(cheeks)

mustache stripe
(moustachial)

malar stripe
(lateral throat
stripe)

malar

eyeline
(pre-ocular)

Fig. G9. Head markings (Lark Sparrow fledgling)

BABY BIRD IDENTIFICATION

INTRODUCTION

This guide is written with the wildlife rehabilitator in mind; however, other professionals in the fields of biology, ornithology, and avian veterinary medicine may find it a useful addition to their library. Anyone, from those interested in simply knowing what birds are nesting in their backyard to serious bird-watchers, will find this guide helpful, especially when encountering an unknown species.

WHAT IS THIS GUIDE, AND WHY IS IT USEFUL?

Baby bird identification has not been a widely covered topic, nor has it been given as much attention in the rehabilitation setting as has diet or housing, for example.[1] The reasons are largely because identification has been difficult for lack of a method or process for learning and the lack of a comprehensive guide that addresses the bird only from the time it hatches to when it fledges. Wildlife rehabilitators often do not have the time in the crux of the nestling season to sit at the computer or thumb through books and photographs. Moreover, they may not know where to start their search. Some resources are not available without memberships or access to scientific literature. This guide significantly contributes to the fields of avian research and rehabilitation because it provides a readily available resource as well as a method to lessen the difficulty.

Naked baby birds may appear unappealing or too fragile to handle, but with a gentle hand, they can easily be studied up close without harm. Overcoming some misguided perceptions about them allows one to discover their uniqueness. Some myths about birds have become deep-seated

Fig. I.1: Anna's Hummingbird chick

beliefs in our culture. One of the most persistent is that we shouldn't touch a baby bird or the parents will abandon it. This fable may have been handed down from parents who simply didn't want their children touching them. In early and sometimes not so early descriptions, nestling birds are likened to tiny prehistoric monsters. A well-regarded ornithologist, D.R. Dickey (1915), described 7-day-old Anna's Hummingbirds (Fig. I.1) as "loathsome black worms" that turn into "tiny porcupines" as their pin feathers emerge. Naked Common Raven hatchlings with a few tufts of grayish down were described as clumsy and looking like "grotesque gargoyles" (Tyrrell 1945). These adjectives imply ugliness or something to be feared and do not help us appreciate baby birds for the amazing beings they truly are. Fuzzy ducklings and goslings and other precocial babies with their soft downy coat seem to have more cuteness appeal than altricial young. Altricial chicks become more charming when they are fully feathered, with wispy remnants of natal down poking through their juvenile feathers. A little bird that flutters its wings, begs with its mouth wide open, and shows little fear tugs at our heartstrings. A baby bird that is too young to be separated from its family needs help. The first step is to get it warm, and then, depending on conditions, the bird can be replaced in its nest or be taken to a wildlife rehabilitator.

WHAT SPECIES ARE INCLUDED?

Species included in this guide are the young of birds that have nesting records in the contiguous United States, Alaska, and Canada. However, much of the information, especially the process of identification taught in this book, could be applied to many birds in other parts of the world. Common names of North American birds are used in the chapters; scientific names for North American birds are referenced in the species accounts. There is an emphasis on altricial species, especially those of the woodpecker family, and species in the order of birds called Passeriformes (passerines, or perching birds) because they are the least represented in any similar resource and the most challenging to identify. Many preco-cial species are included, with emphasis placed on those most commonly encountered, such as Mallard ducklings, Canada Goose goslings, Wild Turkey chicks, and various quail species. Several waterfowl species are already well-covered elsewhere (Nelson 1993; Baicich and Harrison 2005) and therefore are unnecessary to duplicate in this book. Several predatory birds (hawks, eagles, owls, falcons, and vultures) are included.

Not every North American bird species has been illustrated or other-wise represented because of time and space constraints and the complexity and scope of this project. To create a guide that would include hundreds of species and subspecies in all their life stages from hatchling to juvenile would be an insurmountable task. Illustrated in this guide are only those species for which adequate reference materials were available: (1) live and preserved specimens for study and (2) photographic documentation. As mentioned, data for some species are either lacking or fragmentary. For example, many wood warblers are so secretive that their nests are nearly impossible to find. The chicks of species that nest high in the tree can-opy, in cavities, or on cliffs may be completely inaccessible. Some species excluded are those that are highly unlikely to end up in a rehabilitation setting. These include nonpasserine birds that are closely associated with water, particularly oceans, seashores, or salt marshes. Some of these spe-cies, such as those in the family Laridae (gulls and terns), nest in colonies or on islands where identification would be easy because adults are pres-ent. Exotics, or nonnative birds that have escaped into the wild, may be mentioned but are not discussed to any extent. A companion website to this guide, at www.BabyBirdIdentification.com, provides additional spe-cies accounts, rehabilitation notes, photos, illustrations, and more.

HOW TO USE THIS BOOK AND WHERE TO BEGIN

Identifying baby birds to species may seem impossible at first, but on looking closely, they reveal features that can at least place them in a taxo-nomic family or group of birds. The most difficult to identify are altricial birds (Fig. I.2), those that hatch in an undeveloped state and are con-fined to their nest until they fledge. Precocial young (Fig. I.3) are capa-ble of walking, running, or swimming soon after hatching. Their beaks and feet bear a close resemblance to those of adults. These features and the dorsal pattern of down can quickly identify a precocial chick. When identifying adult birds, three main structures provide clues: wings, beak, and feet. These parts reflect birds' method of flight, how they forage, the kinds of foods they eat, and even the kind of home they prefer. Attributes

Fig. I.2: Altricial hatchling, Northern Mockingbird

Fig. I.3: Precocial chick, Ring-necked Pheasant

(otherwise known as "field marks") such as facial pattern, eye rings, wingbars, or spots on the tail can distinguish one species from another. In altricial nestlings, until some of these features appear, the most conspicuous characteristics that can be used are arrangement of toes, shape of the beak, color of the mouth and gape flanges, and form and color of the natal down. As the nestlings develop, size of legs and feet, emerging juvenile feathers, and behavior help to reveal their identity. Even newly hatched altricial birds with transparent skin and only a few tufts of natal down, stubs for legs and wings, and enormous heads with bulbous eyes and rubbery beaks have unique characteristics that define "who" they are and distinguish them from other species.

The step-by-step process of identifying a baby bird is explained in Chapter 4. How you approach identification may depend on your level of experience. If you are new to bird watching or a new volunteer in wildlife rehabilitation, you should begin by learning the birds in your area. Use this guide to learn anatomy, how to assess an age, and why identification is important. General characteristics of the taxonomic family, followed by species descriptions, are provided in the species accounts section. Illustrations are organized in plates with one or more stages of life for the species. An identification worksheet, to help when first learning how to identify hatchlings, nestlings, fledglings, and precocial chicks, is provided in Appendix B. Growth and development data are included when they could be found and for a few representative species and are presented in Appendix C.

Bird identification requires being a bit of a "super sleuth," developing a keen eye, and sometimes resorting to using a magnifying glass. Clues are collected and examined. Through a process of elimination, comparison, and deductive reasoning, you will narrow down the prospects to "bring the case to a close." You may need to look at the bird as if you are an artist creating a painting. Speaking from experience, to create a realistic rendition of a bird forces you to study its finest details. If a bird is found with no story or natural history facts to help, some of the features on the bird may reveal what type of nest and habitat it came from, for example, in a bush, on a ledge in a barn, on the ground, or inside the cavity of a tree. Once you have studied all visible details, you may be able to envision a complete picture of your bird leading to an accurate identification.

The purpose of this guide is not to make anyone an expert at identification but rather to provide a comprehensive and accessible resource to identify baby birds from the time they hatch through a few weeks after they depart from their nest.

ONE

THE IMPORTANCE OF CORRECT IDENTIFICATION

Many reasons exist for why we should not be too casual about quickly identifying a baby bird. When a young bird is found on the ground, the finder must make a decision to leave the bird alone or to intervene. It will need to be identified to know if it is normal for the bird to be out of its nest. If a healthy young bird has been brought to a wildlife rehabilitation center, the first consideration should be the possibility of reuniting it with its natal family. Even under ideal conditions, there is no substitute for wild animals being raised by their parents. The process of rehabilitation can last a few days or several weeks, depending on the age and condition of the chick. All birds, regardless of age, must receive species-specific care in the artificial setting. Aside from physical development, caregivers also need to be concerned with psychological and social development. In wildlife rehabilitation, every measure must be taken during a bird's development to keep a young bird on its ancestral path.

PREVENTING A MISIDENTIFICATION

A misidentification may be more harmful than no identification. Caregivers should always remain attentive to developmental changes that do not resemble the species it is believed to be. Many misidentifications result from relying on only one feature and overlooking more definitive characteristics.

Identifying young birds requires much more than matching a picture or using a phone app. You must have an inkling of what your bird is before thumbing through photos or pictures in field guides. Online photos can be deceptive for many reasons. Assuming the bird in the photograph has been identified correctly as to species, important information about a bird's age and possible condition is rarely included. Color in a photo may be far from the true color of the live specimen and varies given the different devices used to take photos as well as monitors used for viewing. Lighting affects all the aspects of color. Color and color interpretation is explained in Appendix A. Illustrations have their own set of drawbacks. Many depictions are a generalized representation of the species and, owing to space limitations in most guides, cannot include all possible variations, races, or subspecies. In some cases, where there are no specimens or reliable photos available, an illustration may be based on a description only. Illustrations and photos in this guide are intended to support the descriptions in the text and not necessarily stand on their own.

The most common misidentifications by wildlife rehabilitators occur within the passerine order of birds. Nestlings of goldfinches are often confused with such other yellow birds as orioles and some warblers. Mistaking these species is an example of the identifier relying on color only while ignoring beak shape, size, and weight. Birds in the finch family have dietary needs very different from those of all other passerines. Even a few days on an incorrect diet can cause serious issues and setbacks in a bird's growth and development. Learning to identify species in the finch family first will reduce the chances of mistaken identity.

Networking with experienced rehabilitators, local birders, ornithologists, and biologists is a good way to get help with identification. If not experienced with identifying young, they would at least be knowledgeable of the nesting birds in your locale. For guidelines on taking a photo for identification purposes, see "A Note to Readers: Ethical Considerations Regarding Wildlife" near the beginning of this guide.

WHEN TO INTERVENE

Encountering a baby bird in the wild can be a wonderful experience, or it may pose some concerns based on the reporting person's ability to assess and describe the situation. What kind of bird is it? How old is it, and should it be out of its nest? Is it in imminent danger? Are the parents still around? Other concerns of finders are if they can touch it, if it can be put back in the nest, if *they* can take care of it, and where they can take it. Injuries, overexposure to the elements, and contact with a cat or a dog are conditions that require a bird to be taken to a wildlife rehabilitator. These chicks will not survive without human intervention, but sometimes it takes a few minutes of discreet observation to see that all is well.

During the spring and summer months, wildlife rehabilitation centers receive hundreds of calls regarding baby birds that may need a helping hand. Rehabilitators have state and federal permits that allow them to take in injured or orphaned wild animals. The only time a healthy nestling should be "rescued" is if it cannot be reunited with its family. Even if the species is not yet known, being able to assess the stage of life of a young bird (see Chapter 3) can help determine if intervention is necessary. Newly fledged birds are the most common age group brought to wildlife centers. Certainly, this is a very vulnerable time for young birds, but although they appear alone and helpless, they have many innate abilities to help them survive. Their parents may be off finding food but close enough to hear distress calls of their young. If you have found a healthy fledgling or precocial chick, it is possible that others are around. Observing from a distance will hopefully reveal siblings or parents tending to the young.

Determining if a bird is ready to be out of its nest depends on many factors, beginning with nest type, if known. In general, open-cup nesters leave their nest before their wings and tail are grown. Most can run fairly well, but their flight capability is limited. On the other hand, cavity-nesting chicks that spend a longer time in their nest are more developed when they leave and should not be on the ground. Each case is unique and usually requires someone experienced to evaluate the situation. Identification is imperative to know the appropriate course of action. Simply asking the right questions may prevent a needless rescue, which in turn will save a wildlife rehabilitator many hours of work, not to mention the cost of care.

One of the most valuable assets to wildlife rehab groups is the phone volunteers who answer calls from the public. They should be well-trained super sleuth–type people who can interface with the public in a diplomatic manner while artfully extracting pertinent information. Making the decision to bring the animal in or leave it alone is one that requires patience, tactfulness, experience, and knowledge about the species. It is extra challenging to identify nestling passerines over the phone. Phone volunteers must be able to evaluate the finders' knowledge level and whether their description of the bird can be trusted. Photos are helpful, but in many cases, until the bird is in hand, absolute certainty of its identity or age may not be possible. The entire situation must be evaluated before giving advice over the phone

Assessing the Situation: Do Parents Abandon Their Young?

Sometimes finders "rescue" an entire nest because they believe the parents have abandoned their young. Further investigation is warranted for many reasons. Merely accepting a claim of parent abandonment without asking more questions or attempting to identify the species under consideration may be a great disservice not only to the parents of the birds but also to the wildlife rehabilitator who receives the birds. There is an enormous investment when it comes time to raise young, so although abandonment can occur, it is uncommon and usually the result of extreme circumstances. Birds may desert a nest that may or may not contain eggs, the most likely causes of which are discovery, continual disturbance, insect infestation, egg failure owing to harsh weather conditions, or poor location. There is some evidence that a parent will abandon late-season nestlings when the urge to migrate becomes overwhelming (Welty and Baptista 1988). The most likely reason a parent has not returned to feed its young is because of injury or death, in which case the young become orphans. Unless the caller witnesses the death of a parent bird or finds a dead parent (assuming it is actually the parent), careful and diligent observation is necessary to determine if something has actually happened to one or both parents. Knowing some life history of the species will determine if an intervention is necessary. For example, in some species, such as hummingbirds, the female alone tends to the young. In other birds, such as the House Wren, House Finch, and Western Bluebird, the male continues feeding the first brood of the season while the female begins incubating eggs in another nest. As nestlings begin to thermoregulate, the parents may be gone for longer periods of time, and when they do return, time at the nest is brief. As larger insects are fed, time between feedings increases. Additionally, parent birds are very stealthy with their comings and goings so as not to reveal the nest to a predator. Though it may seem to the observer that parents are not tending to their young, the opposite may be true. Observations for an hour or longer may be necessary. If the nest is accessible, cold and lethargic chicks are an indication that something has happened to the parents.

REUNITING OR FOSTERING

Reuniting a young bird with its parents or finding a wild foster family that will accept an orphan reduces rehabilitators' cost and workload and puts the bird back into the wild, where it can be raised naturally and normally. However, even some healthy chicks may not be candidates for nest replacement depending on why they became displaced from the nest. When a young bird is first received, a medical evaluation can uncover obvious issues such as wounds, broken bones, or ectoparasites, but it doesn't reveal the entire story. Unfortunately, sometimes a reunion is not possible because inadequate information was collected from the finder. The exact step-by-step process for every species is beyond the scope of this guide, and experts should be consulted.

When a nestling is found on the ground, a tendency exists to state that it "fell from its nest" whether or not the event was witnessed. Assuming this claim to be true excludes consideration of other possibilities. Therefore, advising a finder to replace the bird in its nest may not be the best course of action. It is quite possible the chick's leaving was intentional. Open-cup nests are not the safest place to be, especially once the nest has been discovered. There are a multitude of other reasons nestlings are found on the ground: removal by the parent of an unhealthy nestling, sibling competition, weather, insect infestation, removal by competitive species, and premature departure owing to lack of food or disturbances.

When considering a reunion with the natal family, the bird's age and health must be assessed. Several conditions must be met, including the accessibility of the nest and whether the chick can be monitored following replacement. It is inadvisable to place a single passerine nestling in a separate nest, especially if there are siblings still in the original nest. In general, passerine parents will not brood or feed babies in two separate nests, thus the single chick will not survive. Healthy fledglings should not be placed back in a nest but in a nearby bush or tree if parents are still in the vicinity. Fledglings have left their nest because they are ready to leave so will not stay if renested. Attempting to replace a fledgling may cause siblings still in the nest to leave before they are ready. For many species a reunion with the natal family can still be considered even if the chick has been absent for a few days or so. Within the corvid family, there is a variable window of time that young can be returned to the natal family and still be accepted. Magpies may not recognize their own young after a short absence and may attack them if reintroduced into the colony. On the other hand, juvenile ravens can be returned to their natal family several weeks or months after hatching.

If a reunion with the natal family is not possible, fostering with a conspecific should be the next consideration.[1] Orphaned nestlings and precocial chicks can be placed with a wild foster family if the chick can be monitored to ensure it has been accepted as well as other criteria: the species should be known to accept orphans, the orphan chick should be at the same stage of life as foster siblings, and there should be enough room to not overburden the parents. American Robins, for example, accept foster nestlings of a similar age as long as clutch size does not exceed five (or fewer if the second or third brood) and two parents are attending.

Precocial chicks—ducklings, goslings, coots, loons, snipes, grebes, and Killdeer—and gallinaceous species—quail, pheasant, and turkey—

uninjured and otherwise healthy, can be reunited with their family if it is still in the vicinity. Some waterfowl species, such as the Canada Goose, are accepting of conspecific young that are not their own, whereas Mallards will not accept orphaned Mallard chicks that are not theirs. Recently hatched chicks are probably not good candidates because if the family is frightened off, the chick may get left behind before having a chance to join the group—it will have a much better chance if it can fly. Gallinaceous birds sometimes share nests and families often blend together, therefore successful fostering in these groups seems likely.

Reuniting nestling and fledgling raptors has proven to be very successful for many species. Although many raptors will feed grounded juveniles, whenever possible an attempt should be made to return a nestling to its original nest. If the nest has been destroyed, a substitute nest basket can be secured near the original nest location. A nestling still in down is not able to thermoregulate and should not be placed in the substitute nest alone. Older nestlings in juvenile feathers can be placed in a nest basket in a nearby tree. There are some exceptions to reuniting raptor babies. Any species with a notably high siblicide rate, such as in some hawks and eagles, may not be candidates for reuniting. When food is scarce, older juveniles compete aggressively for food, therefore attempting to reunite a nestling that has been driven out of the nest should not be considered. Careful consideration must be given to species with large broods and asynchronous hatching, such as Barn Owls and Burrowing Owls. The last to hatch may be several days younger than the first hatched. Nestlings should always be monitored for a few days following a renesting attempt to make sure parents are feeding.

PROPER HUSBANDRY

Correct identification is imperative to know how to care for the species. Successful rehabilitation of wild baby birds requires techniques that duplicate as closely as possible their diet and home. Developing birds are sensitive beings with digestive systems specifically adapted to age-appropriate foods they receive in the wild. There is little room for error in providing a proper diet. The consequences of a young animal not receiving a balanced diet during its development may not manifest until weeks after release, which has been documented in zoos, when chicks are fed too much protein and develop "airplane wing."[2] An imbalance of minerals such as calcium and phosphorus can result in metabolic bone disease. The bird may seem fine in the pre-release aviary, but just because it demonstrates some flight ability and eats on its own does not ensure that it will be strong enough to survive the elements, evade predators, and find food. Knowing the identity of the bird will help you prepare housing that replicates its wild home as closely as possible. Housing enrichment that provides hiding places and natural foraging opportunities helps to reduce stress in captive-raised birds.

LEARNING, BEHAVIOR, AND SELF-IDENTITY

Another important reason to identify a young bird as soon as possible is with respect to learning, behavior, and self-identity. A releasable bird is not just one that is physically fit to survive; it must also be psychologically fit. In some species there may be a critical window of time where a young bird needs to learn its identity, future mate preferences, habitat and food preferences, and important behaviors such as song learning. Some behaviors, such as bathing and preening, are wholly or partly innate (inborn), whereas others can be learned only from a related individual. Exactly *how* some birds learn is not well understood, whether by stimulus enhancement (getting a reward) or by a more demanding kind of social learning called imitative learning. Formal studies on behavior and learning began in the 1920s when pioneer ethologist Konrad Lorenz discovered that incubator-raised goslings followed him wherever he went (Welty and Baptista 1988). Subsequent studies showed goslings formed a bond with the first moving object they saw immediately following hatching. Lorenz called this type of learning *imprinting*. Since the early 1920s, scientists have found that imprinting is much more complicated than originally described, and the degree of imprinting varies in different species.

Birds are social animals by varying degrees. Even solitary species must be social enough to secure a mate and raise young. With few exceptions, when a bird breaks out of its shell, its first experience involves social behavior by interacting with its parents and nestmates (Welty and Baptista 1988). Strong social bonds hold families together and help chicks survive. Chicks have already formed their identity when they depart from the nest. The basis by which they will select a future mate has been formed by their family relationship and recognition of their parents and siblings. Some have learned their natal birthplace and food preferences. Slagsvold and Wiebe (2011) indicate that many birds depend on social learning to acquire feeding sites, food items, and foraging skills. While in the nest, others will learn celestial orientation, which will become important for navigation during migration. Those that will eventually sing may have heard their father's song at a certain age or they may eventually learn their own song by hearing an unrelated singing adult. A naive young bird just out of its nest must immediately learn survival skills. Parental care most often accompanies learning, and chicks' survival depends on their parents' behavior, not just their own. The period of dependency after a bird leaves its nest varies by species. Some family groups such as Acorn Woodpeckers, crows, and bluebirds retain young from the previous year who become helpers with future broods. Corvids, especially ravens and the more social crows, possess exceptionally high intelligence rankings within the animal kingdom and have longer dependency periods than most birds. In his close work with ravens, Bernd Heinrich (1999) discusses their intelligence, problem-solving ability, ability to pass information to others about the location of food, and how they play.

A bird's behavior is usually a response from internal and external stimuli from its environment and is largely directed toward self-survival. Some of these behaviors appear spontaneous (innately driven), while others must be perfected. Differences between learned and innate behaviors are not always easily distinguished. Behaviors that are highly resistant to being modified by experience are said to be innate. Instincts are more complex forms of innate behaviors and are more difficult to define. As it turns out, the brainpower of some birds shows a much higher capacity for learning and memory than previous analysis has shown. Birds have an area in the forebrain, called the hyperstriatum, that can carry out complex functions. Some birds have highly specialized learning abilities. Nutcrackers and chickadees, for example, have extraordinary abilities to store spatial information in their brains that help them remember where they have hidden food.

Possibly the simplest form of learned behavior is *habituation* or learning not to respond to meaningless stimuli such as sights and sounds heard repeatedly, thus causing a decrease in the reaction to the stimulus. Habituation, often confused with imprinting, is a very different form of learning with different outcomes. Tameness can be the result of habituation. Some species are very prone to habituation, especially when a bird associates a human with provision of food.

Imprinting and Self-Identity

Konrad Lorenz (1935) proposed that most birds must learn their identity by visually imprinting on their parents and then learn to recognize their own species through early experiences with others of their kind. Bateson (1978, 2017) described imprinting as an example of tightly constrained learning necessary to enable the animal to recognize close kin so it can avoid inbreeding. Some ethologists feel that imprinting is a form of perceptual or exposure learning, and some feel it is fundamentally different from other forms of learning whereas others do not. Thus, several underlying mechanisms of imprinting have been proposed. Regardless of the different definitions of imprinting, most agree that imprinting is necessary and occurs during certain critical or "sensitive" early developmental phases (Bateson 1966). In birds, these sensitive periods can be further broken down into two forms: filial imprinting and sexual imprinting (Welty and Baptista 1988). Filial imprinting enables a young animal to distinguish between its parent and other members of its own species (Bateson 1979, 1990). It occurs when a bird first opens its eyes or when the bird can focus on objects. Sexual imprinting is the tendency of young to learn characteristics of their parents or foster parent that will influence subsequent mating preferences (Immelmann 1972; Immelmann and Suomi 1981; Slagsvold et al. 2002). Presently, evidence suggests that filial imprinting and sexual imprinting are likely two separate processes, although they may partially overlap (Bateson 1979, 1981, 1990).

Times of peak sensitivity to various forms of imprinting stimuli occur at different stages in various species and are quite different in precocial versus altricial young. An example of an extremely brief sensitive period is seen in coots. They become wild and difficult to tame 8 hours after hatching. Sexual imprinting in the Mourning Dove extends from ages of about 7–52 days (Welty and Baptista 1988). In many passerines, the sensitive period may be between 7 and 12 days or more after hatching depending on species (Schimmel and Wasserman 1991). The sensitive phase may be longer in more social species that exhibit extended parental care. Because imprinting studies have not been conducted on many altricial species, especially passerines, precise periods of imprinting are largely unknown.

Other Types of Imprinting

Some species demonstrate preferences for a particular habitat or specific types of food. These preferences may be genetically based, or young birds have learned preferences because of imprinting mechanisms. Some juvenile birds must follow others when they migrate, whereas other species are able to return to their exact natal territories the following spring. Studies of Indigo Buntings showed that the young must have personal experience seeing the night sky before they migrate (Emlen 1967; Emlen et al. 1975). If they have not seen natural skies at night until after migration has begun, they will never learn to use the star compass for orientation. Emlen

(1967) concluded that this learning must be a type of imprinting. Apparently, pigeons require magnetic cues and cues from the sun for navigation (Keeton 1969).

In a wide variety of birds, development of food recognition seems to be genetically based, whereas recognition of foods is learned by trial and error in others. Some species demonstrate a preference for a particular habitat. Chipping Sparrows prefer pine versus oak habitats. Hand-reared 2- to 4-month-old Chipping Sparrows raised only with oak showed a decreased preference for pine. In at least two other sparrow species, Ralph and Mewaldt (1975) showed that site fidelity involved a form of imprinting. Some species have a fairly strong genetic predisposition for particular food types. For example, with no prior experiences, Pinyon Jays have an inborn recognition of Pinyon Pine seeds (Ligon and Martin 1974), and American Kestrels (Mueller 1974) and Loggerhead Shrikes (Smith 1972) recognize mice as their prey. Mueller (1974) proposed that, aside from genetic-based food recognition, food types can be learned from parents and personal experience. It is quite possible that one or more food-learning mechanisms apply to different species of birds. Additionally, there may be imprinting regarding the acquisition of food. Between 20 and 40 days after hatching, Loggerhead Shrikes must learn how to impale their prey. Certain perches that allow for "dabbing" and "dragging" and eventually snagging the prey are required. Birds that were raised (in captivity) on smooth perches never learned how to impale (Smith 1972).

Song learning, to be discussed later, has been well documented in passerines and hummingbirds as a skill that is acquired only during a critical period and shows a high degree of persistence.

What Species Are Susceptible to Misimprinting?

In a reassessment of the evidence of imprinting in birds, ten Cate and Voss (1999, 6) tentatively concluded that "sexual imprinting is present wherever it has been looked for." Imprinting seems to be the rule rather than the exception, although more systematic studies on variations of imprinting are necessary (Hess 1973). "Sexual imprinting has been demonstrated in a great variety of birds with precocial young—ducks, geese, chickens, turkeys, pheasants, quail, bitterns, and coots—and even in such altricial species as owls, doves, ravens, and finches" (Welty and Baptista 1988, 195). In the few passerine birds that have been studied, imprinting was found in Corvidae, Passeridae, Icteridae, Fringillidae, and Cardinalidae (ten Cate and Voss 1999). Examples of captive-raised species known by experience to have imprinted on a heterospecific (a human or other bird species) include woodpeckers, roadrunners, shrikes, all corvids, the Pygmy Nuthatch, House Finches, and other songbirds.[3] Nearly all species with precocial young are susceptible to misimprinting.

Can Imprinting and Habituation Be Reversed?

Behavioral problems can arise with birds where humans are the surrogate parents or when birds have not been raised with their own kind. The degree to which a bird can be influenced by human behavior (as the surrogate parent) and how long-lasting the effects of human influence are, seems to be a gray area with respect to some species. In the wildlife rehabilitation community, imprinting is typically addressed with waterfowl and some raptors. However, with passerines the idea of imprinting is believed to be inconsequential, nonexistent, or occurring only in corvids. Evidence suggests that imprinting should not be ignored. The sooner a positive identification can be made, the better. Precocial chicks and altricial chicks that arrive with eyes fully open will have already imprinted on the parents and siblings. Older fledglings or juveniles are usually quite wild because they have been with their family during the sensitive period. If possible, nestlings should be exposed to an adult conspecific during the critical period of imprinting for that species, especially when focused vision first occurs. If an adult is not available, other conspecifics should be provided: a juvenile, fledgling, or nest buddy (sibling or foster sibling). Other devices are a taxidermy mount, a life-size photo of the adult, the use of a puppet (bird-skin or other simulation) of the adult, and a mirror. For some raptor species, use of a hack box has proven to be successful.[4] Birds inside the box cannot see humans providing food.

Under the traditional view of imprinting, one would believe it is not reversible, that imprinting is like a "stamp" on the brain one does not forget. However, studies in recent decades have shown that imprinting may be more flexible or forgiving than once believed (Bischof 1979; Bolhuis et al. 1990). Although it may be possible to reverse incorrect imprinting, the least costly method is to take preventive measures. Behavioral issues can be minimized or are preventable if wildlife caregivers have properly identified and understand the biology of the species they care for. There are several suggestions on misimprinting prevention in *Hand-Rearing Birds* (Duerr and Gage 2020) and more information on this guide's companion website.

Reversibility of imprinting or habituation depends on the species, the individual, the age of the bird, which type of imprinting took place, how long the bird was without its own kind, and many other factors (Welty and Baptista 1988). Because of the universal misunderstanding and out-dated ideas about imprinting, every bird that has been subject to possible misimprinting or that has become habituated deserves a chance to be evaluated and properly socialized with its own kind if resources are available. It is certainly worth the effort to try to correct the imprinted bird by presenting it with a conspecific. For example, Great Horned Owls raised improperly by humans are most often assumed to have imprinted on the humans, whereas it may be more likely the owls are habituated to food being brought by humans. These owls display a lack of fear and approach humans expecting to be fed, which can result in a dangerous situation. I have found that placing these habituated owls with conspecifics to test acceptance by the wild buddy, disassociating food-bringing by humans, creating experiences that allow the normal fear reaction to develop, and live-kill experiences can result in releasable birds. However, this process may require an extended period of testing and time for the owls to learn what they normally would have learned with their wild parents.

COMMUNICATION

Survival for any bird requires being able to communicate with others. Some types of communication are innate, and some must be learned. Sounds and body movements produced by the young of some species, especially while begging, can be useful to assist with identification. This section focuses on the importance of identification with respect to song learning. "Because the typical young songbird must learn his songs, the strategy for acquiring the proper songs must include how to learn them at the right place, at the right time, and from the right adults," states Kroodsma (2005). He discusses how some species learn dialects (variations of a song), including the Black-capped Chickadee, Tufted Titmouse, (eastern) Winter Wren, Marsh Wren, Chestnut-sided Warbler, and White-crowned and Song Sparrows. Song acquisition is an important reason why some captive-raised songbirds should be returned to their exact natal territories. Another question that may arise, if singing is solely a behavior of the male songbird: Does a young female need to hear the song of her father? Because song plays many roles, such as species recognition, mate selection, and territory maintenance, the answer is yes. Additionally, a number of (temperate-zoned) female songbirds sing songs they learned during their first year, and their songs may serve different purposes from those of the male (Yamaguchi 1998; Kroodsma 2005). Yamaguchi (1998) found that songs of Northern Cardinals are sexually dimorphic and that differences in songs may be due to physiological and/ or morphological factors.

How Songbirds Learn Songs

There has been much research on the inheritance and learning of song, revealing that song may be genetic, learned, or an integration of both. Suboscines (Tyrant Flycatchers) have one to three songs hardwired in their brain. In oscines (songbirds) that must learn their songs, some must do so when young, whereas others acquire new songs throughout their life. Most species learn only from their own kind but some mimic other species' songs. Most songbirds hatch with a basic template consisting of only the simple components of the song. Some must learn the complete song by hearing singing adults. Others acquire the complete song by practicing and developing their own song or repertoire of songs that develops into full song when they hear the adult song. Young songbirds do not master songs they heard as nestlings until the following spring. Juvenile birds burble and jabber just as humans do, and they need to hear others to produce adult sounds. In humans, different languages are more difficult to learn beyond childhood. The same is true for birds. Becoming an accomplished performer requires a focused period of song learning, called the sensory period. In some species this would be before they are six months old, whereas for others it may not be until their first spring. Then there follows an intensive period, called the sensorimotor period, that lasts 4–6 weeks during which the bird practices its song (Alcock 1979; Marler 1990). For most songbirds there are two distinct learning phases: in phase 1, a nestling memorizes the songs of the neighborhood, then in phase 2, it fledges, moves to a new territory, and practices those songs until it can masterfully defend its territory.

On another extreme, cowbirds (members of the blackbird family) have an entirely unique method by which they learn their songs. The Brown-headed Cowbird is an obligate brood parasite.[5] The female lays between one and seventy eggs in a season in a host species nest, never raising her own young. How, then, do cowbirds, raised by parents of a completely different species, recognize others of their own species when they become independent? Originally it was believed that their identity was hardwired in their brain because male cowbirds could sing their father's song without ever hearing it *and* female cowbirds could recognize the male song without having heard it when young. Recent studies have discovered a more sophisticated mechanism going on: a species-specific vocalization as a "password" that triggers conspecific recognition. When young cowbirds hear the "chatter" call from adult cowbirds in the vicinity, their brains respond by rapidly producing a protein known as ZENK. Apparently, this protein may trigger "learning of additional aspects of the password-giver's phenotype" (Hauber et al. 2001).

FINDING A COMPANION CAGEMATE

Once a nestling has been evaluated and positively identified, it should be placed with a companion, preferably a conspecific buddy of similar age. With few exceptions, baby animals are not typically raised alone in the wild, nor should they be in captive care. Companionship reduces stress and helps nestlings keep each other warm. Many precocial species tend to exhibit a high degree of stress in captivity, so they need to be identified immediately and placed with a similar-size buddy. A mirror in the cage with single chicks may or may not help to reduce stress until a companion can be located. Sometimes quail or Killdeer seem to be more agitated (incessantly peeping) when they cannot make contact with the "sibling" they see in the mirror. As mentioned previously, conspecifics help birds form appropriate self-identity and prevent sexual misimprinting. When heterospecific species are put in the same enclosure for buddy purposes, considerations should be given to compatibility, susceptibility of imprinting on the wrong species, and degree of socialness. Some species are aggressive by nature and may not be good cagemates. Even among their own species, there may be food bullies, especially in more social birds such as corvids. In highly social species, where learning takes place over weeks to months after fledging, careful consideration should be taken when placing different species together. In general, a bird past fledgling stage has most likely formed correct identification with its own species. Songbirds that demonstrate fear reaction by hunkering down in the nest or attempting to flee have a strong sense of identity.

NONNATIVE AND INVASIVE YOUNG

Currently in the United States there are three nonnative species commonly found by the general public that are not protected under the Migratory Bird Treaty Act: Rock Pigeon, European Starling, and House Sparrow. These species can be among the top ten intakes in terms of numbers at rehabilitation centers that still accept them. They produce many young in a season and nest in, on, or near dwellings and businesses. Some US states have strict rules regarding the legality of rehabbing and releasing nonnatives because of the threats they pose to native species. Rehabilitation centers have had to establish policies about not taking nonnatives to avoid overburdening their facility and to save resources for native species. Inevitably, centers can still end up with nonprotected birds. They require correct identification, and challenges exist in dealing with the general public's lack of understanding or feelings about saving all birds. Fortunately, all three species are fairly easy to identify as each has its own set of unique characteristics.

RARE OR ENDANGERED SPECIES

Many threats face birds today in all parts of the world. Vanishing habitats and climate change are the most pervasive. In North America, according to Audubon's climate report (Waters 2019), of the 604 species modeled, 389 are vulnerable to extinction. By 2080 it is possible that more than half of birds' current range would become too inhospitable to survive. Partners in Flight (PIF) has placed 86 species (requiring immediate action) of the nearly 450 breeding landbirds in the United States and Canada on the PIF Watch List.[6] Wildlife rehabilitators and other stewards of the earth can help by reporting a correctly identified rare, endangered, or threatened bird to the agencies that monitor them. Some of these species may require additional permits, or the bird may need to be transferred to a facility that is permitted and experienced.

RECORD KEEPING FOR WILDLIFE REHABILITATORS

Wildlife rehabilitators are required to keep records on each patient and submit annual reports to the permitting agencies. Keeping accurate records is important for several reasons. Wildlife biologists are increasingly turning to rehabilitation intake records to understand national impacts of threats to wild birds (some examples include impacts from feral cats, collisions with glass, and diseases). Records are useful to persons who are interested in what species are nesting in certain areas. Researchers may be interested in following up on a reported nest location for birds not known to have previously nested in the area. For all the above reasons, accurate information is important.

TWO

⌣

ANATOMY

The *key* anatomical characteristics of nestlings, fledglings, and precocial chicks in this guide that aid in identification are features of beaks
and nares, legs and feet, plumage, eyes, and head shape. Some altricial
chicks (mainly passerines) have colorful mouth interiors and gape flanges
that help narrow down choices. Colors of the skin, beak, and feet are not
as reliable because of variability but are helpful supporting features.

THE MOUTH

The oral cavity of a young bird is perfectly engineered to get food into the
crop or stomach so that it can be broken down into micronutrients, which
enables the bird to grow at an extremely fast rate. Figure G1 (Illustrated
Glossary) shows the basic parts of the mouth of a gaping bird. The palate
(roof) of the mouth is hard anteriorly and somewhat softer posteriorly. A
median slit, called the choana, separates the palate into two folds with
small horny papillae that project backward. The choana is divided by a
nasal septum that leads into the right and left nares (nostrils) in the nasal cavities. The tongue on the floor of the mouth conforms generally to
the shape of the lower mandible. Four small pairs of salivary glands are
located near the floor of the mouth. The posterior end of the choanal slit
leads into the pharynx, which opens into paired eustachian tubes con-
necting it with the middle ears. Two laryngeal folds located on the middle
of the floor of the pharynx are bound by a narrow slitlike opening called
the glottis, the opening to the trachea (windpipe). When a bird closes its

mouth, the glottis fits neatly into the choanal slit that closes the connection to the trachea.

The Tongue

The anatomical shape, length, and structures on the tongue are adaptations to the types of food, foraging methods, habitat, and lifestyle of the species. The back of the tongue is forked and sharply pointed and also lined with backward-projecting horny papillae. The base of the tongue contains at least twenty-five to sixty taste buds (more depending on species) with a few more in the softer part of the palate and some at the inside tip of the upper and lower mandibles. Spots or structures on the tongue are unique features in a few species where present.

Mouth Colors

Colors of the interior mouth linings of nestlings vary according to species. With the exception of hummingbirds and cuckoos, nonpasserine species do not have brightly colored mouths. According to Wetherbee (1961), mouth colors in birds are influenced by at least three different factors: (1) the horny yellow covering on the bill that gives the yellow color to young of such species as flycatchers, swallows, wrens, titmice, starlings, and thrushes; (2) carotenoids, pigments that are derived entirely from diet; and (3) the extent of capillary vascularization. Brewer's Blackbird chicks have bright red mouth interiors (see Photo 61), and Rose-breasted Grosbeak chicks have reddish-orange mouth interiors. These species may receive pigments from their egg yolk that influence mouth and skin colors because their parents feed on phytophagous (plant-eating) insects that are rich in carotenoids. The same may be true for frugivorous birds, such as waxwings, that have linings in the pink-red range (see Photo 37) from seeds in the fruit they consume. Many insectivores have colors in the yellow range, and many omnivores and granivores have colors in the red range; however, not every species fits nicely within these general correlations of diet and color. There are several exceptions within the trophic categories (Ficken 1965).[1] Wetherbee's conclusion does not explain why some insectivores have red mouth linings when most have yellow. Whereas carotenoids certainly play a role with colors in feathers, the extent to which they affect mouth colors in nestlings is uncertain in some passerine species. Wood warblers have the most varied mouth colors within their taxonomic family (Parulidae). Although they are primarily insectivores, colors range from pink to red and orange. Red Crossbills have two colors,

Table 2.1. Mouth colors of passerine chicks

Color ranges	Family, group, or species
Reds and pinks	Corvids, Ruby-crowned Kinglet (bright red), waxwings, Phainopepla, House Sparrow, finches, longspurs, sparrows (New World), Yellow-breasted Chat, blackbirds, most warblers (New World)
Oranges	A few Tyrant flycatchers, Bushtit, Golden-crowned Kinglet (orange or orange red), Wrentit, pipits, some warblers, Cardinalidae family (most fall in orange range, some fall more in red range, one is yellow)
Yellows	Most Tyrant flycatchers (some may fall in orange range), larks, swallows, chickadees, titmice, Verdin, nuthatches, Brown Creeper, wrens, gnatcatchers, American Dipper, thrushes, mockingbirds, thrashers, European Starling, a few warblers, Summer Tanager

pink and yellow. Sylviids, such as the Wrentit, are insectivorous and have orange mouth linings. Motacillids (pipits) consume a large variety of arthropods, and their chicks have orange mouth colors.

Mouth color immediately narrows down choices. In most cases it can reliably place a passerine bird in a taxonomic family or group. In general, fewer than 6% of chicks of passerine species in this guide have orange mouth colors. Of the remaining 94%, about half fall in the red range and the other half in the yellow range. Table 2.1 and the color chart in Appendix A list the wide variety of colors found in mouths of nestlings within the visible spectrum of reds, oranges, and yellows. Colors that begin to venture out of the red range are vinaceous and crimson. Some colors in the mouth interior and gape flanges are in the spectrum of light in the ultraviolet (UV) range and are not visible to humans.

A very close look inside a nestling's mouth may actually show a mixture of colors rather than a single uniform color. The roof, throat, and corners of the mouth are usually a deeper shade than the tongue; areas that are often thinner tend to be more transparent or pink because of blood vessels. In some hatchlings, there may be darkened areas on the roof of the mouth at the corners toward the back of the throat. Tissues in these areas are thin and may be influenced by external structures, such as the dark bluish skin color of eyelids. These darkened areas are not to be confused with distinctive markings in some species of sylviids or motacillids, for example.

Describing mouth color of a nestling is to describe the gape (Fig. G1,

Illustrated Glossary).[2] The colors of the palate and tissues on the inside
at the rictus that contain blood vessels are the colors that stand out. The
tongue and back of the throat are generally not included when describing
mouth color. The intensity of mouth color changes within a few days after
hatching, becoming more intense or brightening as the birds are fed. As
nestlings become fledglings, the mouth color may fade and darken. In
some juvenile birds the entire interior of the mouth can remain fleshy
and more brightly colored. For example, first-year crows and ravens retain some of the pink they had as nestlings. In a few species, the gape
of the adult may remain distinct and can serve as a useful field mark. In
general, mouth colors of adults are dark shades (mostly gray to black),
may be completely different from nestling mouth colors, and may change
during breeding season. A nestling European Starling, for example, has a
yellowish-orange mouth color, but the adult's is pinkish.

The Function of Mouth Color

The evolution of the various signals of nestling birds designed to solicit
parental investment has been widely studied. Altricial nestlings communicate to their parents by performing begging calls, stretching their necks,
shivering their wings, and exhibiting their gapes with wide open beaks.
Traditional views hold that these behaviors are an adaptation to increase
detectability to feeding parents (Ficken 1965). In passerines, a brightly
colored open mouth, surrounded by swollen tissues of a contrasting color,
provides a target toward which parents direct their feeds. Kilner and
Davies (1998) investigated nestlings from thirty-one species and found
flanges of nestlings in darker nests were more prominent and less densely
colored compared with chicks in nests that were better illuminated. This
finding may suggest that gape color contrasts with skin coloration and the
nest background, particularly in covered nests, which may enhance the
efficacy of parental feeding.

Aside from enhanced visual conspicuousness, mouth color is also believed to serve as a signal to parents about the condition and quality of
the nestling (Saino et al. 2000; de Ayala et al. 2007). Pale or grayish color
may indicate poor health. Studies on many species of passerine young
have shown that brighter gape color advertises good condition and therefore may bias the parents' investment toward offspring that show the best
fitness. Not only do carotenoids affect colors, they may also play a central role in immunostimulation, which has been demonstrated in House
Sparrows (Loiseau et al. 2008) and Barn Swallows (Saino et al. 2000,

2003). Parents of sparrow and swallow chicks preferentially fed offspring that showed the brightest gapes.

Clotfelter et al. (2003) found evidence that mouth color in nestling Dark-eyed Juncos reflected thermal state. In the first week after hatching, when ambient temperature was decreased, the juncos increased blood flow to their mouths, thus increasing the saturation of redness in their mouths. As the nestlings achieved homeothermy, after about 6–7 days of age, mouth saturation decreased. This may be an indication to parents that chicks no longer need to be brooded. Because other passerine species are able to thermoregulate around the same developmental stage as juncos, additional studies of the relationship between homeothermy and mouth color in different species would be informative.

GAPE FLANGES

Gape (rictal) flanges are the soft tissues that become swollen and are largest near the base of the beak at the rictus (Fig. G2, Illustrated Glossary). In some species the tissues that border the entire beak edge or mouth lining become swollen. The tissues are supplied with tactile nerve endings called Herbst corpuscles. When the flanges are touched, the nestling's beak springs open. Other stimuli can also elicit gaping. Even a slight shaking of the nest will cause younger nestlings to gape. Altricial non-passerines do not have swollen gape flanges, except woodpeckers with enlarged "knobs" on the lower mandible at the corners of the mouth. Hole-nesting birds develop relatively larger flanges than open-cup nesting birds (Ingram 1907, 1920; Kilner and Davies 1998).

The surrounding parts of the mouth lining the edge of the bill, egg tooth, and gape flange are predominately white, yellow, or cream-colored (Clark 1969). When the beak is closed, the flange may be whitish, but when the bird opens its mouth, it may be brighter or darker, especially at the corners of the mouth (Dugas and Dillow 2013; Dugas 2015). The visual contrast between the interior color of the mouth and the gape flange may help the parents' perception of color and health. Prominence or thickness of the flanges varies by species and throughout the stage of nestling growth. Nestlings hatch with small beaks, and by the end of their first day of life, the flanges have swollen in size. They may continue to increase in size, becoming more prominent during the fastest period of growth; by the time they leave the nest, the flanges may be reduced to thin "flaps" evident only at the rictus. In some species the flanges are so swollen that they appear as lips, such as the European Starling, which has

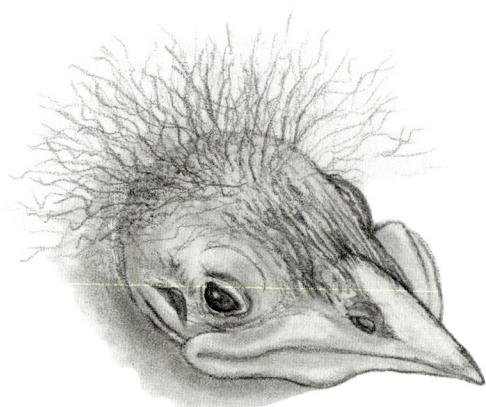

Fig. 2.1: Beak and gape flanges (starling nestling)

been described by some rehabilitators as having "clown lips" (Fig. 2.1). The gapes of the lower mandible of the nestling starling are noticeably larger and protrude much farther than the upper mandible. At the opposite end of the spectrum, some species of blackbirds and corvids have relatively thin gape flanges throughout the nestling growth period.

The degree of prominence of gape flanges can be a key characteristic in some family groups, but its reliability is sometimes in question because the thickness can vary among siblings and is continuously changing as the bird grows. Sometimes differences appear even in a single day. Color of gape flanges may vary slightly within a species especially if influenced by diet.

Gape Size and Shape

The gape is measured across the width of the mouth. The size and shape of the gape relative to head size of a begging nestling passerine can help with

Fig. 2.2. Gape of Northern Mockingbird nestling (photo by Vonda Lee Morton)

identification. Some birds have tiny gapes (e.g., Bushtits and hummingbirds), whereas others have enormous gapes where the head can barely be seen when looking straight at an open mouth (Fig. 2.2). The "big gapers" are robins, bluebirds, grackles, blackbirds, mockingbirds, thrashers, starlings, corvids, and waxwings. The gape may have a particular shape to it that resembles another familiar object, such as a heart shape or calla lily flower (Fig. 2.3). Gape flanges that are turned down at the corners may give an *impression* that a bird looks "sad" or "angry," as with some sparrows (e.g., Chipping Sparrow or Dark-eyed Junco) and especially thrushes (Fig. 2.4). Looking at the side of an open mouth, the commissure of some

Fig. 2.3. Calla lily flower

Fig. 2.4. "Sad or mad?"
(Western Bluebird nestling)

Fig. 2.5. Angulated commissure
(finch nestling)

Fig. 2.6. Straight commissure (nuthatch)

conical billed birds (finches, buntings, sparrows) appears to be angulated similar to a crescent wrench (Fig. 2.5) or straight (Fig. 2.6).

BEAKS AND NARES

One of the most distinguishing attributes among the external features of a bird is its beak. *Beak* and *bill* are used synonymously, but some prefer to use beak when referring to birds with more pointed beaks and bills for such species as ducks. Even without feathers a bird could be placed in a particular taxonomic family based on characteristics of the beak.

Beaks are three-dimensional structures that show tremendous diversity in shape, size, color, and function. Although the design of a bird's beak is largely considered a reflection of the foods it forages for and eats, beaks are highly modified for a variety of other activities that involve nest construction, preening, courtship, grappling with rivals, and feeding young. The pointed beak of an oriole is designed to catch insects but also to weave very elaborate nests. The Hooded Oriole attaches its nest by stitching palm fibers to the underside of the palm frond. Some birds, such as parrots and birds of prey, have a cere, located at the base of the upper mandible, made of a tough horny material. The cere gives way to bare thick skin adjoining the forehead. It is supplied with touch corpuscles and is often brightly colored according to species. A thin, keratinized layer of epidermis, called the rhamphotheca, covers the beak. The rhamphotheca is continuously replaced as it becomes worn.

Figure G2 (Illustrated Glossary) shows the basic structures and regions of the beak: the lower and upper mandible, culmen, commissure, gape flanges, rictus, and nares. Some family groups have features of the bill that pinpoint an identification. Tomia are the cutting edges of a beak present on each mandible that can range from rounded to sharp. Fish-eating birds have serrations in the tomia, seedeaters have ridges for slicing seeds, and falcons have a "tooth" to help kill their prey. The commissure is a line formed when mandibles are closed from the tip of the beak to the rictus, or, depending on usage, this term may refer to only the point of union (at the rictus) of the mandibles. Table 2.2 describes beaks by size. Table 2.3 and Fig. G3 (Illustrated Glossary) describe adult beaks by their shape and features; in many species beak shape is evident even in hatchlings. The size of a beak is described *relative* to the size of the head. A bird may have more than one type or characteristic of beak described (as shown in bold in Table 2.3). Species within a family of birds typically have similar beak and foot structure; in certain family groups, however, there may be some variation in beaks within species that have adapted to different modes of life. Length of a beak may even be different between sexes of the same species, for example, females and males of both Downy Woodpeckers and Hairy Woodpeckers have slightly different bill lengths (as well as other anatomical parts) that also vary geographically within each species. These differences between the sexes are believed to assist birds in foraging for different types of insects in the same area.

The beak of a newly hatched chick, except for the egg tooth (if present), begins as a soft structure, then becomes dry and hard in many birds, such as passerines and gallinaceous species. In other birds, such as waterfowl, the beak is relatively soft and supple.

Table 2.2. Beak description by size

Length	Description	Function	Family, group, or species
Short	Bill much shorter than head	Picking up seeds and grains	Gallinaceous birds and small finches
	Small, with wide gape	Catching flying insects	Nightjars, swifts, swallows
Medium	Same or not longer than bird's head	"Generalized" or multipurpose for eating a large variety of foods; picking up small objects or grasping larger food items	Killdeer and other plovers, many shorebirds, some woodpeckers, thrushes, Northern Mockingbird, Gray Catbird
	Medium to medium-long		Corvids, blackbirds, grackles, orioles, cowbirds, meadowlarks; some thrashers (may be decurved) and nuthatches (lower mandible slightly recurved)
Long	Longer than bird's head	Stabbing and catching prey	Bitterns, herons, kingfishers
		Probing in ground	Woodcock, many shorebirds, large rails
		Chiseling wood	Some woodpeckers
		Probing into bark crevices for insects and spiders	Wrens, Brown Creeper, some thrashers (decurved), Greater Roadrunner
		Feeding on nectar	Hummingbirds

Table 2.3. Beak description by shape

Shape/feature	Description of beak	Function	Family, group, or species[a]
Acute	Delicate, fine, tapers to a sharp point, like tweezers	Catching and gleaning small prey under bark and leaves	Most **insectivores**, warblers, **wrens**, chickadees, titmice, Verdin, **Bushtit**, kinglets, gnatcatchers, **tanagers**
Bent	Deflected at an angle at the middle		**American Flamingo** (very young birds do not show this)
Chisel-like	Tip is beveled (underside curves up)	Boring into wood to enter tree cavity or search for food	**Woodpeckers**, nuthatches
Compressed (laterally)	Most of bill length is higher than wide		Puffins, kingfishers, **thrushes**

[a]Names in boldface are in more than one category.

Table 2.3. Beak description by shape (*continued*)

Shape/feature	Description of beak	Function	Family, group, or species[a]
Conical	Shape of a cone, widest at the base	Stout, large: crushing seeds	Grosbeaks (in the finch and cardinal families)
		Medium or small: cracking seeds	Finches, sparrows, **blackbirds**, members of Cardinalidae
Crossed	Tips of mandibles cross each other	Cracking open specific kinds of seeds	Crossbills
Decurved	Curved downward toward the tip, used like forceps	Digging up prey items in deep mud or pulling insects out from bark of trees	Ibises (very young birds may not show this), curlews, cuckoos, **Bushtit**, Brown Creeper, thrashers, **wrens**, Pyrrhuloxia
Flattened (also called depressed)	Wider than high, flattened from top to bottom		Some ducks (shoveler), **flycatchers, swallows**
Gibbous	Has a pronounced bump on top of bill near the face		Scoters, pigeons, doves
Hooked	Upper mandible longer than lower; upper tip bent over tip of the lower	Shredding and dismembering	All raptors
		Grasping flesh of decaying or recently killed animal	Vultures, ravens
	Hook at the very tip; see wide shape for aerial insectivores	Crushing shells of nuts	Parrots
		Capturing flying insects	Aerial insectivores (some flycatchers, swifts, swallows, nightjars)
Lamellate or sieve-billed	Series of transverse toothlike ridges at cutting edge of bill	Filtering or straining tiny plants and animals from water	Swans, geese, some diving ducks, **American Flamingo**
Notched	Slight nick in the tomia of one or both mandibles	Killing prey	May be difficult to see: vireos (two notches), **thrushes** (slight, near tip of upper mandible), **tanagers**
	Large notch in upper mandible	Killing prey	Falcons
	Several notches	Killing prey	Trogons
Recurved	Curves upward	Swishing end of bill on surface of water to search for food	American Avocet, godwits

Table 2.3. (*continued*)

Shape/feature	Description of beak	Function	Family, group, or species[a]
Serrate	Sawlike tomia or "teeth," looks like a sawblade	Catching fish	Mergansers (also may have a hook at tip)
Spatulate	Spoon shaped—much wider or flattened toward the tip than at the base	Sweeping in water for aquatic food	Northern Shoveler, Cinnamon Teal, Roseate Spoonbill (very young birds do not show this)
Stout	Conspicuously high and wide	All-purpose; for eating a wide variety of foods	Gallinaceous birds, partridge, grouse, corvids
Straight	Line (along the commissure), when mandibles are closed, is in line with the axis of the head	Catching and holding prey before swallowing	Herons, bitterns, egrets, **woodpeckers**, European Starlings, **blackbirds** (orioles and meadowlarks)
Terete	Generally circular or cylindrical either in cross section or when viewed anteriorly	Probing flowers for nectar	Hummingbirds, Gray Catbird
Wide	Wider at the base than it is long	Common in birds that catch insects while flying	Aerial insectivores: flycatchers, swifts, swallows, nightjars

The beak of a passerine hatchling is typically short, usually pinkish or pale yellow, and can even resemble the family group to which it belongs. The tip of the bill and egg tooth are predominately white or yellowish white in all species. The actual beak itself may be tiny compared to the swollen fleshy gape flanges. By the end of the first or second day of life, the beak will have remarkably changed in length and color. At 4 days and older, the rhamphotheca hardens. Typically, by day 7 the beak takes on mature shape and color of the family group. Distinctive beak contours and features (such as hooked tips) may reveal how a bird will eventually capture and prepare its food. However, in birds such as spoonbills, pelicans, flamingos, and ibis, the beak of the hatchling may not resemble the adult beak at all.

Color of the beak may not always be a reliable feature for identification until the bird reaches juvenile or adult age. Size of the beak can sometimes cause confusion. Occasionally, the beak of a nestling (such as a robin or thrasher) may appear abnormally large compared with the rest

of the body. A limited food supply restricts energy and can result in stag-
gered growth of body parts that are most important for survival (Lepczyk
and Karasov 2000): the tarsus (for escaping) and the beak (for feeding).
These parts may continue to grow normally while other components can
slow or stop growing.

Nares (Nostrils)

Location and shape of the nares can be helpful features and are generally consistent within a family group. The distance from the base of the
beak (where feathers begin) are diagnostic features for identification. As
a bird's beak grows, the nares may be positioned farther from the base,
or the feathers at the base may grow longer to cover the nares. Nares
may be oval, round, linear, mere slits, in a soft cere, or partially or fully
covered by feathers, or they may consist of a soft operculum that can
close completely. In some birds such as nightjars, the nares are long, soft,
flexible tubes. In most species the nares are imperforate (separated by a
septum). Less common are perforate (continuous with each other), as in
the Turkey Vulture. Operculate (a fleshy partial covering) may be raised
to some degree.

SKIN

The integumentary system, or skin, in all vertebrates serves as a sensory
organ and protection. Like the skin and fur in mammals, skin and feathers in birds help with thermoregulation. In contrast to mammals, though,
bird skin does not have sweat or sebaceous glands, and the epidermis
(outer layer) is much thinner and drier than mammal skin. Owls and
members of the nightjar family have extremely thin skin. Avian skin is
thinnest in areas covered by feathers and thicker in areas that are exposed
and featherless, such as the beak and feet. The areas of skin that produce
and support feathers are thickest. Underneath the epidermis is the der-
mis, a thicker layer of skin that contains blood vessels, fat deposits, nerves
and free nerve endings, several kinds of neuroreceptors, and smooth muscles that move the feathers. Bird skin has truly amazing capabilities to
help birds regulate body temperature, retain water to facilitate evapora-
tive cooling during daily and seasonal temperature fluctuations, and store
food reserves for times of limited food availability and migration.

Skin Color and Assessment

Skin color, with some exceptions, is considered a supporting feature along with other characteristics used for identification. Most altricial birds hatch with an overall pink, light apricot, or yellowish appearance, darker above (especially on the head) than below. North American cuckoos and road-runners have black skin, some blackbirds have reddish skin, and some thrushes have an orange tint to their skin. In passerine hatchings, skin color changes within a few days after hatching as pin feathers develop just under the dermal layer and darken feather tracts. Because most precocial chicks are completely covered in down, skin color of the body of a precocial chick is not as obvious and not a particularly helpful feature for identification. Table 2.4 provides a list of skin colors based on the overall appearance to the naked human eye.

Many factors can affect an altricial nestling's skin color. The epidermis, especially the areas that bear feathers, is usually nonpigmented or very pale yellow. Skin is so thin and transparent that colors of underlying tissues, including muscle, fat, and blood vessels, influence the colors being observed. The viscera (internal organs) in hatchlings show through the skin of the abdomen: a yellow yolk sac, a dark pink liver, and a dark green gall bladder. In areas where skin is wrinkled or folded, such as around the neck or joints of the legs and wings, true skin color may be evident. Some thrushes, such as robins and bluebirds, will show a yellowish-orange hue to their skin, more intense where it is wrinkled. Highly concentrated yellowish areas may not necessarily be skin color but fat deposits that can vary from one individual to another depending on food consumption. Diet and other factors can influence skin color. Birds that are injured or sick will show colors that are slightly to completely off what would be normal. An overall yellowish tone may indicate liver issues. A greenish tone in most birds can indicate bruising or internal bleeding. In the family Ardeidae, however, it is normal to have greenish or greenish-gray skin, legs, and feet. A grayish or ashen tone may indicate that peripheral circulation has shut down because of dehydration or hypothermia. Naked chicks can become sunburned if they have been exposed to the sun too long, resulting in dark red or bright pink skin.

Scales

Scales of birds are found mainly on the tarsi and toes or farther up in some birds. The pattern of skin or scales on the tarsus, described in Table 2.5 and Fig. G5 a–e (Illustrated Glossary), can place a bird in a particular

Table 2.4. Skin color appearance of nestlings

Color: variations in species accounts[a]	Family, group, or species (examples)
Apricot	Hutton's Vireo, Red-eyed Vireo, Mountain Chickadee, Western Bluebird, Bullock's Oriole, Scott's Oriole, Common Grackle, Northern Cardinal, Black-headed Grosbeak
Black, grayish-black	Cuckoos, American Roadrunner, some hummingbirds, Vermilion Flycatcher (on back), Phainopepla, Mourning Dove, Virginia Rail, Black and Turkey Vultures (on face)
Brown or brownish, cinnamon	Usually a variation with another color: Cormorants (brown-black), Canada Jay, Blue Jay, California Scrub-Jay, Pygmy Nuthatch, Bronzed Cowbird, Prairie Warbler, Townsend's Warbler, Swainson's Thrush, Chestnut-sided Warbler, Western Tanager
Gray	Ash-throated Flycatcher, Eastern Kingbird (d2), shrikes (with orangish appearance), House Wren, Gray Catbird, Brown Thrasher, Green-tailed Towhee, California Towhee, Brewer's Sparrow, Yellow-headed Blackbird (gray-blue)
Green or greenish	Some herons and egrets
Orange or yellowish-orange	Many flycatchers have orange variations, a few vireos, many thrushes, Sage Thrasher, Pine Siskin, many sparrows, many blackbirds, some warblers, members of Cardinalidae
Pink (light tan, pinkish-tan or pinkish-yellow), coral-pink, salmon	Woodpeckers, Common Raven, Carolina Chickadee, Oak Titmouse, Tufted Titmouse, Carolina Wren, Lark Sparrow, Red-winged Blackbird (dark pink to red), Great-tailed Grackle, Yellow-headed Blackbird
Red or reddish, vinaceous, scarlet	Northwestern Crow, Bank Swallow, Cliff Swallow, Carolina Wren, Hermit Thrush, Curve-billed Thrasher, Sage Thrasher, House Sparrow, American Goldfinch, Chipping Sparrow, Lincoln's Sparrow, Dark-eyed Junco, meadowlarks, Tricolored Blackbird, Worm-eating Warbler, Louisiana Waterthrush, Prothonotary Warbler, Magnolia Warbler, Yellow Warbler, Prairie Warbler, Summer Tanager
Yellow or yellowish-orange	Bluebirds (may hatch pink, quickly change), Hermit Thrush, American Robin, Northern Mockingbird, Pine Siskin, Sagebrush Sparrow, Savannah Sparrow, Nelson's Sparrow, Saltmarsh Sparrow, Song Sparrow, White-crowned Sparrow, Yellow-headed Blackbird, some warblers, some members of Cardinalidae

[a] Color may be in part or in combination.

Table 2.5. Types of covering on the tibia, tarsus, and toes, and tarsus shapes

Covering type	Appearance	Family, group, or species
Booted or ocreate (smooth)	Skin of tarsus continuously smooth without scales or plates	Dippers, thrushes, kinglets
Feathered, wholly or in part		Grouse, ptarmigan, frigatebirds (partly), nightjars (partially), some swifts, some hummingbirds
	Feathers on back of tarsi point upward	Family Tytonidae (Barn Owl)
	Feathers on back of tarsi point downward	Family Strigidae (typical owls)
Reticulate (plated)	Skin cut up into irregular (polygonal) plates	Whistling-ducks, swans, geese, Wood Stork, ibises, Roseate Spoonbill, loons, plovers, oystercatchers, pelicans, American Avocet, Black-necked Stilt, vultures, Osprey, falcons
Scutellate (shingled)	Horny skin cut up into overlapping scales (resembles shingles on a roof)	Gallinaceous birds (unfeathered parts), American Flamingo, kingfishers (irregularly scutellate), vireos, corvids, swallows, finches, buntings, titmice, Verdin, Bushtit, Brown Creeper, some wrens, gnatcatchers, wagtails, pipits, waxwings, Phainopepla, shrikes, European Starling, warblers, tanagers, grosbeaks, sparrows, House Sparrow, blackbirds
	In front	Ducks, herons, American Woodcock, snipe, sandpipers, phalaropes, nuthatches, some wrens
Scutellate-booted	Skin of tarsus scutellate in front, booted behind	Flycatchers, Gray Catbird, larks, some thrashers, mockingbirds
Scutellate-reticulate	Plates scutellate in front, reticulate behind	Gulls and relatives, pigeons, doves, dowitchers, woodpeckers
	More scutellate than reticulate and feathered to the toes (booted)	Eagles and hawks
Serrate	Skin of tarsus has serrations on rear edge	Grebes (posterior edges)
Spurred	Rear part of tarsus modified to form a spur	Pheasants and domestic roosters

Table 2.5. Types of covering on the tibia, tarsus, and toes, and tarsus shapes (*cont'd*)

Covering type	Appearance	Family, group, or species
Compressed	Very flat, side to side, with sharp edges in front and behind	Loons, grebes, pelicans, auks, and a few other aquatic species
Rounded in front (most common shape)	Somewhat flattened sides that converge to fairly sharp ridge behind	Thrushes
Rounded in front and behind	Rounded on both sides	Flycatchers, Horned Lark

family, thus scales are considered a key characteristic for identification. Scales are composed of keratin, layers of epidermal cells that are very dense and compact. The feet of birds may be booted (entirely without scales), such as in the thrush family, or may overlap significantly, as in kingfishers and woodpeckers. In the embryo, the skin is smooth. After hatching, the corneum (outermost layer of the epidermis) thickens and keratinizes on the legs, forming scales.

LEGS AND FEET

The anatomy of a bird's legs and feet (Figs. G4 and G5, Illustrated Glossary) is very diverse and designed to perform a wide variety of functions. All birds are bipedal. The major bones of the leg and foot consist of the femur, tibiotarsus, tarsometatarsus (tarsus, *pl.* tarsi), and metatarsi (toes). By many accounts, including this guide, the foot of a bird is defined as the tarsus and the toes, as birds actually walk on their toes (Van Tyne and Berger 1976). The tarsus is the prominent part of the leg between the "ankle" joint (immediately above the toes) and the knee joint. It is the part of the foot that bears the toes. The tarsus is covered with a scalelike skin, called the podotheca or tarsal sheath, which extends to the tips of the toes. A newly hatched altricial bird has legs and feet that are smooth and "stubby," but within a few days, the skin changes and the legs begin to show characteristics of the species. As with beaks, where size is relative to the head, the size of feet and length of legs are compared relative to each other or to the overall size of the bird. For example, towhees have very big feet in relation to their leg length, which is especially obvious in nestlings and fledglings. These birds forage on the ground in leaf litter by hop-scratching to uncover bugs or seeds. Bushtits have long delicate legs. The legs of swallows and House Sparrows are short and stocky.

The feet of most hatchling birds are the same color as the body. As the bird grows, the feet change, but they may not show the adult color until well into the juvenile stage. In most species, legs and feet are various shades of tans, browns, grays, and blacks. However, in a few groups of birds, leg color is unique. For example, the legs of many orioles and vireos become blue or blue-gray as they mature. Legs in many New World sparrows are pink to orange. Legs may be feathered all the way to the toes, such as in certain grouse and owls. Therefore, until the bird becomes a juvenile or adult, the *colors* of legs, feet, and nails of baby birds should be viewed as supporting features, unless unique to that particular species.

Arrangement of Toes

The position of the toes, or how they are arranged on the foot, is a diagnostic feature for an order or family of birds (Figs. 4.1 and 4.2). Toe arrangement is one of the first characteristics to note on a baby bird. If the arrangement is anything other than anisodactyl (one hind toe and three fore toes), you may be able to immediately place your bird in a family or species. All precocial and semi-precocial birds are anisodactyl (Botelho et al. 2015), but they may have adaptations such as webbing or lobes that can further narrow down a bird to a family. In all birds the front toes are inserted on the tarsus at the same level. However, the hallux (hind toe or first digit) varies in position or may be absent. An incumbent hallux is inserted on the metatarsus at the same level as other toes. An elevated hallux is inserted high on the metatarsus and the tip does not touch the ground. Table 2.6 and Fig. G5 a–h (Illustrated Glossary) show types of bird feet and how the toes are arranged. Each type may have variations.

Nails, Claws, and Talons

Nails of birds are found at the distal end of all toes and, depending on species, are called nails, claws, or talons. Claws are more curved and acute than nails of ground-dwelling species, and talons are the strongly hooked claws in birds of prey. Claws are not just for feet; they can also be found on the wings of some species, such as falcons, ducks, and rail. The dorsal portion of the nail is highly keratinized and very hard. Nails or claws vary by the manner in which members of a family or species forage or capture prey. Nails vary in relative length and pointedness. They can be acute, obtuse, lengthened, flattened, or pectinate, as described in Table 2.7 and Fig. G5 a–e (Illustrated Glossary). Claws are curved because the top grows faster than the underside. The amount of curvature increases for birds that are increasingly arboreal in their foraging habitats.

Table 2.6. Types of feet

Type	Arrangement of toes	Family, group, or species
Anisodactyl	Three toes in front, one toe (hallux) behind and incumbent (same level as front toes)	Pigeons, humming-birds, herons, cranes, nightjars, raptors (except Osprey and owls), all passerines, and others
	Palmate: forward three front toes connected by webbing for swimming; hallux elevated	Geese, most ducks, gulls, terns, puffins, murres
	Semipalmate: half or partial webbing between anterior toes; hallux reduced or absent	Quail, pheasant, grouse, turkeys, rails, coots, shorebirds, and plovers (3 toes)
	Totipalmate: all four toes connected by webbing; hallux pulled medially	Cormorants, pelicans
	Lobate (lobed): Anterior toes separated from each other, but each toe edged with a lobe of skin, for swimming; diving ducks have webbed front toes with a lobed hallux	Some diving ducks, grebes, phalaropes
	Semilobate: similar to lobate but with separate lobes on each joint of the toes	Coots
Zygodactyl	Toes arranged in pairs: two toes in front (2nd and 3rd toes), two toes behind (4th and hallux), for grasping and climbing trees	Woodpeckers, cuckoos, owls, parrots
Pamprodactyl	All four toes in front (hallux turned forward)	Swifts (Chimney Swift hallux rotates between anisodactyl to pamprodactyl)
Syndactyl	Third and 4th toes fused together for most of their length with a broad sole; hallux partly connected to inner toe	Kingfishers
Heterodactyl	Toes arranged in pairs: two toes in front (3rd and 4th toes), two toes behind (2nd and hallux)	Trogons

Table 2.7. Characteristics of toenails

Shape	Description	Examples of birds
Acute	Extremely curved and sharply pointed (talons)	All raptors
	Curved and sharply pointed	Climbing birds, such as woodpeckers and nuthatches
Flattened	Extremely flattened and broadened	Grebes
Lengthened	Rather straight, elongated, sharply pointed	Hallux nail of Horned Lark, pipits
Obtuse	Less curved and rather blunt	Grouse
Pectinate	Serrated edges (Fig. G5 i, Illustrated Glossary)	Herons, nightjars, Barn Owl

Birds that climb on vertical surfaces, like woodpeckers and nuthatches, have sharply curved and pointed claws to help grip the bark of the tree. Ground-dwelling birds have longer, slightly curved nails.

FEATHERS, PLUMAGE, AND MOLT

Feathers are unique to birds, setting them apart from all other living beings. Like the hair and nails of mammals, feathers are dead structures. They are keratinized growths of the skin that enable flight, protect the skin, and make warm-bloodedness possible by providing insulation from heat and cold. Therefore, feathers are replaced regularly according to a process called molt, which can show unique species-specific strategies. There are six main types of feathers (Welty and Baptista 1988) and various kinds of modified feathers, such as those forming horns, crests, ruffs, pinnae, ear tufts, rictal bristles, the speculum on wings, or a facial disc. Some birds are adorned with highly specialized feathers for attracting a mate (such as the ornamental plumes of egrets that are used for display), making sounds, or carrying water to their young. Diving and wading birds have feathers that are water repellent. The first feathers of most newly hatched birds are infantile downy feathers. Within a few days, juvenile (sometimes spelled "juvenal" in older literature) feathers emerge. There are various definitions for "plumage" of a bird. In this book, plumage is a full coat of juvenile or adult feathers that cover the bird's entire body at one time.

Feather Types and Function

Vaned or *contour feathers* are the feathers you can see that cover the bird's body (excluding remiges and rectrices). These include the "coverts" of the upper- and underwing surfaces. Contour feathers are found everywhere except the beak, legs, and feet. They consist of horny shafts that generally lack distinct coloration (whitish to brownish) on which variously colored and patterned barbs and barbules are attached (the only part we see). At its base, a contour feather can become downy, which helps insulate the bird. Contour feathers provide protection from vegetation, airborne particles, and the harmful rays of the sun.

Flight feathers are those of the wings and tail. They are long, have large vanes, and have strong barbules to give them more strength for flight. Their strength also results from being anchored by connective tissue to the bone.

Down feathers help insulate birds by trapping body heat next to their skin to stay warm. They have a short or absent rachis, giving the bird its soft and fluffy appearance. Natal down, described below, is different from adult down.

Semiplume feathers are intermediate in structure between down and contour feathers. Unlike down, they have a well-formed rachis, but the vanes are soft. Semiplumes are found underneath contour feathers, and their loose structure helps with insulation.

Bristle feathers are very stiff with a few barbs at base; however, they are highly sensitive, acting as sensory organs. Rictal bristles are found around the mouth of insect-eating birds, where they act as a funnel to help catch prey. Bristles can also be found around the eyes, where they work like eyelashes to protect the eyes. They also occur on the feet of Barn Owls.

Filoplume feathers have a tuft of barbs at the end of the shaft. They are incredibly small and are scattered throughout the plumage next to contour feathers. Because they are attached to nerve endings, they give a sense of touch within the plumage. Filoplumes send messages to the brain giving information about the placement of feathers for insulation, preening, and flight. Sometimes modified filoplumes are used for display, as in the breeding plumes of cormorants.

Powder down feathers are special feathers in some birds, such as herons, where the barbs disintegrate into a fine powder as the feather matures. The bird then spreads the powder all over its body to help repel water.

Molt

Molt is the replacement of some or all feathers of a bird. As will be discussed in Chapter 3, molt is useful in aging birds. In first-year and adult birds, as feathers become worn or damaged, they fall out in a symmetrical pattern that is consistent with the species. Feathers develop from tiny follicles that in their embryonic stage form little bumps on the skin. Newly growing feathers are called pin feathers or blood feathers because they contain blood at the base as the feather emerges from the follicle. New feathers are encased in a sheath (or quill) that breaks away as birds preen and scratch. Altricial young go through their first molt, called the prejuvenile molt, starting at a few days old when pin feathers emerge. Natal down becomes worn by abrasion of the nest and jostling by other nestlings and the parent. Feather tips break and barbules become stripped, removing the down entirely, although in some cases natal down can cling to the tips of emerging juvenile feathers. By the time the birds leave the nest, only a few filaments may remain attached to the juvenile feathers. The duration of the prejuvenile molt can vary from about 15 to 30 days in small passerines to much longer in larger nonpasserine birds, in which some down may persist for up to two months or longer (Beason and Franks 1973). Some larger birds may grow feathers continuously during their first year, during which time they can be difficult to identify. First-year and older birds (in some species) will pass through one or more distinct plumages before reaching full adult plumage or definitive basic plumage. Adults go through a complete molt, whereas the preformative molt out of juvenile plumage is variable, often partial or incomplete, with many species retaining their flight feathers for the first full year of life. Juvenile feathers are grown quickly and are often weaker in structure than basic feathers; thus, damage to flight feathers in first-year birds can affect survivability. All birds have a complete or near-complete molt once per year, while some undergo a second partial or incomplete (rarely complete) prealternate molt between prebasic molts. The more brightly colored birds (such as warblers, tanagers, buntings, and some sparrows) may have a partial prealternate molt to add brightly colored feathers for the breeding season, whereas the prealternate molt in other species (such as flycatchers and wrens) can simply serve to replace feathers without a distinctive change in coloration. Migratory species often molt at faster rates than nonmigratory birds. Most species acquire adult (definitive basic) plumage in a single year, whereas others may take up to three or four years (e.g., gulls) or even five years (e.g., eagles) to reach full adult plumage.

The color of new juvenile feathers encased in sheaths may not be evident because the sheath reflects light, often casting a "bluish" appearance. Use of a magnifying glass can help to see the feather color. As the pin feather continues to grow, it begins unsheathing at the tip, resembling a tiny paint brush. The outer layer of the new feather sloughs off and allows the two sides or vanes to straighten out on either side of a center shaft. This is the first coat of feathers a bird wears when it fledges and will be retained until the next molt. In passerine birds this next molt may occur a few weeks after fledging or not until the following spring when they go through an entire body molt. Fledglings are fully feathered with some feathers partly in shafts at the base. Generally, the wings and, especially, the tail are shorter when fledging than in adults, giving them a stubbier look. In altricial young, the tail grows out last and does not reach its full length until after the chick fledges. This may be because wings and tails can take up too much space in a crowded nest. Growing feathers may be unable to grow straight or may be damaged from abrasion on nest edges or cavity walls. In fledgling birds, feathers on the breast, flanks, abdomen, and particularly the undertail coverts (crissum) are often fluffier and more filamentous, which gives a loosely textured and less organized appearance compared with adults. The plumage of fledglings is typically drabber or duller, often resembling the adult female. Fledglings may also have speckles, spots, streaks, or other disorganized colors and markings to help with concealment. The upperwing coverts may have buffy to white edges, forming either one or two wingbars that do not appear in the adult.

Feather Tracts

A feather tract is called a pteryla (*pl.* pterylae), the area(s) of a bird's skin where feathers grow. Pterylosis is how the feathers are arranged in the feather tracts (Fig. G6 a–c, Illustrated Glossary), and pterylography is the description of the pterylae of birds. The follicles of contour feathers are densely concentrated in the pterylae and separated by bare areas called apteria (*sing.* apterium). The apteria are hidden underneath downy or overlapping contour feathers. Within a few days after hatching, nestlings begin to show pin feathers visible under the skin (Fig. G7, Illustrated Glossary), which makes these areas look pigmented (bluish or blackish), within distinct and symmetrically arranged tracts. Basic regions of feather growth are shown in Table 2.8. Boulton (1927) further divides the basic areas into subregions. The dorsal or spinal feather tract is variable by species or family group. In some species, such as woodpeckers, swallows, and some corvids, the spinal tract forks on the middorsal region.

Table 2.8. Feather tracts

Name	Regions of feather growth based on general location on bird
Alar	Lower wing: lower edge of wing, including remiges (primaries and secondaries) and associated upperwing and underwing coverts
Capital	Head: includes all feathers on the head, which can be further divided into regions: frontal (forehead), coronal (crown), and occipital (back of head)
Caudal	Tail area; includes rectrices and uppertail and undertail coverts
Crural	Lower portion of leg from "knee" to tarsus; sparsely feathered
Femoral	Uppermost part of leg (along femur) on dorsal surface of thigh; slightly crescent-shaped patch of feathers
Humeral	Upperwing: leading edge of wing where the wing meets the body, over the dorsal surface to the trailing edge of wing
Spinal	Upper regions on the back (that extends along the middorsal line); includes cervical, interscapular, dorsal, and pelvic regions
Ventral	Side and underside of bird; includes cervical, sternal, and abdominal regions

Table 2.9. Birds that hatch without down

Hatch completely naked; develop down within a few days	Cormorants, hummingbirds, some vireos (see species accounts), most corvids (except possibly magpies)
Hatch completely naked; no down stage, first feathers are juvenile plumage	Swifts, kingfishers, woodpeckers, some vireos, Yellow-billed Magpie, Common Raven, Verdin, Bushtit, Wrentit, House Sparrow, Cedar Waxwing, Yellow-breasted Chat

Natal Down

Although it seems a tradition to refer to altricial hatchlings as "naked," most birds hatch with some down or the down grows within a few days after hatching (Table 2.9). Nestlings that never develop down begin growing juvenile feathers very quickly. These species should be easier to identify because there are fewer in this category. Downy feathers of newly hatched chicks are called neossoptiles or "true down." The generations of definitive feathers that succeed neossoptiles are called teleoptiles (or definitive feathers). Neossoptiles are variously colored and may follow a general trend within family groups. They are shorter and less distinctively shaped than adult down feathers. The most common type of natal down is a terminal spray of unhooked barbs on a short shaft (Ricklefs 1983), which gives birds their fuzzy appearance. Variability in the details of the neossoptiles depends on the kind of bird and the juvenile feathers that the neossoptiles precede. Pigeons and doves have hairlike down because of the absence of a central vane. Newly emerging feathers of cuckoos are "spinelike" in appearance before they burst from their sheaths. Grebe chicks have a primitive form of natal down that lacks a separate quill; rather, there are downy barbs and hooked feather barbs on the same shaft. Some birds have more than one coat of down that can vary in color with age. For example, certain species of raptors hatch fully covered in whitish down and grow a second coat of down that is darker or lighter than the first coat.

Pterylosis occurs in a fairly definite pattern of neossoptiles arranged in rows. There are endless variations in the distribution and number of neossoptiles in passerine birds (Fig. G6 b–c, Illustrated Glossary). The number and distribution of neossoptiles can be used to help with identification of some birds. For example, some nestlings show an arrangement on the head in the shape described as a "star cluster." House Finches have four "cornrows" on their heads. Unfortunately, pterylosis of young songbirds is a field that has not been thoroughly studied, and the pterylography is not available for every species. Access to information is made difficult unless one is associated with an institution. Some early works that described natal plumages and pterylosis of passerine birds include those by Rudyerd Boulton (1927), Aretas Saunders (1920, 1956), and David Wetherbee (1957, 1958, 1960, 1961).

The amount of down on nestlings can be sparse or plentiful, long or short; it may appear on the head only or the entire body and any combination in between. There seems to be a close correlation between the amount of down of young birds and their physical ability at hatching, the kind of nest they hatch in, and the care their parents provide. Nestlings with thicker down are logically those that are stronger at hatch and require less brooding by the parent. Along with their own thick downy coat, siblings of such species help to keep each other warm. Thickly coated babies would tend to be in open nests that are more exposed to the elements than would those in enclosed cavity-type nests. Wrens and thrashers that nest in arid desert environments are reported to have relatively thicker down than conspecifics in less arid climates (Wetherbee 1958). The length of time true down lasts may be a short interval or several weeks, such as in many hawks, owls, and falcons that have two successive coats of down.

Wing Feathers

Remiges (*sing.* remex) include the primaries, secondaries, and tertials. Birds vary greatly in their power and method of flight. Variations of flight stem from differences in the structure of the wing and the muscles that move them. The main disparities result from feather length in regions of the wing. For example, long-winged birds (e.g., swifts, swallows, falcons, gulls, and terns) have graceful flight that appears effortless. They spend much of their time in the air. The outermost primary feathers in these birds are the longest. Also, the distance from the head to the tip of the wing is longer than the trunk of the bird. Shorter and broader-winged birds (e.g., herons, woodpeckers, thrushes, thrashers, jays, and crows) have more of a "utility" type of wing. In these birds, the second and third primaries from the outside are longest rather than the outermost primaries. Short-winged birds, such as quail and rail, have wings that are short in proportion to their trunk. The fourth or fifth primary from the outside is the longest. The breast muscles are unusually large. They are capable of flapping their wings at high speed. When startled, quail can rise straight up into the air and then fly off. However, the wings beat only for a short moment and then, like a plane landing with stiff wings, the bird sails to the ground again. Once wing feathers of nestlings unsheathe, wings begin to take their shape even though length won't be attained until after fledging. Wing and tail feathers are usually similar to the adult and the best feathers to consult for identification. The number of primary and secondary feathers can help narrow a species down to a family. For example, many passerines have 10 primaries, but some have only 9. Primary wing feathers are numbered p1 to p9 (or so) from the innermost outward; secondaries are numbered s1, s2, etc. from the outermost inward.

Tail Feathers

Rectrices (*sing.* rectrix) have the same structure as wing feathers, featuring an interlocking branched structure. Rectrices are numbered (r1, r2, etc.), starting from the centermost pair (r1 and r1) outward in both directions. Most birds such as passerines have 12 rectrices (or 6 pairs); hummingbirds, swifts, and cuckoos have 10; and rail and grebes only have 8. The Ring-necked Pheasant has 18 rectrices, and the American White Pelican has 24. In flight, the tail of a bird is used as a rudder for steering left or right, to assist the wings in lift, and as a brake to slow down or come to a stop when landing. Some birds use their tail as a prop when climbing trees or pecking holes in wood. Movements of the tail in some species

Table 2.10. Characteristics of tail feathers (of adults and juveniles)

Shape	Characteristics	Kinds of Birds
Forked	Rectrices increase in length successively from the middle to the outermost pair, in abrupt graduations	Terns, some swifts, some flycatchers, Barn Swallow
Graduated	Rectrices shorten successively from the inside to the outside, in abrupt graduations	Cuckoos, magpies
Notched	Rectrices increase in length successively from the middle to the outermost pair, in slight graduations	Finches, some swifts
Pointed or acute	The middle rectrices are much longer than the others	Ring-necked Pheasant, Mourning Dove, terns
Rounded	Rectrices shorten successively from the inside to the outside, in slight graduations	Crows and other corvids
Square	Rectrices are all the same length	Accipiters, Cliff Swallow

can be used as identifiable features, even in young fledglings. Wrens cock their tails up. Some flycatchers jerk their tail as they call or wag their tail when they are being fed. Many birds that live near waterways bob their tails; wagtails wag, whereas pipits, waterthrushes, and dippers bob. Some birds swing the tail from side to side or cock their tail at various angles when they appear excited or agitated. Tail length and shape of the tail tip (Table 2.10) are very useful characteristics for identification. A long tail is one that is longer than the trunk of the bird, such as with pheasants, cuckoos, and grackles. A short tail would be approximately the length of or shorter than the trunk of the bird, such as in many quail, rail, water birds, shorebirds, dippers, and nuthatches. The tails of fledgling altricial birds are not of full length when they leave their nest. Tail markings such as bars or spots are also very defining features for identification, as is the color of the rectrix edges.

Feather Color

Feathers come in every color of the rainbow and in varying shades of white, gray, brown, and black. Feather color comes from pigments and feather structure. Pigments are colored substances found in plants and animals, and there are three different kinds: carotenoids, melanins, and porphyrins. Carotenoids, derived from foods eaten by the bird or from prey that consumed a plant, are fat-soluble pigments stored in birds' feathers, skin, flanks, eyes, and such organs as the liver. Carotenoids are responsible for producing colors of yellows, oranges, and bright reds that we see in warblers, goldfinches, and the Northern Cardinal.

Melanins, the most common pigments in skin and feathers, produce colors that range in pale yellows, red-browns, and black. The colors of gray, brown, brick red, dull yellow, and tan contain black melanin. Green colors are produced by the interaction of carotenoids and melanins. Dark-colored feathers and flight feathers contain more melanin. As hemoglobin is broken down by the liver, amino acids are modified and produce porphyrins. These are what make up brilliant greens, reds, and another range of colors of pinks, browns, reds, and greens. Red and yellow colors have particular significance in the bird world. The color red (in beaks, feathers, and skin) is used by many species to attract mates or deter rivals. Red mouth linings in altricial songbirds may signify a condition of health, that is, the brighter red being very healthy, the paler pinkish or grayish linings being in poor health. Two teams of researchers have independently identified an enzyme-encoding gene, allowing some bird species to convert yellow carotenoids from their diets into red. Apparently, these same red pigments that are deposited into feathers also accumulate in one type of cone photoreceptor in the retina that enhances color vision. The researchers' discovery led to the "redness gene" being found not just in birds with red feathers but also in the genomes of many bird species (Lopes et al. 2016).

How color appears to a human eye depends on the amount of light and the viewpoint of the observer. Some feather colors are a result of the structure of the feather as light is reflected or refracted. The color blue is created by reflection of white light (that has all the colors of the rainbow). When white light hits tiny air pockets in the barbs of feathers, it is scattered back so only the blue light is reflected. If light is shined on the underside of a blue feather or it is held up to strong backlighting, the blue color disappears. Red appears red regardless of which direction light is passing through. Green feathers are made from a combination of pigments and reflection of blue wavelengths. White-colored feathers are created by a feather structure that reflects back the entire spectrum of colors. Iridescence in feathers is also caused by structure of the feather. Incident light is refracted on microscopic structures on the barbules.

Hormones also play a role in plumage coloration. The underlying mechanism of how hormones control feather coloration is not well understood. However, in many adult birds there is a marked visible difference in both size and plumage between males and females, which is called sexual dimorphism and sexual dichromatism, respectively. Males are usually larger (though in some species such as raptors, females are larger).

Males in their nuptial plumage are most conspicuous. The less colorful and even drab plumage of the female offers her some protection against predators as she is performing incubation or domestic duties confined to the nest area. Sexual dichromatism occurs in variable degrees. In adult woodpeckers, the difference in sex coloration may only be the presence of a small patch of red feathers on the head or nape of the male, with no red on the female. In other kinds of birds, the adult male is brilliantly colored, has lengthened plumes, sings distinctive songs, and displays bizarre behaviors and theatrical effects, all in pursuit of a mate. Some families that show pronounced sexual dichromatism in adults are ducks, hummingbirds, warblers, and tanagers. Other species, such as loons, grebes, raptors, gulls, terns, flycatchers, corvids, thrushes, and wrens, show little or no dichromatism. Some juvenile males usually do not begin resembling their fathers until they go through later molts.

Natal Down Color

Natal down color is an important characteristic but, with the exception of black or white, it cannot be solely relied on due to variation and color interpretation (Table 2.11). Also, there are notable differences in descriptions in the literature. Color of natal down is typically in shades of white, cream, gray, brown, and black. Down may be described as one overall color or the dominant color, but on close examination it is often a combination of colors. Down color descriptions in previously published literature are from historical accounts by early field studies, anecdotal observations, or preserved museum specimens. Passerine species described by Wetherbee (1957, 1958) are largely of museum collection specimens, although he provides references of previous descriptions and found that "on the whole, color determinations seemed futile because of variability and the almost hopeless problem of precision in measurement" (Wetherbee 1957, 355). In the search for more current descriptions of down color of neossoptiles in birds, I found the same issues with the lack of a precise method or a single standard reference to describe color. Several reasons may contribute to variations or contradictions in the literature: (1) interpretations by different describers, (2) geography, (3) subspecies, (4) individual variation, (5) altered color in preserved specimens (Ingram 1920), (6) downs on live birds may change with exposure to light, air, or dirt/dust (Saunders 1920), or (7) abrasion or staining from preening or movement in the nest.

Table 2.11. Natal down feather colors

Dominant color	Examples of family, group, or species
Black	Some sparrows, Brewer's Blackbird (blackish-gray to grayish-white)
Brown	Great Blue Heron (light brown with gray), Black-crowned Night-Heron (brown with gray or rufous), Rough-legged Hawk (1st down, grayish-brown), Harris's Hawk (2nd down, light brown), Bald Eagle (2nd down, light brown), Osprey (brown and white), thrashers, and some warblers. Some jays, crows, and some sparrows are grayish-brown
Buff	Least Bittern (dorsal only), Black Vulture, some kites, Harris's Hawk, Barn Owl (2nd down buff-cream), poorwill, Ruby-throated and Costa's Hummingbirds
Gray	Western and Clark's Grebes, some herons and egrets, swans, White-tailed Kite (2nd down), some hawks, Bald Eagle (1st down), Peregrine Falcon, Merlin, Elf Owl, Burrowing Owl, Great Gray Owl, some hummingbirds, Tyrant flycatchers, ravens, most swallows, chickadees, titmice, Bushtit, nuthatches, creepers, some wrens, kinglets, bluebirds (bluish-gray), California Thrasher, mockingbirds, starlings, pipit (brownish-gray), some warblers, some sparrows, tanagers (can be in white range also), some blackbirds (Brewer's Blackbird varies blackish-gray to grayish-white), most finches, and House Sparrow; some jays, crows, and some sparrows are grayish-brown
Sepia	Color found in many bird families more commonly mixed with other colors or as markings (precocial chicks)
Tan	Longspurs
Tawny	Yellow-headed Blackbird, pheasant, turkeys, some nightjars
Umber	Brown Thrasher
White	Some herons and egrets, Turkey Vulture, White-tailed Kite (1st down), some hawks, Golden Eagle, falcons, many owls, Horned Lark, Tree Swallow, some wrens, American Robin (as hatchling, then turns gray), Phainopepla, tanagers (range to pale gray), cowbirds (Brown-headed Cowbird has long snow-white down on head), Bullock's Oriole, House Finch, Evening Grosbeak
Yellow	American Bittern (yellow or greenish), some geese, some pigeons

Feather Color Abnormalities

Normal variations of color in birds have been discussed above, but certain variations or conditions exist that are considered abnormal. These are albinism, leucism, melanism, xanthochroism, pied, and diluted plumage (faded colors). Albinism is the lack of melanin pigment caused by a genetic mutation. Leucistic birds that show abnormal white in the feathers may still have color elsewhere because mutation can prevent production of melanin in some but not all feathers. Leucism can result in faded, pale, or reduced colors in all kinds of pigment. The absence of pigment of some feathers causing irregular white patches may be called pied. Melanism, caused by the deposition of too much melanin pigment, results in darker-than-normal feathers. Another variation is xanthochroism, which involves the color red being replaced with yellow or orangish pigments. This unusual condition would show up only in birds with red, such as the Northern Cardinal, some woodpeckers, and tanagers. Some male House Finches have yellow or orange coloration; this variation is not an abnormality but believed to be mostly the result of reduced intake of carotenoid pigments in the diet or a combination of diet and genetic predisposition.

Plumage Coloration, Patterns, and Field Marks

There are two general types of plumage coloration in birds. One involves advertisement: recognition used in courtship displays. The other, which makes identification challenging, enhances concealment, such as camouflage or cryptic plumage. Many ground-dwelling birds have inconspicuous colors in shades of brown, tan, and gray as well as spots, stripes, streaks, mottling, or other patterns that help them blend in with their surroundings. The Common Poorwill and Western Screech-Owl have patterned plumages that closely resemble the bark they roost on. In fact, these two species so closely resemble each other that adult poorwills are often mistaken for baby owls. The Hermit Thrush has spotted underparts that mimic leaf litter where it forages. Countershading, such as in the Black Phoebe, refers to birds that have darker colors on the back and lighter colors underneath. Disruptive coloration involves contrasting patterns that make the bird appear to be in pieces. Examples are rings on the head, neck, or backs of birds such as in Killdeer and some ducks.

EYES

Vision in chicks is one of the last senses to develop. In precocial species, vision is fairly well developed at hatch. Precocial chicks must be able to see so they can follow the parent once all in the brood have hatched. In altricial species, complete maturation of vision does not occur until several days after hatching (Nice 1962). The eyelids are closed; however, the bulging eyeballs are visible through the lids. Eyelids begin to open at predictable ages of altricial young, but the time it takes for lids to *fully* open varies significantly between species. Also useful for identifying and aging a species is the colored part of the eye called the iris (*pl.* irises or irides). Many birds have blackish or brown irises, but a variety of other colors occur. In some species, iris color changes as birds mature. For example, irises of male Brewer's Blackbird fledglings change from brown to grayish-white 12–21 days post-hatching. Iris color can vary geographically, by sex, with age, by season, and within individuals, although the underlying reasons for iris color variation of birds is poorly understood (Bortolotti et al. 2002).

HEAD SHAPE

The shape of a bird's head is determined by (1) the skull, including the bill, and (2) the feathers that adorn the skull. When observing birds in the field, identification at least to a group can often be made by a bird's silhouette. In some species, a bird's head shape, including the beak, can be a dead giveaway, for example, kingfishers, herons, oystercatchers, ravens, and grosbeaks. Other features that help with the identification are conspicuous crests or plumes. The shape of a bird's skull is diagnostic to a family group, whereas feathers within a family group can vary among species in the family. Figures 2.7 and 2.8 show the differences between owls in two different families. In most altricial baby birds, the shape of the skull is fairly obvious through sparse down or pin feathers. As feathers unsheathe, some feathered head shapes begin to take form. Various nestling and fledgling birds show crests that, although smaller than the adult, are distinctive enough in shape and placement on the head to define the species. Certain crests, for example, on quail and the Phainopepla, are on the very top of the crown. In other birds, such as Blue Jays and some flycatchers, crest feathers lie down in a relaxed position, but when the bird becomes excited or alarmed, the crest raises up on top of the crown. In some birds there appears to be no distinction between the beak and the forehead, giving the head a flattened appearance. Vultures and Barn

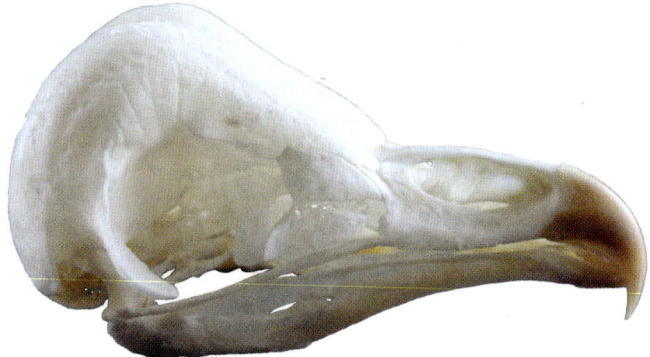

Fig. 2.7. Barn Owl skull

Fig. 2.8. Great Horned Owl skull

Owls may be the most extreme example of this shape. The Brown-headed Cowbird, some sparrows, many ducks, and some geese have flattened foreheads. Other birds have higher crowns, with no slope to the forehead, making the head appear rounded—thrushes tend to have nicely rounded heads. Jays and many shorebirds have high crowns. Some foreheads are high then flatten out on the crown, giving the head a squarish appearance. Bushtits seem to have more forehead and flatter crowns. Creepers and small wrens also have flatter crowns. Within the vireo family, there are slight variations of head shape appearance due to feathers. Some vireos, such as the Red-eyed Vireo, have a gradual slope to the forehead with a long and slightly oval crown. Others, such as the Hutton's Vireo, have a more rounded crown.

THREE

GROWTH, DEVELOPMENT,
AND ESTIMATING AGE

The terms *altricial* and *precocial* describe the type or condition of development of a young bird at hatching. The Latin root meanings of the terms are "to nourish" for altricial and "precocious" for precocial. In birds, a simple difference between the two is that "the precocial bird undergoes stages of development within the egg that the altricial bird undergoes after hatching" (Welty and Baptista 1988). Altricial hatchlings are unable to regulate body temperature and are wholly dependent on parents for warmth and food. Precocial chicks are well developed at hatch with a full coat of down and are able to walk, run, swim, and feed themselves. Nice (1962) further divided the precocial-altricial classification into different grades, describing various intermediate states between altricial and precocial (Table 3.1). Semi-altricial young are down covered but unable to leave the nest on their own. Semi-precocial birds hatch with eyes open and are down covered, and although capable of mobility, they stay at or near the nest. Two additional terms sometimes used to describe a young bird or mammal are *nidifuge* (nest fugitive) and *nidicol* (nest dweller) that refer to leaving or staying at the nest (Welty and Baptista 1988). Nidifugous birds are precocial young, such as quail and shorebirds, that leave the nest shortly after hatching and most can run or swim. Swifts, woodpeckers, hummingbirds, and all passerines are nidicolus young that stay in the nest until they fledge. There are substantial differences in growth rates within the altricial-to-precocial spectrum, varying as much as thirtyfold (Ricklefs 1983). Altricial species can grow at three to four times the rate of precocial species.

Table 3.1. Condition of development at hatching

Type	Condition	Family or group
Altricial	Eyes closed. Nidicolous. Immobile. Naked or partly covered with down.	Pigeons, doves, cuckoos, Greater Roadrunner, swifts, hummingbirds, cormorants, pelicans, kingfishers, woodpeckers, falcons, parrots (and relatives), all passerines
Semi-altricial	Eyes open. Nidicolous. Immobile. Down in all tracts, including ventral.	All members in Ardeidae, ibises, vultures, Osprey, kites, eagles, hawks, falcons
	Eyes closed. Nidicolous. Immobile. Down in all tracts.	Owls
Precocial	Eyes open. Nidifugous. Self-feeding. Down in all tracts.	Geese, swans, ducks, quail, grouse, pheasant, stilts, avocets, plovers, sandpipers, phalaropes
	Eyes open. Nidifugous. Fed by parent at first. Down in all tracts.	Turkey, grebes, rails, gallinules, coots, cranes, oystercatchers, woodcock, snipe
Semi-precocial	Eyes open. Nidicolous for short period although able to walk. Fed by parents.	Nightjars, gulls, terns, loons (unable to walk, fed for 10 wks post-hatching)

Sources: Nice 1962; Baicich and Harrison 2005.

LENGTH OF NESTLING PERIOD

The growth rate of young birds, especially of altricial species, is incomparable to any other vertebrate. Why do baby birds grow so quickly? One obvious reason, especially in birds with high predation rates, is that the sooner they are fully mobile, the better chance they have to escape attack. For all wild animals, birthplace may not be the safest place to live. Thus a bird's chances of survival increase with its ability to leave the nest as soon as possible. Precocial young are already ahead of the game, leaving the nest within a few hours of hatching. The length of time in the nest of an altricial bird varies by species and by individuals. Individual variation of the nestling period is dependent on growth rate. In general, altricial nestlings grow rapidly for a short period, then reach a peak and level off in a sigmoidal shape. Most songbirds leave the nest between 9 and 15 days from hatch, with wings and tails still in sheaths at the base. They usually attain full adult mass within 10–20 days of hatching. Corvids are the exception in songbirds, with ravens having the longest nest period of

28 to 49 days before fledging. Some sparrow species and cuckoos fledge as early as 8 or 9 days. Nestling period is correlated with length of incubation period and with size of species (Welty and Baptista 1988): a long incubation equals long nestling period, and larger species have a longer nestling period than smaller species. The exceptions are cavity-nesters and long-winged species, for example, aerial foragers (swallows and swifts), whose nestling period is about a week longer than most open-cup nesting passerines of comparable size (Skutch 1945). Because their wings and tails are almost fully developed when they depart, they are more capable of flight. As soon as a nestling can thermoregulate and most feathers have unsheathed, it is capable of leaving the nest if there is a danger or perceived danger, even before it can fly. Hiding in brush and shrubs, where parents can still find and feed it, may be safer than staying in a nest that has been exposed. Many sparrow species in the Passerellidae family nest on or near the ground, and the young are capable of running before flying.

ALTRICIAL YOUNG

Nestling growth and development for altricial young has been extensively studied since the mid-1930s. Biological information from these studies is important to avian management, conservation, and rehabilitation. However, despite this history of studying nestling development, many questions remain unanswered. Basic information is limited or absent for many species and is site specific, and in some cases the literature is contradictory. It would be prudent for wildlife rehabilitators to create their own growth and development charts for species they care for (see Appendix C for examples).

Determining the stage of life of a young bird is important when considering not only identification but also reuniting and fostering, appropriate diet, housing, and looking for a nestmate. Three terms—*hatchling*, *nestling*, and *fledgling*—are used to describe early stages of life of altricial young still dependent on adults. Generally speaking, *juvenile* and, in some cases, *first-year* are used to describe young birds that are independent but have not attained their adult plumage.

- A *hatchling* is a newly hatched nestling that is naked or with varying amounts of natal down depending on species. Hatchlings are fairly weak, rather ungainly, can barely raise the head to gape, and evacuate in their nest. They have "stubs" for wings, and the legs and toes are short with clear or pale nails. They are referred to as hatchlings until the pin feathers (on the body) poke through the skin. In passerines, this usually happens by the third day after hatch.

- A bird is considered a *nestling* from the time of hatching until its normal departure from the nest. This is a period of very rapid growth and feather development.
- A *fledgling* is a young bird that has left the nest but is still dependent on its parent(s) for food. It is a fledgling from the time it leaves the nest until independent of all parental care. Fledglings are the most common age of birds brought to rehabilitators for care. They are targets for free-roaming cats and natural predators. Nestlings may leave the nest prematurely if disturbed, but they are not yet considered fledglings. In this guide, a *near-fledgling* is occasionally used to describe a bird that is very close to fledging.
- A *juvenile* is a young bird that has fledged and is able to care for itself. However it has not completed its preformative (post-juvenile) molt, and it has not attained sexual maturity and is incapable of breeding. It is considered a juvenile until the preformative molt commences. The term *juvenal* was formerly used as an adjective to describe a feather or plumage, but to avoid confusion, *juvenile* is used in this guide as both adjective and noun (Howell and Pyle 2015).

A young bird that has not acquired its full adult plumage can be in formative, first alternate, or predefinitive basic plumages, depending on the species and maturation rates. Most songbirds and many other species attain definitive (adult) plumage when a year of age, prior to which they can be identified as *first-year* birds. However, some species, such as young Brown Creepers or Bushtits, are indistinguishable from adults by the time they are few months old. Such larger birds as gulls, hawks, and eagles may not attain definitive plumage for several years.

The Passerine Hatchling

A newly hatched passerine bird has an enormous head compared with the rest of its body. Its eyes are bulbous, closed, and undeveloped. The openings to the ear canal are also closed at first. The skin is thin, appears oily, and is transparent. Internal organs are visible through a protruding abdomen. The beak is rubbery, short, and light in color. An egg tooth may be evident then disappears within a few to several days. The mouth interior and gape flanges, natal down, juvenile feathers, and other anatomical features are described in Chapter 2. The temperature of a hatchling corresponds to ambient temperature, as it is unable to thermoregulate. Hatching weight is approximately two-thirds that of the fresh egg from which it hatches (Welty and Baptista 1988). Wetherbee (1960) found most passerines hatch at 75% of the egg weight and 70% of the egg weight in warblers. Weight gain is slow for the first few days until the digestive system becomes more developed. Then they may double their mass several times during the first 10 days following hatching. Sounds, if any, are faint, high-pitched peeps, pips, or chirps.

The Passerine Nestling

Table 3.2 describes general passerine development from hatchling to fledgling stages.

The Passerine Fledgling

Fledglings retain some of their nestling features for a while after fledging. Compared to adults, the bill is lighter and may be slightly smaller. Remnants of gape flanges may be present. Colors of mouth and iris are generally lighter. Legs may be fleshier with fewer feathers. Juvenile plumage is filamentous (looser in structure) owing to lower barb density and may be streaked or spotted. There may be one or two wingbars that are distinct, indistinct, or broad, whereas adults may not have wingbars. Eye rings may be lacking, incomplete, or duller. Flight feathers are shorter and

Table 3.2. Stage of life and development of small to medium-size passerines

Stage of life	Age in days	Description
Hatchling stage	0 (hatch day)	Skin: thin, transparent, can see internal organs. Down: most are sparsely covered, some with more in one area, few with down on undersides. Some hatch bare and do not develop down or develop in few days. Eyes: undeveloped, closed, bulbous. Unable to thermoregulate, require brooding. Weak, lie prone, barely raise head to gape.
	1	Some pin feathers visible (pigmented) beneath skin (dorsal areas). Beg for food. Still weak, evacuate in the nest, sounds (if any) are faint.
	2	Most feather tracts, especially primaries, visible under skin. First food calls.
	3	Beak darkens, begins to take shape. Gape flanges thicken, color intensifies. Mouth color intensifies. Egg tooth may disappear. Feather tracts continue to darken, some pin feathers poke through skin. Down may thicken. Eyes begin to open as slits; vision is poor. Stronger: most can raise head, gape, use wings as props. Evacuate at edge or over nest.

Table 3.2. Stage of life and development of small to medium-size passerines (cont'd)

Stage of life	Age in days	Description
Nestling stage	4–5	Beginning of rapid-growth stage; weight increases. Pin feathers emerge in various tracts. Ventral feather tracts first appear; belly and chest area up to neck are bare. Primaries emerge. Eyes still opening. Begin to use legs. Evacuate at nest edge or over rim. Vocalizations stronger.
	5–6	Some temperature regulation. Beginnings of motor coordination: grasp with feet, stand, stretch, some preening.
	7–8	Eyes fully open (cavity-nesting species may take longer). Nearly fully feathered, belly and cloacal area still bare. Primaries unsheathing. Down may still be attached to feather tips. Rapid growth of motor coordination: scratch, shake head, stretch legs, flutter wings when begging. Freeze, cower, or crouch when alarmed or in response to alarm call from parent.
	9–11	Ground-nesting birds may fledge (day 10). Fully feathered, juvenile plumage covers apteria, wing and tail feathers still unsheathing. Thermoregulation nearly established; siblings keep one another warm. Preening continues, move about, stretch, flap wings, sleep less, respond to parent's alarm calls, alert to sights and sounds outside nest, may peck at objects on or near nest rim. Fear response strong: may leave nest if provoked and will hide in surrounding vegetation if accessible. Nest replacement may not be possible.
	12–13	Most species fully feathered; downy feathers may still protrude from head. May be antagonistic toward siblings. May flutter wings while perching on nest rim or may venture out (walk or hop) onto nearby branch. Peck at objects outside of nest.
	14–15	Most open-cup (arboreal) species fledge. May still have a few downy feathers protruding from head. Area around cloaca still absent of feathers.
Fledgling stage	9–24	Many cavity-nesting and long-winged species fledge day 20. Fledgling mass is 70–80% of adult mass (or more, depending on species). Breast feathers cover apteria on abdomen. Tail: about half grown. Wings: not as long as adult. Can perch, walk, hop, or run well, but flight capability varies. In general, some species do not fly well at first and are not capable of sustained flight first few weeks post-fledging. Cavity nesters are more capable of flight at fledging than open-cup nesters. At first, fledgling remains mostly immobile, apart from nestmates, and quiet except for food calls. Sleeps upright with head under wing, bathes, suns, preens, and begins bill-wiping. Pecks at objects and picks up food. Begins catching insects in the air (swallows, flycatchers) or on the ground, works at grass heads (finches), scratches ground (sparrows). Pursues parent(s) for food. Begging posture horizontal rather than vertical position. May give fear call around day 21.

Notes: Day 0 = hatch day. Day 1 = within 1st 24 hours of life post-hatching. Day 2 = 2nd day of life, etc. Some cavity nesters and larger passerines with longer fledge dates, such as Purple Martin (fledges 27–36 days) and crows and ravens (fledge 28–49 days), do not fit well into this general scheme.

may still be in sheaths at the bases. They are less worn than those of adult feathers that show abrasion on the tips. The primaries and rectrices are thinner and more tapered, especially on the outer two to three rectrices and longest primaries. All less developed features reach full size four or so weeks after fledging.

Estimating Age of Altricial Young

Knowing the exact age of a nestling in terms of days is not possible without witnessing the exact time it hatched; however, it is possible to *estimate* age by using a combination of several growth measurements. The most readily observable factors that can be age-specific are size, weight, and certain developmental events such as feather eruption patterns (Fig. G7, Illustrated Glossary), development of the extremities, eyelid opening, mobility, and behavior. The species accounts for many species contain ages at which certain growth events occur. It is important to keep in mind that nestling growth can be affected by several circumstances related to nest location, brood size, synchronicity of hatching, hatching order, and parental abilities. Limiting factors can be attributed to lack of food or calcium availability, parasitic infections, brood parasites, competition between nestmates, habitat differences, and weather. Beaks and legs in a nestling continue to grow regardless of limited resources. For example, very young American Robin nestlings are sometimes confused with other species because their beaks are enormous relative to the head and body that have not yet fully grown. Contour feathers tend to rapidly grow and unsheathe before the remiges, providing important insulation early in life when young cannot thermoregulate. In finches, the scutella on feet are first visible on the 7th day after hatching (Farner et al. 1983). It is important to look at all the components that affect growth, in combination, to reduce the margin of aging error.

Birds hatch any time of the day or night, and nestlings in a clutch may hatch one or more days apart. Those that hatch first generally weigh more and will be ahead in growth and development of those that hatch last. For example, in the Bewick's Wren, the last hatched can hatch 48 hours later and weigh 50% less than the first hatched. Assessing age is important when considering placing an orphaned chick with a foster family. With altricial young and most precocial species, the orphaned chick should match the size and development of conspecific chicks of a prospective family. In the wild, a chick that is further behind in development may not be able to compete with its nestmates for food. Those that are behind in development and do not fledge within a reasonable time period of their siblings may get left behind. In the rehabilitation setting, however, con-specifics of different ages in *most* species can be raised together because food distribution is controlled, thus eliminating competition among nestmates. Competition for food is much more pronounced in predatory species, although some passerine nestlings can also aggressively compete for food. Even then, these babies can be separated to reduce competition while still within view of their conspecifics to reduce imprinting or ha-bituation issues.

There are many criteria that scientists and bird banders use to age juve-nile, first-year, and adult birds, including skulling, taking measurements, and observing feather condition along with having a basic understand-ing of the bird's molt strategy. However, for the wildlife rehabilitator, many of these techniques are simply not realistic and may be unnecessary. First, the rehabilitator is at a disadvantage because the species may not be known yet. Finders of baby birds most often do not see adult parents or the nest where the baby originated, especially if the baby is a mobile fledgling. Field researchers, including bird banders who are studying any aspect of breeding birds, are often working with chicks that are identifi-able by the adult parent birds.

Taking measurements to determine age or possibly to help with iden-tification *could* be very useful if there were standards for comparison for all species at all their various stages of growth. For those species with such standards (where known and obtainable), the values are provided in the species accounts. Precise methods for aging may involve taking measure-ments of body mass, length of bird (tip to tail), bill length and depth, gape width, wing (chord) length, tail length, and tarsus length. When eyelids open as slits and when they are fully open are fairly consistent measures of age within a species. Ear openings are also fairly consistent measures: (1) when the small orifice first appears and (2) when the covert feathers appear over the ears, as they are usually the last feathers to appear on the bird. In the interest of education and for those who desire more precise methods for aging, a brief explanation of the various techniques used is included in this chapter. However, if you desire to learn the techniques used by bird banders and have the opportunity, you should volunteer at a bird banding station to gain experience. If you have time to do the re-search there are two resources I recommend: *Birds of the World*, compiled by the Cornell Lab of Ornithology (online), and *The Identification Guide to North American Birds* by Peter Pyle, a technical guide in two volumes (1997b and 2008) featuring molt, plumage, and other characters for iden-tifying, aging, and sexing juvenile to adult birds in the field.

Skulling

Skulling is recognized as one of the most reliable techniques for distinguishing first-year from adult passerines, mainly during the fall months (first 4–6 months of life). Skulling is not widely used by wildlife rehabilitators as the technique requires a level of proficiency and, for the most part, is probably unnecessary because other features can be used. As a point of interest, the skull of nestlings and fledglings has a very thin single layer of bone separating the brain and skull. When the bird reaches three to seven or more months of age (depending on species), a second layer of bone develops underneath the outer layer. The layers are separated by small columns that appear as white dots. Until the second layer is formed, age-specific sizes (in certain species) of "windows" visible on the skull (pink areas where columns have not developed) can be used for aging.

Measuring Weight (Mass); Using Weight to Age a Nestling

Weight *alone* is not as reliable as other characteristics and age-specific events used to indicate age because nestling growth is dependent on many genetic or ecological factors that can result in variation. However, combined with other characteristics such as feather development and size, weight is an important component for age assessment.

In the wild, nestling mass may correspond to food availability affected by weather conditions and foraging success of adults. Other factors that affect mass are brood size, nestmate competition, sexual dimorphism, and other environmental stressors. In some species, there can be a distinguishable difference among siblings, especially in the first three or four days of first brood nestlings (Kessel 1957). This may be due to asynchronous hatching and/or sibling competition to occupy a prime position in the nest. For example, "robin nestlings receive a higher proportion of feedings when occupying the central position in the nest" (McRae et al. 1993, 105). Generally, the first hatched, if healthy, will be larger and weigh more than its siblings. In some species the difference between first and last hatched can be significant. This is most apparent in Barn Owls, which hatch several days apart. The runt in a clutch of Acorn Woodpeckers can weigh 30 grams less than its siblings. It is also evident with House Finches, when hatching can take place from 1 to 5 days apart; however, on average young hatch within 3 days of first hatch. The time of day (or night) when the bird is weighed or when it had its last meal can result in weight differences. Mass in a single day can vary by 10%. In some species, such as small owls, mass can vary by up to 40% depending on when they were last fed.

Weights in certain species may be determined by whether the young are a first brood or second brood by parents within a season. In her study on first and subsequent broods of starlings, Kessel (1957) found differences in weight loss of nestlings just before fledging. In the first brood, for the first 10 to 12 days after hatching, weight gain is rapid and sigmoidal in form. In the following stage, energy is allocated to plumage development, and the growth curve may level off or fluctuate. First-brood nestlings may lose weight from a higher peak before fledging, whereas second-brood nestlings usually do not. Kessel also found, compared with subsequent broods, that there seems to be little nestling weight variation in first broods when conditions such as food availability may be most optimum for growth.

Captive-fed nestlings may weigh more or less than wild babies of the same stage of life. Diets and number of feedings per day will vary between wildlife centers. When comparing the weight of your chick with a weight reference, whether on an identification chart or other reference, your chick should fall within a reasonable *range* for the species you are considering. The range of the reference may be a range of individuals within the brood studied or more commonly is a standard deviation. First, and this may seem obvious, you will compare your chick to the highest and lowest adult weights. In general, chicks hatch under 10% of adult female weight, and by the time they fledge they weigh from 70% to 100% of the parents' weight depending on species. If your chick weighs considerably more (or less) than the adult of the species you are suspecting, then consider another species. In larger species, such as crows and ravens, the range of weight at a particular age will be broader compared with the range in smaller species, so other parameters must be used in combination.

Measuring Total Length

Total length is beak tip to tail tip, or in nestlings prior to rectrix growth, the posterior-most part of the body. Place the bird on a piece of paper and mark the tip of the beak, then mark the outermost tip of tail (or body) and measure between the two points. This measurement can be used to narrow down a species but is not as reliable as other indicators for the same reasons as weight. Length varies in individuals and whether the neck is stretched or tucked in.

Measuring the Beak and Gape

Beak length and depth are fairly reliable at certain ages in the bird's growth. Measurements of gape width and gape flanges are not as reliable because of variability from one day to the next. The gape grows quickly during the early stages of the nestling period and can increase or decrease from one day to the next. The gape flanges are very soft and pliable during the growth period. Bill length is taken from two locations (Fig. G2, Illustrated Glossary): (1) from the anterior end of the nostrils to tip of the bill or (2) from the base (where it meets the forehead feathers) to the tip of the upper mandible.

Measuring the Tarsus, Middle Toe, and Hallux

Measurements of leg tarsus and toes can serve as good age indicators because they may grow normally despite food restrictions or brood size. The tarsus grows quickly and early during the growing period, then tapers off at a specific age. For example, in most passerines, tarsus growth tapers around day five or a bit longer in larger species. Tarsus length is measured from the tibiotarsus joint to the distal end of the last leg scale before the toes emerge. The tarsus is often compared to the length of the middle toe with nail. Tarsus length can help distinguish between a family or a group. For example, tarsus and toes of a towhee are considerably longer (and stronger) than those of the majority of passerines, including smaller sparrows.

Measuring Wing Chord

Wing chord measurement also can provide a good estimate of age as wings grow quickly and continuously throughout the nestling stage. This is a length of the wing most often used in published studies for North American species. If not taken properly, the results may be inconsistent with standard measures. Use a thin metric ruler. Holding the bird in one hand, place the wing on the ruler with the wrist at the ruler end and without pressing on the wing (i.e., when the wing is in its natural position, not flattened), and take the reading at the tip of the longest primary. Note that some biologists flatten the wing (e.g., for a shorebird).

Measuring Tail Length

Rectrices begin to develop after the growth of the primary feathers begins to slow down, usually in the latter half of the nestling period. Tail length (using a thin metric ruler) is measured from the tip of the longest rectrix and the point of insertion of the two central rectrices (Pyle 1997b). Tail length in many species can help determine age. For example, the tails of fledgling songbirds are rarely more than half the length of adults when they leave the nest. Some ground-nesting species, such as sparrows and some warblers, barely show tails when they leave.

Estimating Age Using Feather Development

Feather development during the prejuvenile molt is very easy to assess and a reliable component of aging nestlings because it may proceed independently of growth in body size or mass gain. As shown in Table 3.2, following the hatchling stage, nestlings begin to show feathers emerging along specific feather tracts. The pattern of feather growth follows a consistent age-related pattern within a species. A downy altricial baby goes through its first molt when new juvenile feathers replace natal down. The flight feathers will often begin to emerge and develop before the contour feathers; however, contour feathers will often begin to unsheathe before the flight feathers. The stage when the tips of feather sheaths begin breaking on primaries, secondaries, and rectrices is a good criterion for estimating age of the nestling. For example, in the Bewick's Wren, tips begin breaking at around 7 days. In starlings, growth of the first secondary was used as an indicator (Kessel 1957), emerging through skin about 6.5–7 days after hatching. Growth of flight feathers may be affected by factors different from those that affect weight gain. Growth of the remiges is a fairly reliable age indicator and can be measured throughout the nestling period up until about two weeks of age. After that the growth is too variable. Primary feathers grow most rapidly in the first half of the nestling stage, while rectrices start slowly and grow rapidly in the latter half. In some species, the appearance of feathers on ear coverts can be a marker for age; these are the last feathers to appear on finches. Table 3.3 can help distinguish precocial young and altricial nestlings at various stages of life by their feather development.

Table 3.3. Feather development and stage of life of altricial and precocial chicks

Type of natal down or feathers	Stage of life (approx. age: d = days, H = hatchling, N = nestling, juv = juvenile)
No down, skin is entirely naked, does not develop down (Table 2.9)	H passerine, 0–3 d, woodpecker (N 0–14 d before 1st feathers appear)
Down is sparse (can see skin through down); some areas naked	H passerine, 0–3 d
Down is sparse (some areas naked) and looks thick and stringy	H or N pigeon or dove
Down is fairly thick and covers entire body including undersides (but bird would be confined to nest)	H or N raptor, heron, egret
Down thick, covers entire body (able to walk or run)	Precocial chick
Pin feathers starting to emerge, particularly wings	N passerine, 3–4 d
Mixed erupted juv feathers and pin feathers, still has naked patches on back or belly or underneath wings (natal down may still be evident)	N passerine, 4–9 d
Most feathers have unsheathed, appear fully covered, no naked areas (except belly area), many feathers still in shafts at the base, may have remnants of down, would still be confined to nest	Older passerine N, can regulate body temperature
Fully feathered with little or no down remaining; wings and tail feathers shorter than adult	Fledgling

Growth and Fault Bars

Growth bars can be used to assess age but should be used only to support other aging criteria. Growth bars are alternating patterns of light-and-dark bars across the width of the feathers that indicate 24-hour periods of growth. They are easiest to see on rectrices and with backlighting. Fault bars are caused by severe nutritional distress causing growth to be disrupted. They lack pigment and may have loss of structural integrity, which can cause the feather to break. In nestlings and juvenile birds, fault bars usually occur in a similar position across all flight feathers. Adults can have growth and fault bars as well, but because adults molt feathers sequentially rather than simultaneously as in nestlings, bars will appear scattered. Adults show fault bars much less frequently than do juveniles, as they experience trauma much less often while molting. Exceptions include adults that lose their tails and regenerate all rectrices at once or juveniles that molt or accidently lose some of their rectrices so would subsequently appear to have scattered growth bars. Growth bars in owls can be used more reliably for aging as they are more distinct because of the structure of the feathers.

Estimating Age Based on Development and Behavior of Altricial Young

Certain behavioral events coupled with physical development may consistently occur at a specific age. Table 3.2 describes these events, along with feather growth. For example, in altricial young, ages when the eyes begin to open and when they are fully open are fairly consistent; when known, therefore, some of these events can help with estimating age of the young. In larger altricial young, development is a little slower, length of time in the nest is generally longer, and dependency on parents is generally longer as well. Nestling and fledgling birds beg for food by gaping and fluttering their wings, especially when adults are present. They are uncertain in flight. Juveniles after fledging may still flutter and beg, behavior similar to courtship in adults. Other notable behaviors that can give away a bird's stage of life are posture, movement, reaction to the observer, self-maintenance, elimination (defecation), call notes, and feeding behaviors.

Nestlings of a particular age begin displaying a fear response to what could be a predator. It is an innate or learned behavior necessary for survival. Precocial and altricial species will flatten out and lie still. The fear response is so strong in some altricial species that they will leave the nest prematurely (before they can fly) and will not stay in the nest regardless of any attempt to renest them.

Nestling or Fledgling?

Aging altricial young becomes more difficult at the transition time between being a fully feathered nestling and a fledgling that has just left the nest. Some nestlings may leave early if disturbed but are not of fledgling development quite yet, whereas some fledglings of certain species are not able to fly well when they leave the nest. These are the babies that are most often "rescued" by finders. The young of several species, including herons, some owls, and some passerines, venture out of their nest and perch or climb on branches (thus sometimes called branchers) before they can fly. Some return to the nest (especially at night) whereas others do not. If a young feathered bird is out of its nest, on the ground, or in a bush, the situation must be evaluated with a careful eye and knowledge about the life history of the species. It is initially important to observe from a distance far enough away to see if parents are tending to the grounded chick. If the youngster is alert and active, there is no danger from predators, and parents are repeatedly bringing food, it can probably be left alone. Otherwise, *any* baby that is injured or appears sick, is not being supported by adults, or is in the pathway of danger must be captured and evaluated.

Case Example: It Is Imperative to Understand
the Breeding Biology of the Species

Before any advice is given or any decision is made, aside from considering the source and trusting the information given, it may require a trip out to the site by a qualified and experienced volunteer or professional to make an assessment. The following case is a good example of the importance of understanding the breeding biology of the species of concern. In the late spring in the lower foothills of the Sierra Nevada mountains, hikers on a popular trail came across what appeared to be four "fledgling" Belted Kingfishers on a pile of sand just below their nest/burrow on a bank and within inches of a swiftly moving creek. A hiker alerted the local wildlife rehabilitation group, expressing her concern about whether this was normal because the young birds seemed (to her) to be in danger. Another hiker came along and told the first hiker that it was normal behavior that "newly fledged birds cannot fly very well at first." Photos and a video were texted to the wildlife group; however, the birds appeared to be fully feathered and bright blue because of poor photo quality. It was decided not to intervene because the birds "were eating ants" and the parents were nearby (chattering in the trees above). Hours later, the story got to the rehabilitation group's biologist, who hiked out to the site and found one live young kingfisher hiding near the water's edge. Sadly, two others were dead and floating in the water and the fourth was nowhere to be found. Upon close examination of the carcasses, the only feathers out of sheaths were small crest feathers on their heads. Therefore, the birds were still *nestlings*, not fledglings as had been believed. Regardless, even if the kingfisher chicks were of fledgling age, they should not have been on the ground. This is an example of another misconception. Unlike open-cup nesting species, the nestling period of cavity nesters is longer, therefore their wings are a bit more developed, and they are capable of flight straight out of the nest. A kingfisher burrow is a type of cavity nest; therefore, something was very wrong with the young kingfishers being on the ground in such a perilous situation.

This case demonstrates several lessons. Perhaps the main ones are how differently information can be interpreted and that we simply cannot make generalizations about every species. A grounded baby of a cavity nesting species is *not* normal. The two hikers had a different impression or version of what was actually happening. The report that the birds were eating ants was an important clue because kingfishers do not eat ants but primarily fish. The reason the birds *appeared* to have feathers is because sheaths that encase feathers are shiny and reflect light. Feather quills are also transparent, so the bluish-gray color of the feathers that would soon unsheathe was picked up in the photos and video. Because *blue* in birds is caused by light reflection, the photos showed *reflected* color and were also blurry, giving the birds a "fuzzy," fledgling-like look. They were actually older nestlings of the age that would be coming out of their nest deep within the burrow and waiting at the burrow's entrance for parents to bring fish.

The question in this and other cases is, should there have been an intervention? The answer is not simple, and in this instance, it stems from why the young kingfishers were out of their nest prematurely. An additional piece of information, not initially reported by the finders, was that the hiking trail is very popular for dog walkers. The biologist encountered several hikers with off-leash dogs romping freely through the creek. What caused the nestlings to fall out? Perhaps the ground at the nest entrance gave way from (1) a digging predator or dog or (2) high water level from heavy rains during the previous winter. There was no way to ascertain if the birds came out of their nest by natural or unnatural events. However, romping dogs disturbing a nesting habitat is not natural and possibly compounded the event by frightening or "chasing" the chicks into the water. The one live nestling was hiding behind a bush under a bank. There were two small fresh fish lying next to it, and parents were heard and seen nearby, so it was still being tended to. It was entirely possible there were more young birds still in the burrow. Therefore, after the biologist warmed the baby and gave it a brief exam finding no visible injuries, it seemed the best course of action was to place it back into its burrow. The parents were very vocal in the trees above, waiting for signs of safety to tend to their young. The moral of this story is that each situation must be thoroughly evaluated, and one must understand the natural history of the species.

PRECOCIAL YOUNG

In precocial birds, the terms *hatchling*, *nestling*, and *fledgling* are not often used because they leave the nest soon after hatching and learn to fly sometime later. A newly hatched precocial chick is more advanced in development than the altricial hatchling. Hatching weight is about two-thirds of the fresh egg weight from which it originated (Welty and Baptista 1988). About one-third to one-seventh of the yolk in the belly of a precocial chick is retained. Weight decreases between hatching and the first feeding but then increases consistently throughout its development. The newly hatched chick is thickly covered with wet down that becomes dry and fluffy 2–3 hours after hatching. The egg tooth is present, and there are no gape flanges. The lining of the mouth is the same as or often less brilliant than that of the adults. Eyes are open, hearing is already developed, and chicks respond to sights and sounds around them. At hatching they give "peeping" sounds. The wings are small and undeveloped, although some may have feather sheaths; however, the legs are fully functioning. Despite their downy covering, precocial chicks do not have full temperature control until they are 4 weeks old and will become chilled if they are not kept warm by the parent.

Precocial birds have two phases of development: (1) development from hatching through attainment of flight and (2) development after attainment of flight. Precocial young become partially independent of parental care during the first phase but may remain dependent socially on their family or flock during the second phase. The length of time before chicks attain flight ranges widely: 7–27 days in gallinaceous birds, 14–34 days in shorebirds, and 4–12 weeks in most aquatic birds. Precocial chicks may remain on the nest 3–24 hours or more. In Galliformes (including grouse, pheasant, quail, and turkeys), chicks leave the nest usually only with the hen. She directs her chick's attention to food by picking it up and dropping it repeatedly, while giving a food call. The chick responds by taking dropped food and swallowing. Within a very short time the chick is able to recognize and find its own food. In Anseriformes (swans, geese, ducks), the cygnet and gosling leave the nest with both parents. Ducklings are usually with the hen only. Shorebird chicks, such as Killdeer, leave with one or both parents depending on the species. They start feeding without parental assistance by first nibbling indiscriminately at objects and then learning what is edible by trial and error.

There is an evident social bond within a brood. Unlike altricial birds, precocial young typically begin calling before hatching. Pecking, chasing, and play fighting sometimes occur among chicks in a brood. Sudden noises or abrupt unfamiliar actions will cause chicks to scatter or hide under or near large objects. They "freeze" by crouching flat and remaining motionless with their eyes closed. Usually, but not always, they remain frozen until the parent emits a call to reassemble. If one chick becomes "lost," it will utter the loud brood call and behave very distressed until contact with its family is reestablished. The family bond, formed just before hatching, is very strong until chicks can fly well, and it can last sometimes well into fall and winter. Most precocial young give distress calls when being harmed or handled.

FOUR

THE PROCESS
OF IDENTIFICATION

This chapter takes you through a systematic process to make the most accurate identification possible and to eliminate the possibility of a misidentification. Accurate identification begins with good detective work and should be based on a combination of all available characteristics. There is no magic shortcut. Avoid the temptation to start with picture matching. The Identification Worksheet (Appendix B) is useful when first learning how to identify birds until the process becomes routine. The worksheet takes you through a step-by-step procedure from the minute a chick is found. Beginning with the key characteristics (Chapter 2) will immediately reduce many possibilities. Expert birders who have highly developed skills can identify birds in the field often by intuition or what may be referred to as GISS (or gestalt), a general impression of size and shape. Based on birds you are familiar with, you may have an idea of your bird's identity. Investigate that idea first, but keep an open mind. Make your identification based on the evidence in front of you, not what you assume. Along with this guide, you should have a list of breeding birds from your local birding group, a breeding bird atlas of your county, and a few good field guides with maps showing species nesting in your locale as well as sympatric species.[1] Other helpful references have been mentioned throughout this guide and in the Literature Cited.

First and foremost, identification is a process of elimination by making comparisons (see Appendix D for comparisons of species within the same family). Knowing what it *isn't* helps narrow down the field to what it *could* be. When you have such a broad range of choices with so many possibilities, breaking things down into smaller categories will make the task less overwhelming. If you are a wildlife rehabilitator, learn the species most commonly admitted to your facility. Then you will be at an advantage when a different species arrives. The most common and most numerous species admitted into rehabilitation are typical "backyard" breeders, such as American Robins, jays, finches, blackbirds, doves, House Sparrows, and European Starlings. Some species will become easy to distinguish immediately, even as hatchlings. Corvids, for example, have similar characteristics, but their size, weight, colors, behavior, and vocalizations distinguish them from each other as well as from all other avian species. Other species will be much more challenging. Sparrows of the Passerellidae family are the most difficult to identify because many are sympatric and of similar size. In some sparrows, the distinguishing features do not appear until their first fall or winter.

THE ORDER OF BIRDS AND BIRD NAMES

Scientists have used many taxonomic characters to place living things in orders, families, genera, and species. Scientific classification provides a framework by placing similar birds into groups that share fundamental characteristics of their anatomy, physiology, behavior, and genetic makeup (their DNA). If you learn at least the most common orders and families of birds, you can make quick use of most any field guide. Knowing the family groups enriches your knowledge of the birds you encounter. You come to appreciate how every species is a variation—some subtle, others more abrupt—from the general theme that applies to all members of its family. Birds vary in just about every aspect, causing difficulties for authors and artists of identification guides. With few exceptions, nothing in nature studies should be taken so literally that you don't keep an open mind to all the possibilities. No two birds look exactly the same, and no bird looks exactly like its illustration. The differences may be minuscule, but they do exist just enough to throw a monkey wrench into the process. Serious ornithologists and taxonomists use a bird's scientific name rather than (or in addition to) the common name. Although the American Ornithological Society designates official common names, these names can still be problematic to those not familiar with them. For example, many people in California claim they have seen "blue jays" in their yard. Although they are seeing jays that are blue in color, either the Steller's Jay or the California Scrub-Jay, it is highly unlikely they are seeing the official Blue Jay. Note that capitalizing bird names helps to distinguish the Blue Jay from blue jays (jays that are blue) in text.

STEP-BY-STEP: USING THE WORKSHEET

The process always begins with a reporting person or the finder. It is important to be thorough when collecting all available information. Put on your detective hat and begin gathering clues. The bird's natal history (or story from the finder)—where it came from and a description of the parents, if seen—may be definitive to putting a name to your bird. Many people have difficulty describing birds over the phone. While this can be somewhat amusing, it can be an inconvenience to the raptor rehabilitator that has driven 100 miles to find out the chick was not a hawk but a pigeon squab. When asked if the beak and nails are sharp and curved, "oh yes," the finder says positively. But that person's idea of sharp or curved is not relevant unless in comparison with something else. When asking questions, it is necessary to provide a familiar example and avoid questions that are open-ended.

Questions to Ask the Finder

1. Can you describe the toes and feathers? Refer to Tables 2.6, 3.1, and 3.3 and Figs. 4.1, 4.2, and G5 a–h (Illustrated Glossary). Often, we are speaking with a finder who may not know the difference between a baby pigeon and an owl, or even if it's a duckling or a nestling songbird. The person's interpretation of "downy plumage" may not be anywhere close to the definition. However, the finder should be able to describe the down on a bird with little to none (altricial) versus a fully covered fluffy chick possibly with a pattern on the crown and back. Are the feet webbed? If not, are three toes pointing forward or are two toes pointing forward and two pointing back?

2. Are adult birds present? Ask if adults were seen in the vicinity, and make sure they are the actual parents of the baby bird in question. Were they seen feeding the baby? Remember that cowbirds parasitize many songbird species and bear little resemblance to their host parents.

3. What type of habitat was the bird found in? Was there a nest? Tables 4.1 and 4.2 list typical nesting habitats and nest types for various species. General nest type is provided in the species accounts. Unless a baby bird is found in the vicinity of its nest, once the bird is removed from its habitat, a good portion of its identity has been lost. Habitat is as much a signature to a bird as all the other features that make it a species. Birds have unique physical adaptations that are best suited for the habitat they live in. Following are very general categories of habitats: forest or woodland (coniferous or deciduous); water or aquatic (lakes and ponds,

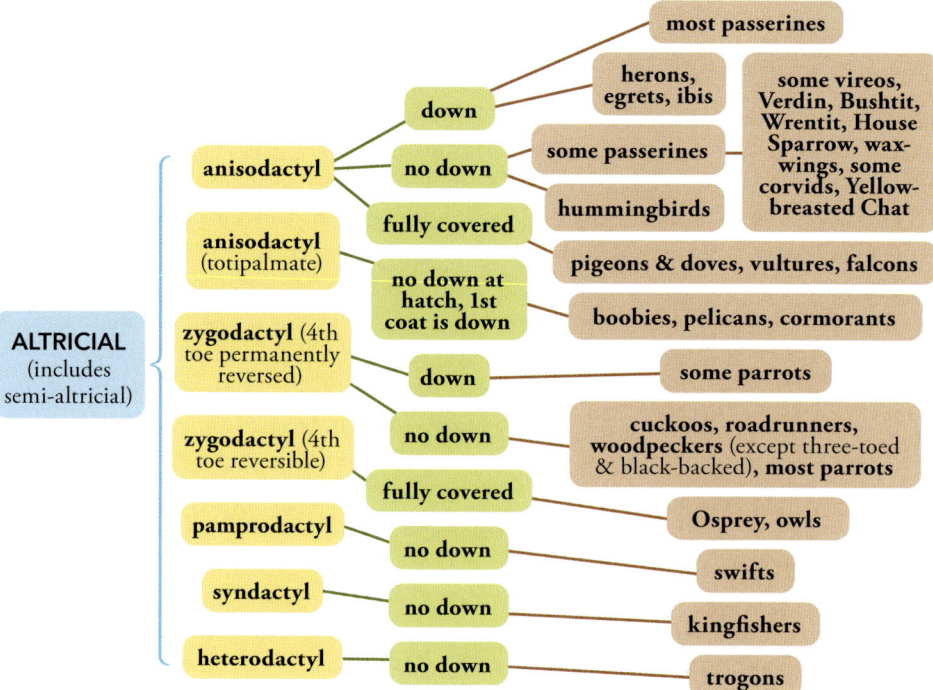

Fig. 4.1. Birds with altricial young by foot type

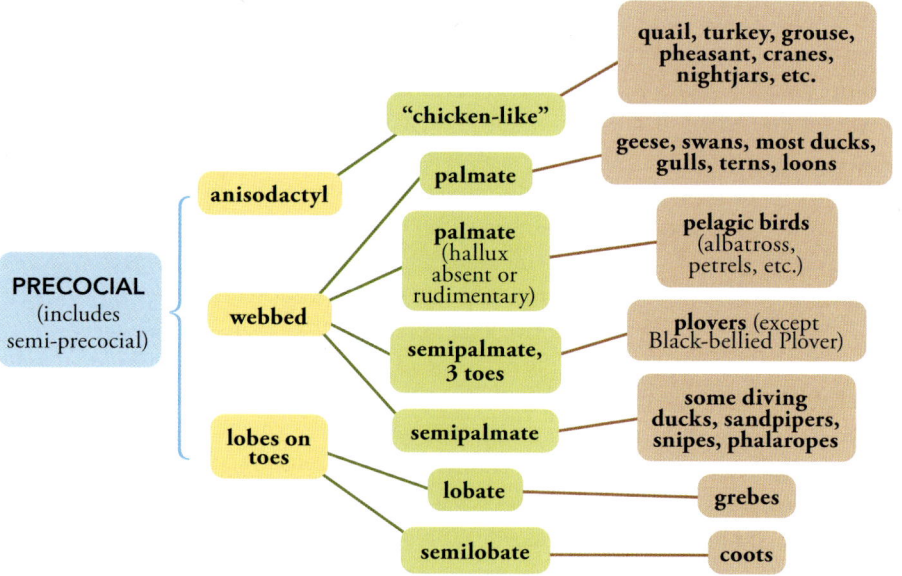

Fig. 4.2. Birds with precocial young

Table 4.1. Nest locations and substrates

Typical location and substrate	Typical family, group, or species
On or in buildings or other man-made structures (such as nest boxes), especially in urban and suburban areas	Pigeons, doves, swifts, some hummingbirds, some raptors, some flycatchers (Pacific-slope Flycatcher, Black Phoebe, and Eastern Phoebe), jays, some swallows, sometimes titmice, some wrens (House Wren, Carolina Wren, and Bewick's Wren), some thrushes, European Starling, House Finch, and House Sparrow
Cave	Cave Swallow
Colony or communally (dense or loose) in trees	Herons, egrets, swifts, Yellow-billed Magpie (loose), sometimes Black-billed Magpie, many swallows (and martins), many blackbirds
Ground	
Little or no nest: near water	Geese, some ducks, some shorebirds, and many ocean birds
Little or no nest: open	Gallinaceous (Chukar, turkeys, ptarmigans), Killdeer (and other plovers), Snowy Owl, Common Poorwill (and other nightjars)
Little or no nest: under trees or shrubs	Some ducks, vultures, Gallinaceous (quail, turkeys), some nightjars, some owls (Long-eared Owl and Short-eared Owl)
In dense clump of vegetation	Northern Harrier, many New World sparrows
Nest: on island (island may even be small one in lake or pond)	Geese (Canada), Great Blue Heron, gulls, terns, cormorants
Water	
Over water: on vegetation growing in water or tree overhanging water	Smaller herons and egrets, sometimes Black-billed Magpie, Marsh Wren, some blackbirds (Red-winged Blackbird, Yellow-headed Blackbird, and Tricolored Blackbird) and grackles, some swallows
Near or next to body of water	Small nest: some flycatchers (Black Phoebe), Marsh Wren, Common Yellowthroat Large nest: large herons, egrets, Osprey
Nest on water: floating platform	Loons, grebes, coots, gallinules

swamps and marshes, open ocean, shoreline); scrub / shrub (short woody plants and bushes); desert (cacti and succulents); open habitats (grasslands, agricultural areas, tundra, fields); and the "urban habitat," which may be in the midst of one or more of the other habitat types. Most baby birds are found in urban areas, where people live, work, shop, and enjoy the outdoors. Bird nests are sometimes found in uncharacteristic places.

Table 4.2. Nest type

Nest type	Typical family, group, or species
Burrow hole in dirt or sand	
In bank	Kingfisher, Bank Swallow, and Northern Rough-winged Swallow
In ground	Burrowing Owl, many seabird species
Cavity	
In tree only	Little or no nest inside: woodpeckers, some ducks (with a little down); nest inside (usually cup): Vaux's Swift
In tree or artificial box	Some ducks (Wood Duck, Bufflehead, goldeneyes, mergansers), some owls, some swallows (Tree, Violet-green, Purple Martin), some flycatchers (Great Crested, Ash-throated, Brown-crested), European Starling, bluebirds, titmice, nuthatches, wrens (e.g., House, Carolina, Bewick's), House Sparrow, Prothonotary Warbler, Lucy's Warbler
In man-made structures (chimneys, air vents, old wells, outhouses, garages, barns, etc.)	Chimney Swift (usually vertical surface with low light), Barn Owl, some wrens (e.g., House, Carolina, Bewick's), House Sparrow, European Starling
Cup or cuplike	
In cavity (see "Cavity")	
Small: in tree or shrub	Hummingbirds, some flycatchers, almost all songbirds that are not cavity nesters have a cup nest or make a cup inside a nest of mud or bulky sticks
Small: behind bark of tree	Brown Creeper
Small: on or near ground or in vegetation (clump of grass or low bush)	Many ground-nesting sparrows (Song Sparrow, Dark-eyed Junco, towhees), buntings, longspurs, Horned Lark, some blackbirds (Dickcissel, meadowlarks)
Large: in tree	Small to medium herons and egrets, hawks, owls, crows, ravens, jays
Very large and bulky: in treetop or top of man-made structure	Bald Eagle, Osprey, Great Blue Heron
Domed with entrance hole at side	Magpies, Verdin, Bushtit, some wrens, some dippers, Ovenbird
Ledge on rock, rocky island, or building edge	
Small cup: glued with saliva to vertical cliff wall or on ledge	Swifts
Large nest	Canada Goose, ravens
Little or no nest	Vultures, falcons, Rock Dove

Table 4.2. (*continued*)

Nest type	Typical family, group, or species
Mud attached to building, rock, or other substrate	Barn Swallow, Cliff Swallow, Cave Swallow
All mud	Cliff Swallow (mud only), Barn Swallow (feathers and grasses incorporated)
Mud used as glue, but not entirely of mud	Phoebes, robins
Pensile (hanging or suspended): usually bound to forked twigs	Bushtit, orioles
Suspended	
Single support	Acadian Flycatcher, Rose-throated Becard
Slung between forked twigs	Vireos, orioles

Many species or individuals may demonstrate some degree of tolerance in adjusting to slightly less suitable sites. For example, the American Robin's nest is usually found in the fork of a tree or shrub, but when these sites are not available the robin may nest on the ledge of a cliff or building or even on the ground. However, the composition and the size of a robin's nest are mostly consistent. Another element of confusion arises when birds use nests of other species. Birds that do not build their own nests, such as owls and falcons, may use old crow or hawk nests. Some species that do build their own nests may take advantage of a nest constructed by another species. If you find a nestling below a colony of Cliff Swallows, you may want to make sure the chick is not a House Sparrow, House Finch, or wren that may have taken over the swallow nest. A cavity in a tree excavated by a woodpecker may have a succession of very different tenants. A more complete explanation of nest types is in the tables.

4. Were any eggs or shells found? What size, shape, color, and markings do they have? Eggs or empty shells can be very helpful clues but are not reliable on their own given a high degree of variability. Even in the same clutch, a few eggs may be smaller, slightly different in shape, paler, or have less of a pattern than the others, and there may even be abnormal eggshell colors. Eggs of two or more species within the same family group nesting in the same area may be indistinguishable. Eggs come in a variety of shapes, sizes, and colors with or without patterns or markings, and there are too many variations to make generalizations. Eggs of more common species are described in the family accounts and some of the plates. More detailed descriptions and illustrations of eggs can be found in Baicich and Harrison (2005) and numerous other sources.

5. How big is the bird? Compare the size of the bird's body to common birds or inanimate objects. In this guide, sizes of passerine and near-passerine birds are defined as follows:

- Small: Adult birds up to and including 6 inches (in.) long (tip to tail). Hatchlings would be very tiny (the size of a dime up to a quarter, or less than 1 in.). A few examples of these would be hummingbirds, swifts, some flycatchers, vireos, most swallows, chickadees, titmice, kinglets, nuthatches, creepers, most wrens, most wood warblers, goldfinches, and some sparrows.
- Medium: Adult birds > 6 in. to < 10 in. in length. These would include some nightjars, small owls, some woodpeckers, catbirds, cardinals, most thrushes, bluebirds, tanagers, martins, grosbeaks, towhees, crossbills, starlings, (medium-size) finches, waxwings, pipits, shrikes, orioles, cowbirds, blackbirds, House Sparrow, and many New World sparrows.
- Large: From 10 in. and up. These would include pigeons, doves, large owls, hawks, ravens, and the like.

When describing a bird over the phone, it is sometimes useful to concentrate on just the size of the bird's body, not including the head, neck, and tail. You could begin with a comparison relative to a familiar bird, such as a hummingbird, sparrow, robin, pigeon, hawk, and so forth. But even then, that may be difficult depending on perceptions or familiarity by the finder with these types of birds. It may be more useful to make a comparison to a solid object that does not vary, such as a golf ball, baseball, softball, or football.

Examining the Bird

Identification requires a set of skills, just like skills that an artist has learned through education and practice. You may want to start by describing your bird in words or making a little sketch of your subject. Divide your bird into sections and work from beak tip to tail. If you have an inkling or "GISS" feeling of what it is already, write it down and look for details that support your feeling. Look at the adult of the bird you are considering. Are you on the right track? Do not make the common mistake of looking at one feature to make the positive identification. Always compare sizes and weights if you are deciding between a few different species. For example, starlings and Oak Titmice are secondary cavity nesters, and both have very prominent, swollen gape flanges. However, there is a distinct difference in size, weight, and down or feather color between these two species. Once you have described your bird, you will want to try to place it into a family (such as thrushes) or group (such as bluebirds). You will then compare the key and supporting characteristics of species in the species accounts in the family you believe the bird belongs. Check a range map to see if your bird falls in the nesting range for the species you have selected. As a reminder, identification study can occur while conducting a health exam or providing treatment. Taking a few photos that can be studied later, to prevent unnecessary handling, may be better for the bird.

As you become more experienced with bird identification, you'll begin to notice that some features are particular to a certain group or species. Bewick's Wren hatchlings have a high concentration of down on their heads, with little to none on other parts of the body. Lark Sparrows, Spotted Towhees, and some other sparrows have white on their outer tail feathers in a specific pattern, area, or size. In a few cases, a supporting characteristic as minuscule as toenail color may be the defining characteristic. In passerine hatchlings, the legs, feet, and toenails are not well developed and generally uniformly the same color (pink, tan, or yellowish). By the time they reach the older nestling stage, toenails of most passerine nestlings have changed to shades in the ranges of browns and blacks. However, many nestling blackbirds and mockingbirds have white toenails (especially when viewing them from the underside of the nail), and cowbirds have black tips on their toenails. Other examples are the yellowish skin and footpads on some thrushes and reddish skin on some blackbirds.

Birders in the field make comparisons of size and shape to other birds they know well. For example, is the bird smaller than a crow or larger than a sparrow? Is the shape chunky, pot-bellied, or long and slender? Birders also compare the individual bird's parts with other parts on its body, for example, beak length relative to head size, or length of tail or wings relative to body length. Although the beaks, wings, and tails of many fledglings may not be fully grown, those of many species will exhibit enough distinguishable characteristics in the wing and tail to support other key features. Once you have narrowed down the field of possibilities to just a few, you will need to make a side-by-side comparison. Table 4.3 is an example of a comparison of two families of long-winged birds with similar traits because of how they forage. It might be useful to make your own tables for comparing family groups or species that have similar characteristics.

You may find the identification worksheet in Appendix B helpful to follow as you continue to collect information.

Table 4.3. Comparison of swifts and swallows

Body part	Swifts	Swallows
Bill	Short, tiny, slightly decurved	Short, wide at base (triangular in outline when viewed from above), flattened, slightly hooked, culmen somewhat decurved toward tip, commissural point below nasal canthus of eye
Mouth	Wide, gape extends back under eyes	Wide, twice the length of culmen
Facial bristles	No	Yes
Legs/feet	Pamprodactyl foot arrangement, legs small and weak, feet strong with strong claws for clinging to vertical surfaces (rocks and chimneys)	Anisodactyl foot arrangement, weak, perch on horizontal surfaces
Tails	10 feathers, shorter, stiffer	12 feathers
Wings	10 primary feathers, long, slender, curve backward	9 primaries, long and pointed

1. Weight and size of the bird (in hand). When comparing your chick with weights provided in the species accounts, you will need to have assessed the age or stage of life. The general size of a bird is measured in two ways: by dimensions and by weight. Dimensions include measurements: the length of the bird, from tip of bill to tip of longest tail feather (or nub on a naked or downy nestling), its wingspan or wing chord, tarsus, and so forth. Weight is measured in grams (to the tenth for small birds). As explained in Chapter 3, size and weight are best used together. After examining all notable characteristics, you can then compare your bird with other birds of similar weights within the general range that are at the same developmental stage. Weight is most helpful when comparing birds in distantly related taxa, such as distinguishing between the American Robin and Northern Mockingbird.

Basic principles to keep in mind:

A lightweight hatchling will be a lightweight adult. Young nestling American Robins and Northern Mockingbirds have similar characteristics and are often mistaken for each other, but at every stage of life a robin will weigh nearly twice as much as a mockingbird (Table D-4, Appendix D).

*Many young birds will weigh more than their parents **before** they leave the nest* (Welty and Baptista 1988). At the time of fledging, they often lose weight to facilitate flight, weighing about 70–75% of the weight of the adult female.

More things to keep in mind:

- Always compare the chick's weight and length to the adult weight.
- A baby bird may arrive weighing less than normal depending on its hydration condition and how long it has gone without feedings. Even during the course of the day, weight can vary by 5 to 10%.
- The smaller the bird, the more reliable the comparative weight in the species accounts will be. In smaller birds there is not a lot of wiggle room. In larger birds, a broader range of weights exists at each stage of life because of differences in individuals (their hatch order and sex) and subspecies.
- If your hatchling weighs less than 5 grams, then you would obviously not consider a hatchling that weighs over 5 grams on d 0 (hatch day).

2. Foot type (arrangement of toes and types of webbing). See Table 2.6 and Fig. G5 (Illustrated Glossary).

3. Mouth. As explained in Chapter 2 (Anatomy), mouth colors are useful in passerines and a few nonpasserines (e.g., hummingbirds and cuckoos). Describe the color of the interior lining and note the size (large or wide) or shape of the gape. Observe gape flange color and thickness and if the swollen tissues are only at the corners or how far they follow the commissure to the beak tip. If the swollen areas resemble knobs on the lower mandible, consider woodpeckers (you should have noted the zygodactyl foot arrangement already). Size of gape flanges can indicate the kind of nest (thicker may be from cavity type). If the chick is a passerine, see Table 2.1 (Mouth Colors). Note any dark markings (like spots) on the upper palate or tongue (not the thin darker shaded areas where the eyes are located). If there are unusual structures on the tongue and upper palate, look at cuckoos.

4. Beak size, shape, and features. See Tables 2.2 and 2.3 and Fig. G3 (Illustrated Glossary). *Relative to the head*, is the beak shorter (short), same as (medium), or longer (long)? Is it tiny, small, thin, or big (like a grosbeak)? What is the shape? Describe any other attributes such as a decurved culmen. Distinguish between a straight commissure (as in a flycatcher) or an angulated commissure (as in a finch or sparrow). Features of beaks in the family descriptions in the species accounts can be diagnostic for young birds once the rhamphotheca has begun to harden.

5. Age assessment. Based on down or juvenile feathers, see Table 3.3. Based on behavior, see Table 3.2. Baby birds in a nest may not seem like they are doing much other than begging for food, but as they develop, they actually perform quite a few behaviors that can help with assessing their age. Predictable developmental events: when they can hold up their head, start preening, stretch their legs and wings, and begin flapping their wings. These behaviors are important functions for one reason or another. Preening and scratching are critical for defense against ectoparasites as well as breaking open feather sheaths to assist with molt.

6. To what family or group does the bird belong? At this point you may be able to place your bird in a family or group of birds. Use the characteristics you've noted, go to the Species Accounts, and find a family, group, or possibly a species or two that fit. For passerine birds, mouth color will eliminate almost half of your choices. Within the selected family or group, continue your exam using additional characteristics below to confirm your finding.

7. Down color and appearance. For precocial species of newly hatched chicks, note the color of down, conspicuous dorsal pattern, beak shape, and details of the tarsus and toes. More dorsal patterns are provided in other references mentioned previously. For altricial species with down, use Table 2.11. When assessing down you will note color, amount, location, areas where down is concentrated, and color pattern of down, if any. Note the texture of down; for example, nestling pigeons and doves have down that appears "stringy."

8. Feather attributes. These would be the juvenile feathers. The pattern and order of feather growth can be an important feature to note.

Colors. See "Molt" in Chapter 2 and the chart of colors in Appendix A. For altricial species: feather color in nestlings is first noticeable at the feather tips. Fledgling plumage descriptions in the species accounts are of fresh feathers that recently emerged from sheaths during the nestling phase. Is the bird uniformly light or dark or strikingly two-toned (e.g., nuthatch or Black Phoebe)?

Patterns or features that stand out. These would be stripes, streaks, spots, speckling, or mottling on the chest, belly, or dorsal areas. Older nestlings have distinctly speckled breasts or some feathers with the same color as the adult of their species. Does the plumage resemble the bark of a tree (see owls or nightjars)? Is the coloring disruptive, such as rings on the neck or back?

Head. See "Head Shape" in Chapter 2. Describe the overall shape and size with attention to the forehead. Is there a conspicuous facial pattern

(Fig. G9, Illustrated Glossary), eyelines (above eye, through eye), and eye rings, arcs, or "spectacles"? Is the crown striped? Head feathers may be crested such as in flycatchers, Steller's Jays, Blue Jays, and titmice. Is the face bald, especially around the eyes (blackbirds, some sparrows, members of the cardinal family)?

Upperparts (dorsal surface). What is the overall color? Start at the crown and work down to the tail. Is there a rump patch?

Underparts (ventral surface). Begin at the chin and work down to the underside of the tail. Include side, flank, and underwing surface. Note markings or mottling. For example, the young of American Robins and other thrushes have conspicuous spots on the throat and chest, and many sparrows have mottling or streaks on the undersides. Are the undertail coverts a different color?

Wings. Are they rounded or sharply pointed? Is the length relative to the body longer or shorter than the tail? Partly or completely in sheaths? Wingbars can help distinguish many warblers, vireos, and flycatchers. Are there one or two; are they thin, bold, defined, or indistinct; and what is the color? What is the color of the edges of the wings? Any spots or barring? Wing patterns are important in waterbirds and shorebirds.

Tail. See Table 2.10 for shapes and characteristics. How long are the tail feathers relative to the body? Are the feathers still in sheaths? Is the tail deeply forked as with a Barn Swallow, square-tipped like an accipiter, pointed like a Mourning Dove, or long and graduated like a magpie? Are the edges notched like finches? Are outer feathers all or partly white, with patches, spots, or patterns (kingbirds, gnatcatchers, wood warblers, and some sparrows)? For tails with stripes, consider wrens. Many young raptors have distinct stripes in their tails (more so than adults).

9. Tarsus, toes, nails, and scales. See Tables 2.5, 2.6, and 2.7 and Fig. G5 a–i (Illustrated Glossary). Colors can be helpful but are variable. What is the length of the tarsus relative to the middle toe with the nail? Are toes small, weak, or strong, and what is the length relative to the leg?

10. Skin color. See "Skin" in Chapter 2 and Table 2.4. Reminders: skin color is highly variable, is influenced by underlying tissues, changes daily, and is subject to age and health.

11. Eyes. The color of eyes of young birds is often different than the color in adults, and eye color may differ by sex (e.g., Bushtit females have pale eyes, males have dark eyes). Are the eyelids closed, partly open (slits), or fully open? In general, for passerines, eyelids are closed the first 3–4 days post-hatching, partly open days 3–5, and fully open days 5–7 or longer for bigger birds. Eyelids of some altricial species are open at hatch (Table 3.1).

12. What is the bird's behavior? Some behaviors are indicative characteristics of the family, genus, or species. These kinds of behaviors would include posture, foraging behavior, and flight style. While perched, does the bird wag, flick, or bob its tail? Does it hold its tail upright like a wren or down like a flycatcher? Does it weave or shiver while begging (like some sparrows)? The way a baby bird evacuates from the nest can be a supporting trait for age and identification: finches create a "poop wreath" of fecal sacs around the edge of the nest, bushtits and chickadees point their bottoms straight up, by nine days old swallow nestlings can back up to the opening to evacuate away from the nest, and phoebes back up over their nest and wag their tails when they poop. Many passerine cavity nesters do a flip or "somersault" when they evacuate. Nestling hummingbirds shoot liquid feces out of the nest but away from the edge so that the nest remains clean.

13. Vocalizations. Nestlings give species-specific auditory clues in the form of feeding calls as they beg for food. At first, they may make no sound when they gape, or the calls may be weak *peeps*. As the birds grow, begging may become very loud and intense as the young compete to be the first to be fed. Like colors, sounds can be subjective to the listener. A *chirp* (single note, melodic, or barking-like) to one person might be a *cheep* to another. Sounds of chicks and fledglings of many species have not been well documented. However, experienced wildlife rehabilitators who have cared for members of the blackbird family, grosbeaks, and swallows are able to identify chicks by their begging calls. A video recording of a chick while vocalizing is invaluable. Descriptions of some sounds are noted in the species accounts. Sounds can also be found online at such sites as allaboutbirds.org and xeno-canto.org.

SPECIES ACCOUNTS

INTRODUCTION TO THE
SPECIES ACCOUNTS

The order and the arrangement of families and species in this book follow the 7th edition of the *Checklist of North and Middle American Birds* (American Ornithological Society), including revisions through the 61st supplement (Chesser et al. 2020). The taxonomic position of birds is constantly being updated, so changes may be expected after the publication of this guide. The 62nd supplement was published in July 2021, as *Baby Bird Identification* was in production. Significant changes were made in that supplement to the order of passerine families that are not reflected here. The accounts in this book are divided into two sections: (1) 172 nonpasserine species and (2) 231 passerine species. With a few exceptions, subspecies are not included.

Descriptions of the family are shared characteristics most useful for identification, summarized with respect to adult size, appearance, diet, nest, eggs, and general features of the young of the family. Sizes (small, medium, or large) are generally relative to other species within the same order or family of birds. Most adult weights are from Dunning 2018 and allaboutbirds.org. Fledgling plumages are described for birds recently fledged with fresh juvenile feathers.

Descriptions of the species include adult mass, length, diet, and nest type, and as many key features of the young as were readily available in published and unpublished resources. There is a need for more data in some cases because the information may represent only a few individuals. In other cases, information may be purely anecdotal, is over a century old, or is missing because some species are highly secretive or nest sites (e.g., cavities or cliffs) are too inaccessible to study. Additional species, photos, illustrations, diagrams, and rehabilitation notes can be found on my companion website: www.BabyBirdIdentification.com.

This is a guide about identification of a bird in hand, therefore not included are details related to habitat, locations, or ranges; specifics on breeding behavior and nesting; or voices of adults. These topics are well covered in many other identification guides and the references listed below. Not every species description has a matching illustration, and some species may have only one stage of life represented. Illustrations on the plates are not life-size; however, sizes are approximately relative to the individuals on the same plate. Colors are defined in the table in Appendix A.

The main sources used for species accounts were the Cornell Lab of Ornithology's *Birds of the World* (BOW) online; the electronic book collection of *Life Histories of North American Birds* by Arthur Cleveland Bent; *Nests, Eggs, and Nestlings of North American Birds* (Baicich and Harrison 2005); *Identification Guide to North American Birds*, Part 1 (Pyle 1997b); *The Sibley Guide to Birds*, 2nd ed. (Sibley 2014); *The Downy Waterfowl of North America* (Nelson 1993), popular field guides, books, and journal articles (see Literature Cited). Other sources used were verifiable photo documentation, unpublished data from colleagues in wildlife rehabilitation, and personal data from preserved and live specimens in the wild, in rehabilitation, or in museums.

KEY TO THE DESCRIPTIONS

Size: Indicated as small, med (medium), or lg (large), generally relative to other species within the same order or family of birds.

Length (of bird): Unless otherwise noted, length is *total* length, given as millimeters (mm) or centimeters (cm), and inches (in.), and is provided as a range, an average, or according to the sex. Total length is beak to tail tip (tip of the emerging pin feather) or in nestlings prior to rectrix growth, the posterior-most part of the body.

Weight (mass): Given in grams (g) unless noted as kilograms (kg), and given as a range, an average (avg = mean), or an estimate (est.).

Diet: A summary of adult diet, and when known, diet for young is provided.

General abbreviations: G&D = growth and development. MC = mouth interior color. GF = gape flange color and description. LF = legs and feet (description of the exposed parts of the tarsus and toes). Voc = vocalizations. sp = species. ssp = subspecies. prom = prominent. mod = moderate.

Age or time period: d = day or days, d0 = hatch day (1st 24 hrs after hatch), d1 = next day after hatch, etc. wk = week, wks = weeks.

Stage of life and sex: H = hatchling. N = nestling. Fl = fledgling. Juv = juvenile. 1st yr = first year. Ad = adult(s). M = male. F = female.

Feather development: PinF = pin feather(s), the sheath (also called feather shaft or blood feather) that contains the juvenile feather. PinVis = pin feathers visible under skin; as they grow, the area becomes pigmented and little bumps (papillae) appear. PinEm = pin feathers that have emerged (poked through skin). PinUn = pin feathers that are unsheathing, usually at the tips first. FF = fully feathered.

Feather tracts: See Table 2.8. and Fig. G6 a (Illustrated Glossary).

Feather terminology: Wings: rem = remex or remiges (primaries and secondaries); prim = primary feather or primaries; p = primary (p6 = primary no. 6); sec = secondaries; cov(s) = covert(s); terts = tertials. Tail: rect = rectrix or rectrices (tail feathers); r = rectrix (r1 = rectrix no. 1).

Plumage descriptions: See also Fig. G8 (Illustrated Glossary). Above = upperparts, dorsal areas (dorsum): forehead, crown, sides of head (eye and facial features), nape, back, mantle, scapulars, upperwing feathers (lesser, median, greater and primary coverts, alula), rump, uppertail coverts, and dorsal surface of tail. Below = underparts or ventral areas: chin, throat, breast, belly (abdomen), sides, flanks, undertail coverts, crissum, vent, and undersides of wing and tail. DefB = Definitive Basic (nonbreeding) plumage; in first-year birds it is the first adult plumage following juvenile plumage.

Location: States of the United States and provinces and territories of Canada are abbreviated. MVZ = Museum of Vertebrate Zoology, University of California, Berkeley, CA. CAS = California Academy of Sciences, San Francisco, CA.

Cardinal directions: n, s, e, w, c = north, south, east, west, central.

Similar Species (sim sp): Mentioned where nesting ranges are sympatric (two related species or populations existing in the same geographic area).

NONPASSERINE BIRDS

GEESE, DUCKS, AND OTHER WATERFOWL

(Family: Anatidae, Order: Anseriformes)

All species are aquatic swimmers. Mallard ducklings and Canada Goose goslings are the most commonly encountered waterfowl species in North America because of their close association with humans (nesting on or near human dwellings). Occasionally found are ducklings of Wood Duck, Common Merganser, and other waterfowl species that have been left behind or separated from their families. The most reliable characteristics for the identification of chicks in the Anatidae family are dorsal patterns of down, eyelines, cheek patches, tibial patterns, bill and foot structure, egg teeth, and vocalizations. Color of down is variable and not as reliable as other features for identification. **Adults.** Small to very large waterbirds. Measurements vary geographically, seasonally, and by sex (males are larger than females). Diet: omnivorous, herbivorous, and piscivorous (some duck species such as mergansers). Nest: varies, some in cavities (including man-made boxes), on cliff ledges, ground, or shallow burrows. Eggs: elliptical (or subelliptical); plain off-white, buff, or pale olive; unspotted. Bill: broadly depressed or narrowly compressed (except two subfamilies), may have a nail-like hook, and either lamellate (filtering ridges) or serrate (jagged) edges. Plumage: well covered with down and oily feathers; lores are feathered or bare, colors vary from brightly colored males to females and young that are in shades of browns and grays for camouflage. Wings: long, pointed, or rounded (whistling-ducks). Tail: usually rounded and short, except the stiff-tailed ducks (Masked Duck and Ruddy Duck). Legs/feet: palmate (or semipalmate in some diving ducks); three toes facing front are webbed; elevated hallux that may or may not be lobate; short legs; tarsi reticulate (swans, geese, whistling-ducks) or scutellate in front (other ducks). **Young.** Precocial, mobile, self-feed shortly post-hatch. Bill: similar shape as adult but smaller, tip rounded, egg tooth may remain several days post-hatch. Down: dense, fully covered, of various colors, patterns define family to which species belongs. Dabbling ducks: down mostly brown above, yellow below, yellow face and dark eyeline. Diving ducks: more uniformly dark than dabbling ducks with more boldly contrasting colors. Cygnets (swans): pale and more uniformly colored. Goslings: lack more clear-cut patterns of ducklings. Feet: variable in color, even within species. Leave nest: shortly after hatching, remain with female (ducks) or in family units (geese and swans) until they attain flight.

Domestic Ducks and Geese: Bills broad, flat, often pink or orange, though not exclusively. Most varieties/hybrids of domestic geese and ducks, except the Muscovy, originated from the Graylag Goose and Mallard duck.

Plumage colors of domestic ducks and geese vary but usually solid yellow ducklings are domestic. Domestic Mallards and Muscovy may become feral if they escape or are dumped.

Cackling Goose (*Branta hutchinsii*)

Cackling Geese raise their young on northern tundras & winter in parts of the US. **Adult.** Small goose (short neck, short bill), 1398–2380 g, 63–65 cm (24.8–25.6 in.). **Gosling.** Bill: H pale blue-gray to blackish, egg tooth light colored & prominent. Down: thick; above pale & bright yellow with olive pattern, crown patch round & dark; below yellow. Down appears grayish with incoming 2nd downy coat. LF: vary, pale grayish to olive-gray, blackish-gray, or black. Iris: blue-gray. G&D: d0: 68 g (*B. h. minima*), mass varies according to Ad body size of ssp. Flies wk 6–7.

Canada Goose (*Branta canadensis*) Plate 1

Adult. Large to very large goose, 3000–9000 g, 76–110 cm (30–43 in.). **Gosling.** Leaves nest 1–2 d post-hatch. Survives on yolk sac 2 d before becoming self-feeding. Bill: H blue-gray to blackish, gradually changing to black (in fall), tip rounded; egg tooth remains several days. Down: little patterning, lemon-yellow to light olive with darker brown patches on back; older goslings have yellowish-green hue. LF: vary, pale grayish to yellow-green, olive-gray, blackish-gray, or black (after 1st fall). G&D: mass varies according to Ad body size of ssp: *B. c. maxima*: range 80.5–134.8 g; d20: rem & rect PinEm. Voc: single strong *peeps* or warbling *trills*. **Juvenile.** Dirty gray; cheek patch evident; downy crown feathers disappear just before 1st flight, 6–7 wks in smaller races, 8–9 wks in larger. Mass: 1st year juv 1.5–3.0 kg. Remains with family 1st year. *Can be reunited with own family or introduced into foster family if gosling is close in size to foster siblings. Observe to make sure gosling is eating & swimming with foster family.*

Mute Swan (*Cygnus olor*)

Introduced, exotic, larger than two native sp, Trumpeter (*C. buccinator*) & Tundra Swans, in N.A. Nesting increasing within urban parks. **Adult.** M 10,200 g, F 8400 g. **Cygnet.** Moves about, feeds, swims soon post-hatch. Down: gray or white morph. *Gray*: juv brownish, LF slate-gray, bill slate. *White*: remains white as Ad, LF pinkish-tan, bill tan. Lores: feathered. G&D: d0: 225 g, d42: juv plumage appears, flies 120–150 d.

Tundra Swan (*Cygnus columbianus*)

Adult. M 7200 g, F 6300 g; 125 cm (49 in.). Eggs: elliptical, ovate & creamy-white. **Cygnet.** Bill: dull pink with whitish or yellowish egg teeth. Down: completely covered, light gray above, whiter below; crown, face & hindneck silver-gray; darker back & rump; 2 white patches on shoulders & wings. Face: tinged pale yellowish-gray or brownish-pink. LF: sides of toes light dull pink & grayish-blue. G&D: (AK) d1: 170.5–189.5 g.

Muscovy Duck (*Cairina moschata*)

Raise young mainly in Mexico, but wild populations have expanded into lower Rio Grande valley & parts of TX. Domestic varieties occur in urban areas throughout N.A. Distinguishing feature: lumpy red facial skin & swollen knob on top of bill. **Adult.** 1990–4000 g, 66–84 cm (26–33 in.). Nest: cavity in tree or box, occasionally on ground near water. **Duckling.** Bill: gray-brown, stout, hooked downward. Down: brown on head & dorsum, yellow forehead (some) & undersides. Markings: boldly patterned; dorsal spots & wing patches yellowish & prom; thin, dark eye-line to nape. LF: heavy tarsi, yellowish- & grayish-brown. **Juvenile.** Wing patches smaller than Ad. Tail: long, fanlike. Also: long body with long down on back of head like a "mane."

Wood Duck (*Aix sponsa*)   Plate 1

Adult. 454–862 g, 47–54 cm (18.5–21.3 in.). **Duckling.** Moves 2.5–4.8 km within 2 d post-hatch. Diet: invertebrates 2–3 wks, vegetation when older. Bill: gray-brown upper, edge light yellow; yellow-pink lower; nail reddish, pale pink toward tip; egg teeth ivory, lower bilobed. Down: dark brown or blackish-olive above, buffy-yellow below & on face; crown deep brown; brown stripe over eye does not continue to bill (as in Mallard). Light yellow spots on shoulders & rump (6–7 mm). Some may have yellow or brownish-yellow wing patch. Tail: very dark. LF: brownish-yellow (or black) tarsus; boldly patterned with black webbing between toes. Iris: M turns red 60 d. G&D: d0: range 19–28 g; d20: 1st rect appear; d30: wing cov PinEm, breast/belly PinUn; d40: prim PinUn, crown PinUn; fly 8–10 wks. Voc: alarm call high-pitched *peep. Chicks can be highly stressed in captivity; need companions.*

Gadwall (*Mareca strepera*)

Adult. 500–1250 g, 46–57 cm (18–22.4 in.). **Duckling.** Bill: narrow with high bridge, upper bluish-gray or bluish olive-gray, lower pinkish-yellow. Down: base color pale yellow or cream-buff, sepia above; sides of head buff. Eyeline may be thin & dark, sometimes wide & grayish. Spots of cream-buff (sizes vary): 2 on shoulder, 2 larger on rump. LF: grayish-brown & dull orange. G&D (M > F): d0: 28–33.5 g; d17: 187; d33: 380–407; d50: fly; (ND) 553–606 g.

Mallard (*Anas platyrhynchos*)　　　　　　　　　　　Plate 1

Adult. Medium-size duck, 1000–1300 g, 50–65 cm (19.7–25.6 in.). Diet of young: < 25 d, mostly animal foods; 20–30 d strains bottom substrate & feeds in vegetation for seeds. Nest: on ground with cover, preferably near water. Eggs: Unmarked, creamy-grayish or greenish-buff. **Duckling.** Leaves nest 13–16 hr after last-hatched; hen leads to water (not always to nearest pond). Can self-feed. Bill: dark olive-brown. Down: long & dense; dorsum sepia with buff-yellow patches; orangish-yellow chin, throat & sides of head; forehead, crown & back of head dark brown; eyeline (from base of bill through eye to back of head) & cheek patch dark brown; below buff-yellow. LF: grayish-brown with orangish patches. G&D: d0 (MI): 32–34 g; d7: 66; d14: 72; weight doubles wk 1, quadruples wk 2, full size 4–6 months; d52–70: flies. Voc: quiet *pips* (contentment), loud *peep* (distress). Care: F only, brooded 14 d post-hatch.

Green-winged Teal (*Anas crecca*)

Smallest N.A. dabbling duck. Bill narrower, shorter than most dabblers. **Adult.** 140–500 g, 31–39 cm (12.2–15.3 in.). **Duckling.** Leaves nest: few hrs post-hatch. Down: dark olive-brown above, yellow below. Head: 2 horizontal dorsal stripes (dark brown) through eye & along cheek; cheek stripe usually well defined but variable, dusky face with darker large ear-spot; dorsal spots small to large or linear. Head: squarish. Distinguish from Blue-winged & Northern Cinnamon teal ducklings: narrow nail, dark forehead & greenish-gray tarsi & toes. G&D: d0–1: ~15 g. Care: F only.

Hooded Merganser (*Lophodytes cucullatus*)   Plate 1

Smallest of three N.A. mergansers. **Adult.** 453–879 g; 40–49 cm (15.7–19.3 in.). Nest: in cavity. **Duckling.** Leaves nest within 24 hr, self-feeds, dives. Bill: brownish-slate upper, yellowish-pink to light orange lower, reddish-brown nail (pink at tip). Down: brown dorsum & upper breast; white, buff, buffy or rufous cheeks; grayish spot on each side of back & base of each side of tail; below: white throat, lower breast & belly. LF: grayish-olive, olive, or brownish-gray webs, except inner web (toe 2) yellowish. Iris: yellowish-brown. G&D: d0 (MO): 31 g, 74 mm (3 in.); d70: flies.

Common Merganser (*Mergus merganser*)   Plate 1

Adult. 900–2160 g, 54–71 cm (21.3–28 in.). Nest: in cavity. Bill: sharply serrated. **Duckling.** 46.2 g. May remain in nest 1–2 d post-hatch. Rides on back of hen. Bill: brownish-gray, slender, relatively shorter than Ad, not ducklike. Down: head dark brown to black, white stripe below eye, white cheeks, rufous "patch" on neck, tawny-brown patch over eye with dark streaks from bill to below eye, white undersides, flanks & dorsum dark brown, white patches on rump & wings. Crown: more rounded & appears larger than Red-breasted Merganser. LF: olive-brown or yellowish-gray. Fly: 65–70 d post-hatch. Care: F alone, independent ~5 wks post-fledging.

Ruddy Duck (*Oxyura jamaicensis*)

Adult. 300–850 g, 35–43 cm (13.8–17 in.). **Duckling.** Leaves nest soon post-hatch, able to swim, dive & catch food within 24 hr. Bill: grayish-brown, spatulate, high bridge, narrow nail. Down: dark crown, dorsum dark gray-brown with pale spot each side of back, undersides white, white line nape to bill, cheeks & throat dotted brown & divided by white line. LF: gray. G&D: d0 (s.w. MB, Canada): 38–60 g; d14: rect PinF evident; d21: wing PinF evident; d42–49: flies. Care: by F only.

QUAIL, NEW WORLD

(Family: Odontophoridae, Order: Galliformes)

   Highly terrestrial and gregarious. Legs and feet adapted for walking and running. **Adults.** Small to medium-size relative to other Galliformes. Diet: pecking & picking of seeds & other vegetation such as forbs, grass, leaves, flowers, fruit, nuts, roots, tubers, and invertebrates (mostly insects).

Nest: usually on ground, occasionally low in bush/tree, shallow depression of grasses and leaves. Eggs: oval to subelliptical or short pyriform, in colors of white, cream, buff, or reddish; some spotted reddish or brown. Bill: chicken-like shape, stout, decurved culmen, sharply pointed, nostrils exposed. Plumage: most with crests called a "topknot" in California Quail and Gambel's Quail and a plume in Mountain Quail. Wings: short, concave, rounded. Tail: short, rounded. Legs/feet: anisodactyl, featherless, tarsi scutellate, nails obtuse. **Chicks.** Distinguishable features of chicks are size (tiny to very small) and the dorsal pattern of down. Precocial, mobile, self-feeding within hours post-hatch. Make short flights within 14 d. Bill: like adult but small. Down: fully covered except legs, pattern varies by genera, crests may appear at age 1–2 days. Juvenile plumage begins emerging within days, functional flight feathers by day 7. Biparental care.

Mountain Quail (*Oreortyx pictus*)   Plate 2

Adult. 210–262 g; 26–31 cm (10–12 in.). Unique head plume: 2 long, black, straight feathers. **Chick.** Young leaves nest shortly post-hatch, self-feeds within hrs. Bill: olive & cream. Down: mixture of buff, cream & sepia above, grayish or yellowish-white below; pattern like Northern Bobwhite but paler & with three buffy stripes on the back. Irregular central stripe running crown to tail is chestnut edged with black & bordered with buff. LF: beige. G&D: d0: 8.5 g (captive); d14: can fly; d40: 100 g.

Northern Bobwhite (*Colinus virginianus*)   Plate 2

Adult. (FL) 161 g, (MA) 233 g; 25 cm (9.75 in.). **Chick.** Bill: H pinkish white, turns dusky or reddish brown & lighter at base of lower mandible. Down above: chestnut streak behind eye, head & back chestnut & blackish-brown with 2 buff-yellow stripes along back; below: drab-gray, chin paler, sides browner. LF: pinkish-white; wk 15 turn dusky or reddish-brown with lighter joints; juv yellowish-brown. G&D: d0: 6–8.6 g, gains ~0.5 g/d until d10 (nearly Ad mass wk 15); d1–2: prim & sec PinEm; d2–3: egg tooth gone, PinEm alar, humeral, femoral & ventral; d11–13: capital PinEm; wk 6–7: FF, sexes distinguishable. May remain with Ad through 1st winter.

Scaled Quail (*Callipepla squamata*)

Adult. 177–191 g; 25.4–30.5 cm (10–12 in.). **Chick.** Head cinnamon-buff to pinkish-buff with wide chestnut stripe from midcrown down to hindneck (edged black & bordered white); small grayish tuft; dark chestnut auricular spot; back mottled buff & chestnut between 2 pairs of blackish lines bordered by 2 whitish stripes. Below: chin & throat whitish, breast & flanks grayish-buff, paler on belly. G&D: d0: 7.5 g; d35: 57 (avg).

California Quail (*Callipepla californica*) Plate 2

Adult. M 140–230 g, F 147–186; 24–27 cm (9.4–10.6 in.). Topknot: 6 plumes, black, "comma shaped," forward facing. **Chick.** Leaves nest immediately, can forage, brooded 2 wks. Bill: gray & very short (chicken-like). Down: fully covered, buff-white background, rusty on back, white below; large dark brown patch back of head with buff border; pair of black stripes bordered by three buffy-yellow stripes running longitudinally on dorsum; dark auricular spot. Prim PinF 1 through 7 emerge at hatch. LF: dark gray, hallux short. Iris: darker brown than Ad. G&D: d0: ~5 g; d2: 6.1 g, gains 2 g/day 1st 90 d, reaches Ad mass 14–15 wks post-hatch; d14: flies short distances. Topknot: raised tuft, then at ~6 wks short, brown plumes evident. Voc: *trill* call after hatch, then high-pitched *peeping, pseu-pseu* when distressed. *Hybridizes with Gambel's Quail where sympatric in s. CA.*

Gambel's Quail (*Callipepla gambelii*) Plate 2

Adult. 160–200 g; 25 cm (9.8 in.). Similar to California Quail (CAQU), ranges overlap in parts of CA & Baja California. **Chick.** Soon after hatch, can walk & self-feed. Bill: pinkish-coral. Down: pattern similar to CAQU but more yellow. Above: dark brown back, flanks & wings; back with three longitudinal yellowish-buff stripes; dark topknot; rufous-brown crown with yellowish-buff sides, cheeks & nape; black mark runs from base of bill through eye becoming wider & curves downward. Below: buffish-white. G&D: d0: 6.7–8.0 g (3 chicks, incubator hatched), mass of wild-hatched chicks may be similar to CAQU; d10: flutter-fly. Biparental care until F begins second brood. Hybrids: see "California Quail."

TURKEY, GROUSE, AND RELATIVES

(Family: Phasianidae, Order: Galliformes)

A very diverse family. Members are highly terrestrial, some are secretive, many form small flocks, some form leks, most are nonmigratory. Legs and feet are well developed for running. Toe arrangement: anisodactyl (except in grouse may be slightly webbed), hallux reduced and elevated. Most distinguishing features of young: cryptic pattern of down and features of the feet in some groups. **Adults.** Small to very large Galliformes: 12.5 in. (Gray Partridge), 46 in. (Wild Turkey). Diet: mostly herbivorous; some take invertebrates from vegetation or ground; vegetable matter is leaves, flowers, fruits, nuts, tubers, etc. Nest: scrape in ground concealed in vegetation. Eggs: subelliptical to oval; white, cream, or olive colors with no markings, or brown with dark spots. Bill: relatively short, stout, not pointed or sharp; nostrils partly feathered (grouse, ptarmigan) or exposed (turkey, pheasants). Plumage: head may be bare or with combs or wattles. Flight feathers: 10 primaries, 18–19 secondaries, and 16 rectrices. Wings: medium length, rounded, with stiff primaries that may be curved. Tail: variable, short and squarish to long and pointed in pheasants, or large and fan shaped in turkeys. Legs/feet: strong, heavy, some entirely feathered (grouse and ptarmigan); hallux elevated, some have spurs (pheasants, turkey). **Chicks.** Precocial; mobile, able to forage soon post-hatch. Bill: like adult but small. Down: fully covered, dense, shorter on head, boldly patterned, tarsus bare. Primaries may be formed at hatch, fully functional flight feathers 7–14 days old; flutter short distances within a few days. Care: highly variable.

Wild Turkey (*Meleagris gallopavo*)   Plate 2

Adult. 2500–10,800 g, (NY) M 6100–10,400 g, F 3200–6500 g; 110–115 cm (43.3–45.3 in.). Bill: orange shading to yellow on lower. **Chick.** Leaves nest soon post-hatch, young fed by hen 1st few days. Bill: chicken-like, bulky, pale pink-buff lower mandible, upper darker pink. Down: various "earthy" colors; background of pinkish-buff, tawny & gray, with various brown blotches. Back: large blotches Brussels brown to Vandyke brown, smaller markings on forehead, crown & nape; sides of head pinkish buff to yellow. Below: near white to pale pinkish-buff chin, upper throat & breast; gray blotches on thighs. LF (Fig. G5 e, Illustrated Glossary): H buffy yellow, turn vinaceous pink wks 3–5. G&D: (*M. g. silvestris*) 46–48 g, wings develop quickly (prim PinF nos. 1–7 are 6–12 mm at hatch); d7–14: short flights, roost in trees; d11: rect evident; d21–27: breast & back PinUn; d42: FF (except head & neck). Fully grown: several months. Voc: soft *purr* call is common; three types of *peep* calls (single, double & multiple notes). Care: F only.

Ruffed Grouse (*Bonasa umbellus*)   Plate 2

Medium-size grouse. **Adult.** 450–750 g, 40–50 cm (16–19.7 in.). **Chick.** Bill & LF: pink. Down: head & nape rust-red with large dark blotch behind eye. Above: reddish-brown to sandy-brown; pale buff below. G&D: d0: 11–13 g, wing feathers have emerged; d5–7: short flights. Juv: plumage similar to Ad F but lacks subterminal tail band, head less distinctly marked.

Spruce Grouse (*Canachites canadensis*)   Plate 2

Small grouse; shares similar habitats to Sooty & Dusky Grouse, but half the size. **Adult.** (WA) M 400–560 g, F 370–513 g; 41 cm (16 in.). **Chick.** Down above: head straw-yellow, rufous crown edged black tapering to nape, irregular sepia eyeline; back mottled rufous & straw-yellow (anteriorly). Below straw-yellow or pale whitish-buff. LF: toes unfeathered. G&D: d0: 15 g (avg); d6–8 flies weakly. Remains with family 70–100 d.

Greater Sage-Grouse (*Centrocercus urophasianus*)   Plate 2

Very large grouse. **Adult.** M 1700–2900g, F 1000–1800 g; 56–71 cm (22–28 in.). **Chick.** Down: mottled & spotted with black, brown, buff & white. Head: 2 brown (edged black) blotches on sides. Below: breast buff with black mottling. G&D: d0: 30–31 g; d10: flies weakly, stronger by 5 wks.

Dusky Grouse (*Dendragapus obscurus*)

Adult. (WA) M 1212 g, F 905 g; 51 cm (20 in.). **Chick.** Bill: tan, black at base. Down: above light brown with black & reddish-brown mottling; below buffy-white to yellowish with black markings. G&D: d0: (MT) 19.4 g, (e. NV) 22.7 g; prim PinF (1–7) present; d8–9: flies short distances; d28: in juv plumage. *Sim sp: Greater Sage-Grouse.*

Sooty Grouse (*Dendragapus fuliginosus*)

Endemic to mountains of w. N.A. from s. coasts of AK & n.w. BC (Canada), s. to the Sierra Nevada Mtns in CA & w. NV. Nostrils feathered. **Adult.** 750–1300 g, 40–50 cm (15.8–19.7 in.). **Chick.** Leaves nest 12–24 hrs, self-feeding. Bill: horn, blackish lower mandible. Down above: black & reddish brown. Undersides & sides of head buff to yellowish, auriculars with black markings. Juv plumage similar to Ad F but crown more reddish-brown with whitish streaks on shafts of back feathers & scapulars. G&D: d0: 25.8 g, PinF of prim (nos. 1–7) evident; d16–17: PinUn; d6–7 can hop-fly; d8–9: short flights; d14: longer flights. Care: F only.

Sharp-tailed Grouse (*Tympanuchus phasianellus*)

Medium-size grouse. **Adult.** 596–1031 g; 41–47 cm (16–18.5 in.). **Chick.** Down above: head & nape sulfur-yellow, pale pinkish-buff tinge on nape, black spots on crown & auriculars, straw-yellow back with black blotches or streaks. Bright straw-yellow below (varies by ssp). Bill: pinkish-horn with dark stripe/spot on culmen base. G&D: d0: 12.8 g; d7–10: flies short distances; fully grown 12 wks.

Gray Partridge (*Perdix perdix*)

Adult. (ND) M 356–483 g, F 298–472 g; 30.5–33 cm (12–13 in.). **Chick.** Mobile within hrs. Down: back brown & buff with 2 black stripes, wings rufous; below yellow buff to gray buff. Juv plumage: dull pale brown, streaks & spots darker than Ad, gray-brown prim & sec, gray spot behind eye. LF: yellow brown. G&D: d0: < 10 g, d14+: short flights.

Ring-necked Pheasant (*Phasianus colchicus*)   Plate 2

Adult. 500–3000 g; 50–70 cm (19.7–27.6 in.) including tail. Nostrils exposed, tail long & pointed, feet spurred in M. **Chick.** Leaves nest with hen shortly post-hatch. Bill: short, chicken-like, pinkish-horn with black patch at base of upper beak. Down above: color varies, pale tawny to yellow-buff; brow cinnamon or tawny; central blackish stripe from forehead to crown (forms small patch) with two lateral black stripes midcrown to nape, small black wavy markings at ear covs, black central stripe down back to rump, black patches on wings & rump on each side; below: buffy-yellow or cream. LF: pink-white, become browner, hallux short. Iris: dark brown. G&D: d0: 18.5 g (avg), 21–25.4 g (5 incubator hatched); d7: ~7.6 cm tall; d12: short flights; d30: 135 g. Stay with Ad F 10–12 wks. Voc: contentment call *ter-it* or *ter-wit*; loud caution call *terreep* or *turreep*.

Indian Peafowl (Peacock) (*Pavo cristatus*)

Some feral populations exist in parts of CA, FL & HI, nonmigratory, mostly dependent on humans. **Adult.** M 4000–6000 g, F 2750–4000 g; M 110–115 cm (43.3–45.3 in.) including tail, F 95 (37.4 in.). **Chick.** Down: darker above, paler & yellowish below, wing feathers with white tips (d1). Eye features: large, round, dark with grayish ring, dark feathers form eye arcs, upper arc continues as semicircle behind eye. G&D: d0: 120 g (avg), 13 cm (5 in.); d7: juv plumage begins to appear, dark brown dorsum, lighter below. Care: F only, leaves nest by d3, up to 5 chicks only (remaining chicks or eggs are abandoned).

Chukar (Partridge) (*Alectoris chukar*)

First introduced in 1893 from India. Game species. Wild populations exist in suitable habitat. Sexes alike in plumage (F slightly smaller than M). **Adult.** 550–675 g, 34–38 cm (13.4–15 in.). **Chick.** Bill: light horn, tinged pinkish. Down above: cinnamon-buff with sepia sprinkles on crown & nape, 2 dark wide streaks down back with 3 grayish-buff streaks (1 center & either side of dark streaks), 2 narrow irregular black streaks on sides of back & thighs; below: pale buff to off-white; sides: sepia streaks from eye over ear covs; narrow pale line over eye. LF: pinkish-tan. **Juvenile.** Plumage above: olive-brown turns dull brownish-gray, outer rect tinged rufous. No black collar. Flies: 7–10 d. Ad size 60 d.

GREBES

(Family: Podicipedidae, Order: Podicipediformes)

Distinguishing features of young: pinkish or red bare areas on head, head feather pattern, and lobate-webbed toes. **Adults.** Small to medium-size waterbirds; males are larger than females. Diet: dive for small aquatic animals, some vegetable matter, and own body feathers. Nest: wet or decaying vegetation floating on water. Eggs: biconical, dull white or pale blue, unspotted. Bill: straight, compressed and acute (except 1 sp). Plumage: satiny and glistening; above brown or black; paler below; throat and neck may be white or rich reddish-brown in some. Wings: short, narrow, somewhat pointed. Tail: rudimentary. Legs/feet: tarsi laterally compressed, scutellate, posterior edges serrate; four lobate toes, extremely flattened nails; short legs on posterior of body designed for swimming and diving, not walking. **Young.** Precocial, fed by parent at first. Down: boldly striped or streaked head and neck (except Western Grebe, gray with no pattern). Leave nest: after last hatched. Biparental care: sometimes young divided between parents; may be fed up to 10 wks post-fledging.

Pied-billed Grebe (*Podilymbus podiceps*)   Plate 3

Adult. 253–568 g, 30–38 cm (11.8–15 in.). Bill: high, laterally compressed, tip decurved with slight hook. **Chick.** Swims shortly post-hatching. Gape pink to reddish. Bill: white egg teeth, black area on both mandibles about same width as white, extending onto culmen; remainder of bill pink. *1st down*: short, thin, back & sides black with 4 white stripes extending down back; crown with 2 V-shaped stripes. Head & neck: black mottling, extends to white chest. Head pattern variable: cinnamon-rufous crown spot, bar on nape, rufous eyeline; lores bare & yellowish. *2nd down* d7: gray feathers grow in between 1st natal down feathers. LF: black, turn grayish or greenish. Iris: black until d7, then dark brown; eye ring rufous. G&D: d0: 13.3–16.2 g (5 captive), weight gain most rapid d10–25, max mass reached 11–14 wks; d10 breast PinUn; d35–37 flies; d42: down mostly gone. Biparental care, young chicks ride on back of Ad.

Horned Grebe (*Podiceps auritus*) Plate 3

Adult. 300–570 g, 31–38 cm (12–15 in.). **Chick.** Bill: pinkish, 2 bands on upper mandible. Down: above blackish, narrow grayish stripes; below white. Head markings: back of crown with pinkish spot, 1–2 white stripes on occiput, white *V* on forehead. LF: H dark gray. Iris: H pale gray; turns grayish-brown. **Juvenile.** Plumage similar to Ad DefB but duller (brownish) with more mottling or striping, tips of scapulars may have downy remains.

Red-necked Grebe (*Podiceps grisegena*) Plate 3

Adult. 800–1600 g; 43–56 cm (17–22 in). **Chick.** MC: red. Bill: buffy pink with black band around base & blackish bar near tip. Skin: bare patches on lores & crown pale pinkish, turn scarlet when excited. Down: dense, short, fully covered; head pattern varies among individuals, white or buff white with bold black stripes & blotches. LF: gray to black. **Juvenile.** MC: dark gray to black. Bill: yellow base, dark grayish-brown tip. Plumage: brownish-red foreneck, bold black striping on head. Iris: yellowish.

Eared Grebe (*Podiceps nigricollis*) Plate 3

Adult. 200–735 g; 30–35 cm (11.8–13.8 in.). **Chick.** Bill: tawny, or light pinkish to coral; tip whitish encircled with 1 narrow black band, base with jagged-edged black band. Skin: bare patches on crown & lores pale, turn coral or pinkish-orange when excited. *1st down*: crown to below eye black & gray with whitish stripes above & behind eye; below: cheeks, chin & throat whitish, neck white with black & gray stripes, spots on neck sides & backside; back & flanks dark grayish to blackish with some whitish intermixed, some appearing as undefined stripes. *2nd down* d7:

brownish-black above, white below; completed by d10. LF: black, turn blackish-green laterally. Iris: black, then gray (d21). **Juvenile.** Plumage similar to Ad DefB except brownish tinge on back & neck, paler flanks & buffy on sides of neck. Sec form white wing patch. Dusky cheeks, white chin & throat. LF: lateral surface olive-green. Iris: tan with orange tint.

Western Grebe (*Aechmophorus occidentalis*)   Plate 3

Adult. 808–1826 g, 55–75 cm (21.6–29.5 in.). Bill: yellowish-green. Black cap extends below eye. Iris: scarlet with narrow yellow ring around pupil. **Chick.** Bill: black; white egg tooth (upper) corresponds with white spot on lower; these gradually merge with pale bill tip as young reach 156 g. Down: dense, velvety, dark gray dorsum, whitish or light gray on face & lowerparts. Crown: bare, skin is yellow or orange, turns red when begging or when alone; d10–15 turns black. White eye ring & lores. LF: slate or near black with greenish lobes, turn olive-green (juv). Iris: black, then pale gray. G&D: d0: mass unknown, d1: 21.7–36 g. **Juvenile.** Plumage similar to Ad DefB, except generally paler, no dark crown, face & back with gray or brown wash. Biparental care: chicks fed directly at first, ride on back of parents. Independent 6–7 wks.

Clark's Grebe (*Aechmophorus clarkii*)   Plate 3

Adult. 718–1685 g, 55–75 cm (22–30 in.). Bill: brighter yellow than Western Grebe. Black cap is above the eye. **Chick.** Similar to Western Grebe but crown & back paler gray in Clark's; feathers appear snow-white d20–50. LF: pink in some chicks.

PIGEONS AND DOVES

(Family: Columbidae, Order: Columbiformes)

Heavy bodied with small heads and short necks, wings, and legs. Most distinguishing features of young: "hairlike" stringy down and features of bill. **Adults.** 30 g (Common Ground Dove) to 360 g (Band-tailed Pigeon). Diet: mainly seeds and fruit; some sp eat insects; squabs fed a seed-based, high-protein fat secretion (called "crop or pigeon's milk") regurgitated from crop of adult. Nest: platform of loose material in tree, bush, or ledge; sometimes on ground or rock cavity. Eggs: elliptical, pale, no markings. Bills: distinct, relatively small and slender with soft skin (cere) at the base; culmen decurved with the middle constricted. Nostrils: lumpy appearance, usually slits overhung by operculum. Plumage: dense, may be bare around eyes. Wings: long and flat, shape varies. Tail: long and square, rounded, or pointed. Legs/feet: toes cleft to the base, sometimes with slight webbing, hallux slightly elevated or incumbent; tarsus scutellate in front (sometimes feathered on proximal portion), reticulate elsewhere. Irises: variously colored; white, brown, and orange to red. **Squabs.** Altricial. Mouth: variable pink to gray; no gape flanges (do not gape for food). Bill: distinctive "lump" on top near face, leathery, pliable. Skin: varies pink to dark, visible through down. Down: stringy hairlike downy feathers. Plumage (juvenile): color variable, stringy down still poking through feathers, fledglings have scalloped wing coverts. Voc: soft *whistling* or *peeping*.

Rock Pigeon (*Columba livia*)   Plate 4

Adult. 265–380 g; 30–36 cm (11.8–14.2 in.). LF: rose-pink. Irises: vary, yellow, orange, or reddish-orange; grayish-blue orbital ring. **Squab.** Bill: smooth, leathery & pinkish (then darker), light at distal tip, cere pink or dark. Skin: pink, visible under down, may be darker on head; turns darker by d2. Down: long, silky, stringy, whitish to bright butter-yellow on head & body, some may have reddish tint. LF: pink to slate, claws grayish-black. Iris: 6–8 months med brown or grayish-brown. G&D: (KS) 15.2 g (avg) or 11–13.3 g (5 incubator hatched), 5 cm (2 in.), gains 4–8 g/day; d4–5: eyes open; d4–6: contour PinEm; d9: PinEm all tracts; d14: well feathered; d18: walks well; d27: feathering 90% complete. Voc: *peeping* call persistent by 1 wk, d7 circular begging "dance." **Fledgling.** Fledges: varies, 25–32 d (longer in winter). *Differs from Band-tailed Pigeon (BTPI): young Rock Pigeons shorter tailed, stockier, white rump patch may show, dark band at tip of tail (vs. at base in BTPI), entire bill dark (vs. yellow & black in BTPI), feet pinkish (vs. yellow in BTPI).*

Band-tailed Pigeon (*Patagioenas fasciata*)   Plate 4

May breed year-round, primarily in spring. **Adult.** M > F, (CO) 226–460 g, (CA) 300–515 g; 33–40 cm (13–15.8 in.). **Squab.** MC: pinkish-gray. Skin: not described. Orbital ring narrow, variably pink, orange, yellow, or brown (most common); becomes dark pink to light red by 120 d post-hatch. Eyelids red. Down: long, fully covered, yellow-brown, orange-brown, or yellowish-buff. Bill: yellow, distal 3rd black. LF: chrome-yellow or orange-yellow. G&D: d0–3: range 12–40 g; d15: may climb onto branch. Iris: pale yellow to brownish nearest pupil, turning bloodred 100 d post-hatch. **Fledgling.** Fledges: 22–29 d, 213 g (avg).

Plumage (CA ssp, *P. f. monilis*): plumage above brownish-gray with pale edgings (scalloping appearance); lacks markings on hindneck. *Differs from Rock Pigeon (see above): relatively long tail, white collar (in Ad), uniformly gray overall, yellow feet.*

African Collared-Dove (*Streptopelia roseogrisea*)

Introduced, likely escaped pet (sold as Ringed Turtle-Dove), some feral populations established in s. US. Survivability unlikely where winters are cold. **Adult.** 130–166 g, 26–27 cm (10–11 in.). Diet: seeds. **Squab.** Down: creamy buff. **Fledgling.** Paler, duller than Ad. LF: gray. Irises: become yellowish, then orange, then red (Ad). Call: 2-note *Ko-kroooo. Sim sp: Eurasian Collared-Dove has dark gray vent feathers.*

Eurasian Collared-Dove (*Streptopelia decaocto*)   Photo 1

Introduced, rapidly spreading, may have negative impact on Mourning Dove. **Adult.** M slightly larger than F, geographic variation, 140–180 g; 29–30 cm (11.6 in.). Diet: seed & cereal grain, some berries, green parts of plants. White undertail covs, dark-tipped wings, 3-note call *koo-KOO-kook.* **Squab.** Bill: pinkish-tan, then slate-gray. Skin: H pinkish-gray; N orbital ring grayish-white. Down: sparse, cream-colored or yellow. LF: gray, turn brownish-red in juv (dark red or mauve in Ad). Tail: squared, white tipped. Iris: gray, turns brown before fledging. G&D: d0 (one chick, Nevada County, CA): 9.4 g (crop full), length 6 cm; egg in nest with chick was 7.2 g & 3 cm long; d7: almost fully covered with down, PinF on wings; d12: FF, prim nearly full length; d14: same color as Ad; d17: black collar may be lacking or not complete. **Fledgling.** Fledges: d17–18. Iris: brown or brownish-red.

Spotted Dove (*Streptopelia chinensis*)

Introduced from Asia into s. CA & islands of HI (late 1800s). **Adult.** Mass varies seasonally; (CA) 128–194 g; 30.5 cm (12 in.). **Squab.** Bill: grayish-pink, tip darker. Down: sparse, hairlike, sandy to sage-buff, paler below. LF: pale flesh. Iris: gray, inner ring brown. G&D: d0: 6.4 g, gains 5.7 g/d; d6: contour PinEm; d12: FF. **Fledgling.** Fledges 15 d (or earlier). Bill: gray, blacker toward tip. Plumage well developed d30–35. Above: light brown; back, scapulars, covs & rem edged cinnamon to pale pink-buff; prim sepia; crown tinged light gray & vinaceous; spotted neck pattern barely evident; tail rounded, broad white tips. Below: paler. LF: grayish to pink. Iris: light brownish-gray, yellowish-gray, or tan.

Inca Dove (*Columbina inca*)

Adult. (AZ) 33–57 g, 18–23 cm (7–9 in.). LF: pinkish-gray, dark brown scutes. Iris dark red. **Squab.** G&D: unknown. **Fledgling.** Fledges 12–16 d. Bill: paler than Ad. LF: dark brown. Plumage similar to Ad but paler above, subterminal bar on tertials & wing covs absent or grayish-buff. Iris: pale yellow.

Common Ground Dove (*Columbina passerina*)

Adult. 28–40 g, 15–18 cm (6–7 in.). Iris: dark red, orange, or pinkish. **Squab.** Skin: dull gray. Down: tan or gray, sparse, hairlike. G&D: d0: 4–6 g; d3–4: mass doubles; d11: FF, can fly. **Fledgling.** Fledges d14. Plumage above: uniformly brown, wing covs edged with buff or cinnamon; below brown with breast scaling. Iris: grayish or pale orange.

White-winged Dove (*Zenaida asiatica*)

Adult. (AZ) 125–187 g, 29 cm (11.4 in.). LF: reddish. Iris: orange or orange-red. **Squab.** Skin: dark. Down: sparse, buff, whitish, or dull yellow. G&D: d0: 7–9 g; d1: 8–12.4; d4: 14–39; d7: 33–67, nearly FF; d8: prim PinUn; d12: 89 g, 13.6 cm. **Fledgling.** Fledges d13–18. LF: pinkish red, then brown. Iris: clove-brown. Plumage similar to Ad but duller & grayer. Above: grayish-brown, scapulars & wing covs edged buff, wings dark, white wing patches (evident ~14 d) appear as narrow strip along lower edge of resting wing. Dependent 4+ wks post-fledging.

Mourning Dove (*Zenaida macroura*)   Photos 2-3

Adult. (M > F) 86–170 g, 23–34 cm (9–13 in.). Gape: "red-wine" in Ad & young. **Squab.** ~5 g. MC: dark neutral gray. Bill: dark gray, sharp egg teeth, dark tip d7. Skin: blackish-gray or olive. Down: woolly, off-white, cream, or buffy on head & body. LF: olive-gray, turn geranium-pink. G&D: d0: 5–6 g (60–65 mm); d1: 11 g; d2–3: wing prim PinEm; d3: 22 g, tail PinEm; d4–7: eyes open; d5–6: body & head PinEm; d6–8: prim PinUn; d7: 31 g (82 mm); d8–9: crown PinUn; d10: 45–60 g; d10–11: ventral PinUn; d12–15: 50–80 g (150–180 mm), PinUn underwings & belly; d16: 85 g. Voc: H weak *peeping*, becomes loud & persistent. **Fledgling.** 50–80 g (60–65% of Ad mass). Fledges: d12–15. Plumage more mottled or scaled than Ad in shades of dark gray, black & brown. Back & wings buffy tipped, no white wingbars. Independent d25–27.

CUCKOOS

(Family: Cuculidae, Order: Cuculiformes)

Six species raise young in N.A. Yellow-billed Cuckoo (YBCU), Black-billed Cuckoo (BBCU), and Greater Roadrunner are the most widespread. YBCU and BBCU are sympatric in e. & c. N.A. Unlike Old World cuckoos, they usually raise their own young and only occasionally lay eggs in other birds' nests. Distinguishing features of chicks: toe arrangement, black skin, coarse "spinelike" pin feathers (cuckoos) or white natal down (roadrunner), and complex pattern of structures (called papillae, spines, or discs) on palate and tongue. **Adults.** Medium-size birds: 65 g (cuckoos) to 380 g (Greater Roadrunner). Diet: omnivorous; mainly caterpillars of most species; larger species catch snakes and lizards. Nest: loosely built cup or twig platform (cuckoos) or compact cup (roadrunners). Eggs: elliptical to subelliptical, white to blue, bluish-green, or greenish-blue. Bill: variable, usually compressed, somewhat decurved. Plumage: mostly brown and gray, occasionally black. Wings: variable form, relatively short and broad. Tail: long (15 cm), graduated. Legs/feet: zygodactyl (with the fourth toe permanently reversed). **Young.** Altricial. Mouth: lining red with white papillae. Gape flanges: grayish-yellow. Skin: black. Down: mostly naked; first feathers of cuckoos resemble quills of a porcupine; roadrunners have stringy whitish down. Juvenile plumage: back two-toned brown, white underside. Develop rapidly; some fledge less than a week post-hatching; capable of flight 17–30 days but cared for by adults several more weeks. Yellowed-billed Cuckoo and Black-billed Cuckoo nestlings are difficult to distinguish until juvenile feathers appear (see Tables C-1 and C-2, Appendix C.). The most apparent difference is dorsal view of wings: in YBCU, inner webs of prim strongly rufous on proximal half; rufous tinge to proximal portion of outer webs of prim & sec; all wing covs with rufous margins. BBCU is described showing white to pale buff on inner webs of remiges and some rufous on outer webs of inner primaries & outer secondaries, but duller than YBCU (Parkes 1984). Anterior underparts of BBCU more buffy gray than YBCU. Tail underside of YBCU has more distinct markings but may not be evident until bird is near full length. Lower mandible of YBCU may be partially yellow. Lower mandible of BBCU varies from black to blue gray (Parkes 1984).

Greater Roadrunner (*Geococcyx californianus*)   Plate 4

Adult. 221–538 g, 52–54 cm (20.5–21 in.). **Nestling.** MC: pink to red with 2 white spots on upper palate & back edges of tongue, black tongue tip.

Bill: H black with pink along commissure, wide & triangular (narrows by fledging). Skin: black, appears oily. Down: coarse, long, white. LF: black, turn blue-gray. G&D: d0: 14 g, d1: 20; d3: 34; d6–8: 69–94, eyes open; d10–14: 130–155 g, FF; d14: orbital area turns orangish; d16: 175 g; d24: 255. Iris: black, then N dark grayish-brown. Voc: whining, growling. **Fledgling.** Fledges ~d11, self-feeding d16, flies d17–19. Skin: orbital area pale blue, orange patch behind eye. Plumage similar to Ad but stripes broader & less defined.

Yellow-billed Cuckoo (*Coccyzus americanus*)   Plate 4

Adult. (PA) 50–84.6 g, 26–30 cm (10–11.8 in.). Eyelids yellow, orbital skin grayish. Bill: yellow lower mandible & portion of upper mandible. Tail: black with large white spots. LF: slate or plumbeous gray. **Nestling.** Mouth features: red, tongue has black edge; 10 large creamy-white papillae on palate & tongue. GF: pink, thin. Bill: dark gray upper, light gray lower, tip darker. Skin: black, shiny, bare area above bill paler than bill, surrounding skin has light gray spot; orbital area pale yellow. Down: white hairlike (at first) neossoptiles, long, pointed, quickly thicken & lengthen. Tail: short. LF: light slate-blue. G&D: d0: 8–9 g; see Table C-1 (Appendix C). Voc: "buzzing hiss like sound of bees escaping from a tunnel in dry grass" (Bent 1940, 57). **Fledgling.** 32–38 g. Leaves nest d7–9, can run, flies short distances. Plumage: similar to Ad but dusky grayish-brown, wing covs with cinnamon-brown tinge, outer webs of prim rufescent; bare just above bill. Below: olive-brown, pale grayish-buff throat & breast. Tail short & less distinct than Ad with dark gray & grayish-white spaces. Bill: all black for at least 3 wks post-fledging in CA sp (Franzreb and Laymon 1993); eastern birds have yellow bills (Oberholser 1974). Eyelids: gray. *Sim sp: Black-billed Cuckoo.*

Black-billed Cuckoo (*Coccyzus erythropthalmus*)   Plate 4

Adult. (PA) 38.3–65.4 g, 28–31 cm (11–12 in.). Diet of young: larval insects, (MI) gypsy moth larvae (81% of the diet). Bill: black. Orbital ring bright red (breeding), yellowish (nonbreeding). LF: plumbeous. Tail: grayish, thin white strips on underside. **Nestling.** Mouth features: red, tongue has black edge; papillae are large, creamy-white, 5 on palate, 2 pairs (elongated) around pharynx, 1 on tongue. GF: thin, pink. Bill & LF: light slate-gray. Skin: shiny, coal-black, dark bare area above bill; orbital ring dull greenish to yellow. Down: gray hairlike (at first) on dorsal, long, pointed; quills become thicker & longer with growth; white & gray on ventral. G&D: d0: 7.4 g; see Table C-2, Appendix C. Voc: begging call resembles buzzing of insects, loud "bark" when disturbed. **Fledgling.** 28.5 g. Leaves nest d6–9, wings long, tail ~¼ in. long, climbs & runs well. Flies well d21–24. MC: upper palate gray through 1st year. Bill: black upper, lower mostly cinnamon (middle portion pale yellow to dull white). Skin: orbital ring: dull greenish. LF: dark cinnamon, nails black. Plumage above: similar to Ad DefB but duller & softer, buff-brown with grayish-white or buff tips. Russet on outer webs of prim & lesser upperwing covs. Below & flanks: silvery white tinged pale brownish-buff, buffy crissum. Tail: olive-brown, small dull white spot at tips, outer edge of outer two feathers with narrow (0.05 mm) margin. *Sim sp: Yellow-billed Cuckoo.*

NIGHTJARS AND RELATIVES

(Family: Caprimulgidae, also called Goatsuckers,
Order: Caprimulgiformes)

Nocturnal or crepuscular (active during twilight). Large heads, large dark eyes, wide gapes, small beaks, rictal bristles (evident in most species) around bill do not cover circular (or tubular) nostrils, small semipalmate feet, some resemble small owls with soft cryptically patterned plumage with a velvety appearance. Aerial hunters; acrobatic flight for following and catching flying insects. Hiss and snap in self-defense but are harmless. **Adults.** Diet: mainly flying insects. Nest: usually on bare ground or litter, no nest material. Eggs: somewhat cryptically colored, elliptical, white to pinkish-buff, dark specks and blotches. Bill: short, small, delicate, depressed, slightly hooked. Plumage: very soft feathers with aftershafts; coloration dull & usually streaked, mottled, or barred. Wings: long, pointed. Tail: long, variable form. Legs/feet: small, weak; outer toes much shorter than the middle, hallux short and elevated; small nails, middle toenail pectinate (see Fig. G5 i, Illustrated Glossary); tarsi partly feathered, short, twice length of hallux. **Young.** Semi-precocial: eyes open at hatch, fed by parent, leave nest within 2 days post-hatching by walking or hopping. Mouth features: toadlike shape, immense gape, tiny tongue. Bill: tiny, hooked, commissure below midpoint of eye. Down: hatch fully covered, fluffy like lambswool, dark to buff (cryptically colored). Juv feathers with markings for camouflage. Fly d14–23 (per species). Care by one or both Ad; young seize food from adult bill; independent 1–2 weeks post-fledge. *Note: Some that are found in a state of torpor and mistaken for being ill should be returned immediately. Warm bird very gradually. A missing tail will regrow in about 2–3 weeks. If a nighthawk feels threatened it may puff feathers into the air or release an oily discharge from a pair of elongated sacs* *(caeca) that branch from the lower bowel. May be hand-fed (even adults) in captivity; force-feeding is stressful. Require spacious, soft-sided enclosure (for feather protection) with log to rest on.*

Lesser Nighthawk (*Chordeiles acutipennis*)   Plate 5

Adult. (TX) 41–64 g (M > F), 20–23 cm (8–9 in.), bill 5–6.8 mm (M > F), tarsus 13–15 mm. Nest: scrape, bare ground. **Nestling.** Down: whitish mottled with buff, paler below. G&D: unknown; d3: can run; d12: FF, natal down gone. **Fledgling.** Flies 21–25 d. Plumage: M similar to Ad M but paler above (less black), smaller buffy wing patches & subterminal tail band; F similar to juv M except throat & wing patches more deep buff, tail band lacking or indistinct. Biparental care. *Distinguish from Common Nighthawk: compare culmen & tarsus length (see "Common Nighthawk" below), fledgling Common Nighthawk should weigh more than Lesser; wing patch more distal on wing of Lesser.*

Common Nighthawk (*Chordeiles minor*)   Plate 5, Photo 4

Lacks rictal bristles. **Adult.** 65–98 g, 22–24 cm (8.7–9.4 in.), culmen 4.8–8.2 mm, tarsus 12.4–15.7 mm. Diet: flying insects. Nest: bare ground. M has white throat patch (indistinct or buff in F) & white tail band (lacking in most F). **Nestling.** Bill: H pale gray, turns darker gray, purplish-gray d17, culmen decurved. LF: dull brown, blackish or plumbeous d17, metal-gray d20. Down above: thick, fluffy, long, sepia to black mixed with grayish-white on dorsum, throat & sides. Replaced with dark gray & buff-yellow feathers, mottled or marbled. Below: pale gray, face at base of bill with buff tint, chin dark, dark midthroat & malar stripe. Iris: H bluish, then black or brown d3, Vandyke brown d15. G&D: d0: ~6 g, 48–55 mm; d2: 11–12; d3*: 15; d5: wing PinEm; d7: contour PinEm; d11: 42 g, wing PinUn; d12: throat PinUn; d14: 54g, prim PinUn (dark); d16: 66 g, can hop; d18: attempts to fly; d20: 78 g; d21–23: short flights; d28: down mostly gone. Voc: soft *peeps.* **Fledgling.** Fledges d10–20, flies d25–30. Plumage highly variable (geographically and individually), some gray, others paler (buffy); similar to Ad F but paler, throat patch mottled or lacking; smaller wing patch; tail lacks white. Covs & rect edged white or buff, prim no. 4 or 5 with mottled white bands. Biparental care, independent d30, d52 may join migrating flock. *Distinguish from Lesser Nighthawk (see "Lesser Nighthawk" above). Young nighthawks sometimes misidentified as Common Poorwill. *Weights d3–20 from one captive, Tullahoma, TN (LouAnn Partington).*

Common Poorwill (*Phalaenoptilus nuttallii*)   Plate 5

Smallest nightjar. Rictal bristles twice as long as bill. No wing patch or stripe. Small white corners on 3 outermost rectrices of M, paler in F. **Adult.** (AZ) 35.3–57.3 g, 19–21 cm (7.5–8.3 in.), culmen ~12 mm, tarsus 15–19 mm. Diet of young: regurgitated insects. Nest: bare ground (or leaf litter). **Nestling.** Down: fine buff to pale burnt umber. LF: light grayish-brown. Bill: small, black, tubular nostrils. G&D: no mass info, H 65 mm long, mobile but nidicolous. Siblings vary in mass & development because of asynchronous hatching. **Fledgling.** Mass 10% more than Ad. Fledges 20–23 d. Plumage: overall buff (black basally), grayish and pale to bright burnt umber with indistinct black barring across dorsal surfaces; wings barred, except tips are mottled. Below: paler, mottled, buff throat patch.

Chuck-will's-widow (*Antrostomus carolinensis*)   Plate 5

Largest nightjar. Large flat head, long rictal bristles. **Adult.** 66–188 g, 28–32 cm (11–12.6 in.). Nest: bare ground (or leaf litter). **Nestling.** Bill: gray. Down: fully covered, long, soft, back golden brown (tawny), head & underparts yellowish-tawny to yellowish-buff, paler on throat & belly. G&D: unknown, d1–3: active, can hop; d4: PinUn; d17: FF, flies short distances. **Fledgling.** Bill: buff with darker tip. LF: amber. Plumage similar to Ad DefB but paler (tawny-olive), brown scapulars, broad buffy wing covs. Care: by F.

Eastern Whip-poor-will (*Antrostomus vociferus*)   Plate 5

Very long rictal bristles present. **Adult.** (KS) M 50.3 g, F 56.5, 22–26 cm (8.7–10 in.). Nest: leaf litter. **Nestling.** Bill: horn, ivory egg tooth. Down: long, soft, silky, orange-buff or cinnamon-buff, center of crown dark, pinkish-cinnamon head & underparts. G&D: d0: 5.4–13.6 g (KS); d4: 28, prim PinEm; d8: 28.2 g; d10: PinUn, very active; d15–16: flies short distances; d18–19: 58 g. **Fledgling.** ~58 g. Plumage: juv M similar to Ad DefB M but crown pale gray with black spots, pale buff wingbars; F similar to juv M but outer rect with buff tips. Below: brownish-buff with dusky barring, indistinct white throat band. Biparental care until F begins 2nd nest, then M only.

SWIFTS

(Family: Apodidae, Order: Apodiformes)

Swifts spend most of their life in the air or clinging to vertical surfaces, therefore features of their wings and feet are unique. Aerial insect hunters with large eyes, tiny beaks, and very wide gapes. **Adults.** Diet: adults carry insects to young in pouch below tongue. Nests: rounded cup of mud or other materials or shallow half cup stuck to vertical surface with saliva; in crevices, in ledges under overhangs, inside chimneys (or other structures) or hollows of trees. Eggs: white, elongated. Bill: very short, small, depressed, decurved culmen. Gape: extremely wide, commissural point below eye. Plumage: compact, uniformly dull but sometimes with white areas; forehead feathers may partially conceal nostrils. Sexes colored alike. Wings: very long (cross in X pattern beyond tail), flat, pointed; extremely short secondaries; short, stout humeri. Tail: form variable, usually forked or notched; rectrices in Vaux's and Chimney Swifts stiffened and spinose. Legs/feet: pamprodactyl; three anterior toes about equal in length with small hallux frequently directed inward (reversible in Apodidae); very short tarsi; toes feathered or unfeathered, strong for clinging to vertical surfaces. Nails: strongly curved, acute. **Young.** Altricial. Mouth: drab or pink, no gape flanges. Down: none, later covered with spinelike quills uniformly covering body, mottled appearance from whitish color of feather sheaths. Nestling wing length same or not much longer than tail. Leave nest: 28–60 days (per species). Biparental care.

Black Swift (*Cypseloides niger*)

Adult. (CA) 41–51.5 g, 18 cm (7 in.). Nest: cup of moss & mud in crevice on rock cliffs. **Nestling.** Bill: blackish, white egg tooth (gone d18–33). Skin: pink & gray. Down: none. LF: flesh pink, nails gray with white tips. G&D (9 nestlings, Riverside County, CA): d0: 3–3.5 g; d7–8: eyes begin to open, (fully open d14–16); d6–7: blackish-brown PinUn head, back, upper chest; d13–14: nearly FF, dark gray down like semiplumes; d8–10: sec PinEm; d13: sec PinUn, outer prim PinEm. **Fledgling.** 64.3 g, fledges d45–49. Plumage above: uniformly blackish with white edging giving scalloped appearance (disappears rapidly); below: white edging broader on abdomen, vent & crissum. Voc: none made while begging. Biparental care.

Chimney Swift (*Chaetura pelagica*)   Plate 5

Adult. (IL) 17–29.8 g, 14 cm (5.5 in.). Nest: twigs & small sticks cemented with sticky saliva; formerly in natural crevices, now almost exclusively in man-made vertical structures, especially chimneys. **Nestling.** Bill: small, weak, light gray. Skin: pinkish-tan. Down: none. LF: tarsus very short; dusky or gray claws, very long, pointed; hallux rotates forward; strong toes, N can cling vertically using wing claw. G&D: d0: 1–1.5 g; d2: 3.5; d4: 5.5, PinVis; d5–6: PinEm; d6: 8 g; d10: 17.5; d10–12: wing PinUn; d12–14: body PinUn; d14–16: eyes open; d15: 21 g; d19–34: 22.5 g. **Fledgling.** Fledges d28–30, flies well.

Vaux's Swift (*Chaetura vauxi*)   Plate 5

Adult. (CA) 15–21 g, 11–12 cm (4.75 in.). Nest: cavity, preferably tree hollow in forest. **Nestling.** Bill: extremely tiny, grayish with prom egg tooth (upper mandible). Skin: light pink. Down: none. LF: short tarsus, tiny feet, nails grayish. G&D d0: ~1.5 g; d3: 3; d4–6: PinVis; d6–8: wing PinEm; d10: body PinEm; d13: eyes open; d14–15: contour PinUn; d18: 21 g; d30: 17–18. Voc: rapid, short *pips* & *squeaks*. **Fledgling.** Fledges d20–22 but clings to cavity wall; d28–30 leave cavity. Plumage above: brownish-gray (some with faint greenish gloss) rect & rem nearly fully grown (but partly in sheaths). Below: paler. *Newly fledged young sometimes miss nest hole & are found on ground. If there are branches to catch them, can climb back to nest hole.*

White-throated Swift (*Aeronautes saxatalis*)

Adult. (CA) 27.8–36 g, (NV) 38.7–33.2 g; 16.5 cm (6.5 in.). Diet: adults carry (in floor of mouth) bolus of aerial arthropods. Nest: saucer shape of a variety of materials cemented with saliva; on shelf in rocky crevice (preferred). **Nestling.** Skin: pink. Down: none, then covered with pale gray semiplumes above, less dense on head & below. G&D: d0: 2 g (d2–17 gains 2 g/d); d7–8: PinVis; d8–11: prim PinEm; d8–15: eyes open; d11–13: PinEm; d22–24: > 40 g, PinUn; d28: juv feathers emerging; d40: FF. **Fledgling.** 30–35 g. Fledges: (CA) d42–43, (BC, Canada) d40–46. Plumage similar to Ad DefB, but crown (appears scalloped), nape & crissum with white edgings, prim with pale edges, sec with narrow to broad white edges.

HUMMINGBIRDS

(Family: Trochilidae, Order: Apodiformes)

Smallest of all birds. In the US the male Calliope Hummingbird is the smallest, weighing 2.7 g or 1/10th of an ounce. Tiny young have short bills and down that resembles minute porcupine-like quills. Hatchling hummingbird species look similar. Identification usually requires seeing the adult, nest details, location, time of year, and possibly down color. **Adults.** Diet: primarily nectar; also tiny insects and spiders by gleaning or flying, and sap. Nest: tiny, cup shaped, deep, tightly woven with spider web and plant down. Eggs: tiny, white, unmarked. Bill: length variable but longer than head, slender, straight (sometimes decurved, rarely recurved), terete (sometimes compressed). Gape: narrow. Plumage: compact, brilliantly colored (iridescent) in both sexes (less so in females). Wings: long, flat, pointed; secondaries extremely short. Tail: variable form. Legs/feet: short, tiny, weak; tarsi not longer than middle toe with claw, feathered or unfeathered. **Young.** Altricial. Mouth: mostly yellow, some with pinkish interior and yellow at corners. Bill: short at first; fine striations or corrugations (in good light with magnifying lens) most obvious on very young birds, extend diagonally along length of bill, more evident on distal end; become reduced as bird matures. Skin: usually dark. Down: naked, then (most species) with two thin rows of hairlike downy feathers along center of back. Juvenile plumage: appears somewhat mottled (different colors interspersed) as feathers emerge on back, lighter below; iridescent feathers on the throat, gorget, or crown can be useful for aging. Fledge: 19–30 days; may perch on nest edge and exercise wings 2–3 days before fledging. Gape response: mother hummingbird elicits gaping of young during days 1–5 post-hatching by touching her bill on the bulge behind chick's eye; after downy dorsal feathers appear, chick gapes when its feathers are moved by the mother hovering above (Welty and Baptista 1988). Care: female only, feeds nectar and insects by regurgitation directly into chick's crop. Female absent from nest for long periods during day so nest may appear to be abandoned. *Song learning: some species must learn songs, some do not.*

Ruby-throated Hummingbird (*Archilochus colubris*)

Adult. (PA) M 2.4–3.7 g, F 2.6–4 g; 9.5 cm (3.75 in.). **Nestling.** MC: varies bright orange-yellow to vermilion. Bill: shorter than Ad, see family description above. Skin: slate-blue, dark slate or dark gray. Down: yellowish or buffy-yellow, thick along middorsum, 4 mm long, 12 pairs in two rows. G&D: d0: 0.62 g; d2: 1; d5: 1.5; d7: 2.5, PinUn; d9: 3 g, eyes open;

d11: 3.5 g; d18: 5. **Fledgling.** ~4.85 g, more than Ad. Fledges 18–20 d. Plumage: see Table D-1 (Appendix D).

Black-chinned Hummingbird (*Archilochus alexandri*)

Adult. (AZ): M 2.3–4.8 g, F 2.5–5 g; 9.5 cm (3.75 in.). Bill: slightly decurved. **Nestling.** MC: yellowish at flanges, pinkish elsewhere. Skin: dark slate to dark gray. Bill: orangish, culmen tip dark. Down: two rows, buffy-yellow to tawny on back, naked elsewhere; d6–9: PinUn on back, wings & tail with natal down still attached to tips. Nails: black, acute. **Fledgling.** Departs 21 d. Plumage: see Table D-1 (Appendix D).

Anna's Hummingbird (*Calypte anna*)   Plate 5

Nesting begins with onset of winter rains; in some areas of CA, egg laying in late Dec. **Adult.** (AZ) M 3.6–6.3 g, F 3.2–5.6 g; 10 cm (4 in.). Length of first-year birds slightly less than Ad, shorter tail. **Nestling.** MC: bright orange-yellow. Bill: short, squat, pale yellow, darker d7. Skin: black dorsally, pinkish throat. Down: "smoky fuzz," 2 lines along spinal tract, head bare, well covered d5–7. G&D: d0: unknown; d3: 0.9 g; d5: eyes open; d7: numerous PinF (resemble quills); d8–10: dorsal PinUn (pale cinnamon); d9: Ad mass; d12: greenish tips appear; d13–14: head PinUn, wing fluttering; d18–19: extends tongue; d19–21: preens, flutters wings on edge of nest. Nails: black, acute. Voc: high *seet* (begging). **Fledgling.** Fledges d23–24, independent 7–14 d post-fledge. Plumage: see Table D-1 (Appendix D). *Song learning: practices song wk 8; other components of song learned through imitation of other Ad by time juvs are 4–7 months.*

Costa's Hummingbird (*Calypte costae*)

Nests in s.w. US Mar. & Apr. **Adult.** M 2.5–5.2 g, F 2.5–3.4 g; 9 cm (3.5 in.). Nest: loose, shallow cup, plant material (flower heads), feathers & spider web. **Nestling.** MC: pinkish interior, yellowish corners. Bill: yellow, triangular, short (slightly longer than width at base). Skin: black above, brownish below. Down: naked except for yellowish filaments along spinal tract. G&D: unknown, d6: PinF appear, more noticeable at d10–12. **Fledgling.** Fledges d20–23, fed by F 7 d post-fledging. Plumage: see Table D-1 (Appendix D).

Calliope Hummingbird (*Selasphorus calliope*)

Adult. M 1.9–3.2 g, F 2.2–3.2 g; M 7 cm (2.75 in.), F 7.6 cm (3 in.). Nest: often built on top of dead cone base (resembles pinecone). Diet: nectar, sap. **Nestling.** Unknown. **Fledgling.** Fledges d18–21. Plumage: see Table D-1 (Appendix D). Care: F only. *Sim sp: juv F Calliope similar to F Rufous & Allen's Hummingbirds.*

Rufous Hummingbird (*Selasphorus rufus*)

Adult. (AZ) 2.8–4.5 g, ~9.5 cm (3.8 in.). **Nestling.** Skin: dark. Down: grayish, two rows along spinal. d6–7: PinEm. **Fledgling.** Fledges 20 d. Plumage: see Table D-1 (Appendix D).

Allen's Hummingbird (*Selasphorus sasin*)

Adult. (*S.s. sasin*): M 2.5–3.8 g, F 2.8–3.5 g; (*S.s. sedentarius*): M 3.5 g, F 3.7 g; 9.5 cm (3.75 in.). **Nestling.** MC & GF: brilliant orange-yellow. Bill: short, flesh-yellow, doubles length d14–19. Skin: dark. Down: naked except 2 rows grayish-yellowish along spinal tract. G&D: d0: 0.26–0.32 g; d5–12: eyes open; d6: able to grasp firmly; d7: spinal tract PinEm; d8: rem PinEm; d9: dorsum & rect PinEm, cinnamon feathers on back unsheathing; d12: 4 g, rect PinUn; d15: FF. **Fledgling.** Near avg wt of Ad. Fledges: as early as d18 and up to d30. Plumage: see Table D-1 (Appendix D). Care: F only, may continue to be fed ~10 d post-fledging. *Song learning: no song development. Sim sp: Rufous Hummingbird; Allen's & Rufous may hybridize. Measurements & notching of rect no. 2 are best way to distinguish (Pyle 1997b).*

Broad-tailed Hummingbird (*Selasphorus platycercus*)

Adult. (CO) M 3.5 g, F 3.6 g; 10 cm (4 in.). Nest: cup, plant materials & spider web, decorated with lichens & pieces of bark & moss. **Nestling.** GF: ochre-orange. Skin: dark. Down: yellowish, other *Selasphorus* sp have some down along spine. G&D (2 chicks): d0: 0.4 g, d5–9: max growth reached (0.34 g/d avg); d10–12: FF. **Fledgling.** Plumage: see Table D-1 (Appendix D).

RAILS, GALLINULES, AND COOTS

(Family: Rallidae, Order: Gruiformes)

Rails are solitary and shy (except coots), some heard more than seen. They have laterally compressed bodies and long slender toes that may be lobed for walking on floating vegetation. **Adults.** Small to medium-size birds. Diet: omnivorous, opportunistic. Nest: shaped like cup or dome, usually on ground, in low vegetation, or over water. Eggs: subelliptical to oval, off-white to brown, spots or blotches. Bill: shape and length varies for probing or pecking. Plumage of rails: black, gray, and browns with contrasting colors of light buff to rich rufous or olive. Plumage of coots, moorhens, and gallinules: purple or black. Wings: stubby, short. Tail: short. Legs/feet: anisodactyl and semilobate; hallux longer than nail of middle toe; moderately long tarsus. **Young.** Precocial, fed by parents at first, very active, can swim. Skin: most have brightly colored bare areas on the head. Down: fully covered, all or mostly black. Wings: wing claw at carpal joint helps chicks grasp & pull through vegetation. Legs/feet: grow quicker than wings. Leave nest: 1–3 days post-hatch; fledge 19–70 days depending on species. Biparental care. Juvenile plumage: first coat of grayish feathers replaces down 6–15 days post-hatch.

Clapper Rail (*Rallus crepitans*)

Adult. (LA) M 180–400 g, F 180–320 g; (FL) M 263–310 g, F 199–314 g; 32–41 cm (12.6–16 in.). MC: orange-red. **Young.** Bill: pied, varies from pinkish-white with black base to blackish with pinkish-white area around nostrils & tip, light gray (wk 5). Skin on belly & cloaca: orange-pink or salmon. Down: black. Iris: dark brown, drab olive (wk 6). G&D: 75–80 mm total length. d4–6: egg tooth gone; 5 wks: all PinUn (except capital, caudal & alar); 6 wks: FF; 7 wks: prim half grown; 10 wks: fly. Voc: faint *cheeps*. **Juvenile.** MC: yellow. LF: gray (8 wks), orange at joint (of tarsus & tibiotarsus) begins to show (9 wks). Plumage: duller & more muted than Ad, 10 wks indistinguishable from Ad.

King Rail (*Rallus elegans*)

Populations declining, threatened or endangered in 12 US states & Canada. **Adult.** (e. US) M 340–490 g, F 253–325 g; 38 cm (15 in.) **Young.** Bill: pied, half (near base) grayish-black, whitish on distal half & nares. Down: thick, black, faint greenish sheen on less dense areas. LF: pale brown-gray.

Virginia Rail (*Rallus limicola*)   Plate 6

Adult. (AZ) 55–124 g (M > F), 20–27 cm (8–10.5 in.). Iris: reddish-brown.
Young. Able to walk & swim shortly post-hatching, fed by parent 1st
3–7 d. Bill: pale pink, narrow black band near center that widens with
growth; completely black at 4 wks, brownish-red appears in lower mandible wk 12. Skin: head bare, bluish-black. Down: glossy black with metallic green gloss, some chicks may have patch of white down below wings.
LF: brownish-black, Ad size wk 3–4. Iris: black, turns dark olive d21.
G&D: d0: 5–5.8 g; d1: 6.2–8.5; d10: 25.5; d14: 30.8; d21: 50; d28: fly;
d42: FF. Voc: young chicks *pee-eep*; then calls of 1–2 syllables, repeat
in intensity when alone. **Juvenile.** Plumage above: duller blackish-brown
than Ad, feather edgings more rufescent; wing covs reddish-brown by
d28. Below: variable sooty black to dirty white; sides of breast fuscous or
grayish olive-brown. Iris: olive-brown to brownish.

Sora (*Porzana carolina*)   Plate 6

Adult. 49–126 g (avg), 20–25 cm (8–10 in.). Bill: Ad M brilliant
chrome-yellow. LF: yellowish-green. **Young.** Weak at hatch, fed by parents, stay in nest 3–4 d post-hatch. Asynchronous hatching extreme: some
chicks hatch 2 wks after 1st hatched. Bill: yellow with reddish base; fleshy
red operculum gone wk 5. Down: black & glossy, orange bristles under
chin (disappear wks 1–2). LF: pink. G&D: d0: 5–7.5 g. **Fledgling.** Flies at
4 wks. **Juvenile.** Similar to Ad except forehead & lores olive-brown, white
& dusky chin & midthroat, breast & lower throat buff olive-brown, buff-
brown flanks & sides. Iris: brown.

Common Gallinule (*Gallinula galeata*)   Plate 6

Formerly Common Moorhen (*Gallinula chloropus*), widespread in the Old
World; taxonomic split in 2011. **Adult.** M 370–456 g, F 310–398 g, 32–
35 cm (12.6–13.8 in.). **Young.** Bill: scarlet with orange-yellow tip & dark
band; frontal shield scarlet (fades to pale maroon by d30). Skin: above
eyes light ultramarine (fades by d8); crown & wings vinaceous. Down:
fully covered blackish neutral gray except head, legs & wings; thicker
below. LF: blackish neutral gray, become grayer, then dull olive-yellow by
d20. G&D: d0: 11.3 g (range 10–13), eyes open; d8: capital PinUn; d26: >
100 g; d26: prim & sec PinUn. **Juvenile.** Plumage: duller than Ad & paler
below (belly whitish), supraloral line buffy-white. Bill & frontal shield:
dull greenish-brown. LF: dull olive-gray. Flies 49 d, independent ~72 d.

American Coot (*Fulica americana*)   Plate 6

Most aquatic of all rails, dives frequently, swims, "runs" over water surface before taking flight. **Adult.** M 576–848 g, F 427–628 g, 32–43 cm (12.6–17 in). LF: semilobate (lateral "flangelike" lobes along toes indented at each toe segment, Fig. G5 d, Illustrated Glossary); green (breeding season), yellow-green (2 yrs), yellow (3 yrs), yellow-orange to red-orange (4+ yrs) (Crawford 1978). **Young.** Bill: scarlet, black tip. Skin: reddish & bare on crown with purplish-blue over eyes. Down: thick, black on back, lighter (mouse-gray) mixed with brownish-gray below; stiff, curly hairlike feathers on forehead; lores & chin flame scarlet & bright orange-red; back & throat light cadmium & pale buff. LF: grayish-green or bluish-gray. G&D: d0: 19–22 g; d19: 56; d36: 223; d71–73: 447–492. **Fledgling.** Flies ~d75. Plumage above: dark ash-brown on head & back; below paler, whitish throat, gray breast & belly. LF: leading edge blue-green, grayish-olive to yellowish olive-green; trailing edge olive-gray or plumbeous. Shield proper varies grayish, or dull flesh color, tinged with olive, or ivory-gray; a "callus" of the shield proper varies from pale red or dark grayish-brown (SC).

Yellow Rail (*Coturnicops noveboracensis*)

Adult. (Canada, 90% of breeding range) 53–70 g; 18 cm (7.25 in.). **Young.** Bill: bright pink. Down: black. Wing claw: 2nd digit. G&D: d18: juv plumage predominates; d24 (3 captive): 39, 39 & 46 g; d35: 1st flight.

Black Rail (*Laterallus jamaicensis*)

Adult. (CA) 29 g; (AZ) M 23–34 g, F 25–46 g; (e. N.A.) 29–44 g; 10–15 cm (4–6 in.). LF: blackish-brown. **Young.** Little info; stay in nest 1–2 d post-hatching. Bill: H sepia, base pinkish. Down: black with greenish sheen. LF: darker than Ad. Iris: dark olive-green or brown. **Juvenile.** Plumage above: crown brown, sides of face gray (auriculars darker); nape to upper back (triangular patch) raw sienna, rufous, or vinaceous-pink; midback, scapulars & uppertail covs with white spots (1–12 mm diameter); spotting on sec (3 or more rows) & covs. Below: white or light gray chin, abdomen pale neutral gray, flanks darker & barred or spotted white. LF: darker & duller than Ad. Iris: greenish-olive (4–6 wks). Parental care: unknown.

CRANES

(Family: Gruidae, Order: Gruiformes)

Very large wading birds with long necks and legs and wingspans of 6–7 feet. Solitary in breeding season, gregarious other seasons. **Adults.** Diet: varied, omnivorous, opportunistic; pecking or probing. Nest: platform of vegetation on ground in marshy areas. Eggs: subelliptical, variously colored creamy or buff to olive-buff with spots or blotches. Bill: as long as or slightly longer than head, straight, compressed. Plumage: most are predominantly pale gray or white, patches of red on face, lores & crown with bristles. Wings: large, rounded. Tail: short. Legs/feet: legs very long, hallux elevated and short (equal to length of nail of middle toe). **Young (colt).** Precocial, fed by parents at first, mobile, fully covered in down. Leave nest within hrs post-hatching. Fledge: d50–130.

Sandhill Crane (*Antigone canadensis*)   Plate 6

Adult. Mass varies by ssp & sex: Greater (southern) 4850 g, 117 cm (46 in.); Lesser (northern) 3350 g, 104 cm (41 in.). Legs: long, yellowish. **Colt.** Bill: narrow, beige-pink, darker toward tip. Down: thick, golden or rusty brown, grayish-white below. Iris: dark umber or gray. G&D varies by ssp: d0: 114.2 g, tarsus 44.2 mm, able to sit on hocks & stand (weakly); d1: leaves nest, begins self-feeding; d2: 125 g; d5: 131; d8: 174; d10: 209; d15: 315; d20: 457; d25: 709; d30: 894, wings & legs nearly half Ad length; d60: bill & wings nearly Ad size. Voc: piping (trills & peeps), contact call high-pitched yelp; loud, rapid slurred notes when stressed. Care: both feed young directly during early period. **Fledgling.** 1st flight varies geographically: e.g., 50 d (Rocky Mtn. population) & 70 d (Boise, ID). Remains with parents until following yr. Plumage mottled brown & gray overall. Above: crown downy orange-brown to salmon. Sec covs, nape & back with cinnamon-rufous edges. Reach Ad mass 10–12 months.

Whooping Crane (*Grus americana*)

Tallest bird in N.A. Endangered status. **Adult.** 6000–7800 g, height 150 cm (59 in.). **Colt.** Can walk & swim few hrs post-hatching; 1st 36–48 hrs chicks are not fed, rely on yolk sac. Bill: pale buffy-brown. Skin: reddish cast from underlying blood vessels. Down: above dull cinnamon to brown, rump russet, neck paler & grayer. Below: pale grayish-buff or dull brownish-white. LF: light brownish. Iris: dark olive. G&D: mass (incubated/captive) 6–10% lower than wild chicks (Ricklefs et al. 1986): d1:

100–150 g, height 19–20.5 cm; d5: 101 g; d10: 173; d15: 261; d18–28: egg tooth lost; d20: 454 g; d30: 1304; d40: 1750–2500 g (65–75 cm); d100: 4300–4950 g (100–120 cm), can fly.

STILTS AND AVOCETS

(Family: Recurvirostridae, Order: Charadriiformes)

Two species raise young in N.A., the American Avocet (AMAV) and the Black-necked Stilt (BNST). **Adults.** Smallish, tall, and thin shorebirds; males slightly larger than females. Diet: small aquatic animals (mollusks, crustaceans, fish, frogs, lizards) and some vegetable matter; AMAV forage by sweeping bill side-to-side, BNST by pecking. Nest: loose colonies, in the open on bare ground near water. Eggs: pale grayish to olive, reddish-brown to black markings. Bill: very slender, long, straight (BNST), recurved (AMAV). Plumage: compact, black or glossy green or brownish-gray above, mostly white below. AMAV with rusty-orange head and neck (breeding), light gray (other seasons). Flight feathers: 10 primaries, 16–17 secondaries, and 12 rectrices. Wings: long, pointed. Tail: short, square. Legs/feet: tarsus very long, slender, shorter than bill; front toes webbed (partially in most stilts, in avocets webbing between outer and middle toes more extensive); hallux much reduced in AMAV, absent in BNST. **Young.** Precocial. Bill: slender, shorter than adult. Down: fully covered, with streaks and mottling. Leave nest: can run and self-feed shortly post-hatch. Legs/feet: long (see adult toe description). Biparental care.

Black-necked Stilt (*Himantopus mexicanus*)   Plate 6

Adult. (NV) 136–202 g; 35–39 cm (14–15 in.) (including bill); wingspan to 71 cm. Bill: black, slightly recurved. LF: extremely long red legs, half web between middle & outer toe, no hallux. **Young.** Leaves nest within 24 hr post-hatching, can run & self-feed. Bill: short, straight, black. Down above: light drab, pale gray & buff mottling with dark blotches, sepia streaks crown & back; 2 sepia lines shoulder to rump (may fuse at midback); grayer on neck, thin dark eyeline. Below: creamy-white, belly white. G&D: d1: 13.6 g, tarsus 28.4 mm; d3: 17–18 g, tarsus 31–32 mm; d5: 28 g, tarsus 38 mm; d17: 68.7 g, tarsus 54.7 mm; d21: PinEm scapulars, back & breast. **Juvenile.** Independent (sustained flight) d28–32. LF: light orangish-pink. Plumage like Ad F but dark areas with fine drab-gray or dull white mottling; back & wings with drab-gray or flesh-ochre edgings; inner prim & sec sepia with thin white edges. Iris: brown.

American Avocet (*Recurvirostra americana*)   Plate 6

Adult. (CA) M 261–362 g, F 215–382 g; 43–47 cm (17–18.5 in.) (including bill); wingspan to 91 cm. Bill: long, recurved. Sexually dimorphic: M mass > F, M bill longer but straighter. **Young.** Leaves nest 1–24 hrs post-hatching, can run & self-feed. Bill: blackish, short, slender, nearly straight. Down: soft, silky, dorsum with broken pattern of pale buffy-gray, orange feathers on head & neck with mottling on crown & back; white below. Eyeline: fine, dark, indistinct or broken. LF: gray with orange edges to webs, webbing between outer & middle toes more extensive. G&D (Lassen County, CA): d0: 17–21 g; d7: 39–49; d14: 74–108; d27: sustained flight. **Juvenile.** Plumage like Ad but head & neck paler. Plumage above: black washed with brown, white edges on inner prim, sec covs with flesh-ochre tips. LF: tinged olive. Independent 6 wks.

OYSTERCATCHERS

(Family: Haematopodidae, Order: Charadriiformes)

Large coastal wading birds. Two species raise young in N.A., the American Oystercatcher (AMOY) and Black Oystercatcher (BLOY). **Adults.** Diet: mainly mollusks and a variety of other invertebrates, sometimes small fish; probe in mud, pry from rocks, stab and hammer. Nest: simple scrape on the beach, in dunes, rocky shore (BLOY) or salt marsh (AMOY). Eggs: oval to subelliptical; spotted or streaked, cryptically colored gray or buffy. Bill: long (twice length of head), strong, stout, compressed laterally, reddish-orange; shape of tip varies by diet and type of foraging. Plumage: black and white or all black. Flight feathers: 10 primaries, 16–17 secondaries, and 12 rectrices; and diastataxic. Wings: moderately rounded. Tail: short, squared to slightly rounded. Legs/feet: sturdy, thick, hallux absent; small webbing between toes; light pink and pale. **Young.** Precocial, mobile soon post-hatch. Bill: shorter than Ad. Down: fully covered. Legs/feet: long. Leave nest: within 24 hours, fledge at 34–49 days. Iris: brown. Care: fed by both parents up to 30 days, may not be independent for 6 months.

American Oystercatcher (*Haematopus palliatus*)   Plate 6

Adult. M 499–657 g, F 568–720 g, 43 cm (17 in.). Nest: shallow scrape in sand, on islands, or on coastal beaches among rocks or in dunes. **Chick.** Bill: dark, pinkish base. Orbital ring: brown. Down above: dull grayish-black, rump & flanks tipped buff & grayish-white; thin black line from base of bill through eye to nape; two black stripes run parallel down spine, one across each wing. Below: chin, throat, breast darker than white belly. LF: light gray. Iris: brown. G&D: d0: 37.3 g; d12: PinEm flanks & ventral midline, prim PinEm; d14–21: sec & sec covs, rect, auriculars & crown PinEm; d42: FF, few traces of down. **Juvenile.** Bill: gray & buff mixed. Iris brown. Orbital ring: dark brown-orange. Plumage above: dark gray or light raw umber; pale feather edges (clay to cinnamon) on dorsum give "scaly" appearance; wings raw umber with white wingbar. Below: white, chin & throat mottled fuscous to buff; white feathers around base of bill. LF: dull grayish. Independent & sustained flight 34–37 d.

Black Oystercatcher (*Haematopus bachmani*)

Adult. (AK) M 582 g, F 619 g, 42–47 cm (16.5–18.5 in.). Nest: shallow scrape on ground, rocky coastal shores, or islands. **Chick.** Bill: sepia, white egg tooth (gone d7–12). Down: pattern like AMOY but darker, less white below. Above: black & drab, tipped buff. Below: med gray, abdomen white. Tail black. LF: drab-gray. Iris: raw umber. G&D: d0: 32–36 g; d3: can walk & swim; d5: pecks at insects; d12: all PinEm; d14–21: attains max growth; d21: capital, prim, sec & rect PinEm. **Juvenile.** Bill: turns chrome-orange, sepia tip. Plumage (complete by d42) above: dark grayish-brown except head & neck jet-black; scapulars, upperwing covs, rump & flank with tawny edges. Iris: d28: cinnamon-rufous. Flies d35–40. Parents may feed several wks.

PLOVERS

(Family: Charadriidae, Order: Charadriiformes)

Rounded heads, large eyes, short necks, and short bills. **Adults.** Small to medium-size shorebirds no larger than 38 cm (15 in.). Diet: mainly invertebrates (small crustaceans and aquatic insect larvae), sometimes berries and seeds. Nest: shallow scrape on ground, on sandy beach, in gravel, by lakeshore, open tundra, or marsh habitat. Eggs: oval to pyriform, usually with whitish, buff, gray, or greenish background; heavily marked with dark spots of variable size and number. Bill: straight, short, moderate length, swollen at tip. Plumage: browns, olive, gray or black, and white in conspicuous patterns. Most are black and uniformly colored; many with broad collar and contrastingly marked tail or rump; boldly marked head and neck. Underparts in many species are white or, rarely, solid black, with one or two broad, contrasting bands. Wings: long, pointed. Tail: short to medium. Legs/feet: legs short to long, near center of body; tarsi reticulate, partly bare; anisodactyl and semipalmate, 3 toes (except

Black-bellied Plover with four toes); toes moderately long, hallux vestigial or lacking. **Young.** Precocial; leave nest area as soon as down is dry, self-feeding but may not feed first day. Bill: slender, resembles adult bill but shorter. Down: fully covered, cryptic, distinctively patterned, mottled with or without white collar. LF: long tarsus, long toes. Crouching fear response. Fledge period varies per species, approximately 21–40 days (species in N.A.) or longer in other parts of the world. Biparental care.

Killdeer (*Charadrius vociferus*)   Plate 6

Common on golf courses, airports, or other open areas with lawns. Adults display broken-wing act to distract predators from nest. **Adult.** Med-size plover. (Great Plains): M 84–109 g, F 87.7–121 g, 20–28 cm (10.5 in.). Diet: primarily invertebrates. Nest: ground, depression often on gravel (parking lots & rooftops). Bill: usually < 1 in. **Young.** Leaves nest (walks): within 24 hrs or less, follows parents to feeding area usually near water, proficient swimmer. Becomes motionless if approached. MC: pink or grayish-pink. Bill: glossy black, short, blunt wedge. Skin: buffy-brown. Down: mottled buff & grayish-brown or blackish-brown on crown & back, black middorsal stripe (broken); crown with black wreath; white forehead, chin, throat, belly & neck; black stripe runs back from eye; conspicuous broad black band around neck, like a collar. Eye ring: bare, yellowish-gray, changes to red by juv. Tail: long tuft of blackish downy feathers ~40 mm long. LF: straw-yellow, lighter at joints; nails black. Iris: dark. G&D: d0: 8–10 g (avg), d7: 15–20; d20–31: flies. Voc: peeping before hatch. Distress call: short, loud piping notes.

Piping Plover (*Charadrius melodus*)

Small plover; endangered status. **Adult.** (MD) 46.5–62 g, 17–18 cm (6.7–7 in.). **Chick.** Bill: black. Down cryptic on pale sand & gravel. Above: head & back with buff & grayish speckles & blackish-brown spots; hindneck with white collar; wings & thighs with dark spots. Below: white. LF: orange. G&D: d1: 6.3–7.2 g; d15–30: acquire juv plumage.

Mountain Plover (*Charadrius montanus*)

Adult. 90–110 g; 21–23.5 cm (8.3–9.3 in.). **Chick.** Leaves nest: few hrs post-hatching, can self-feed. Down: fully covered, above cream or cinnamon-buff with black spotting; below buffy-white. G&D: d0: 7–11 g; d33–34: flies.

Snowy Plover (*Charadrius nivosus*)

Adult. (CA) 35–58 g, 15–17 cm (6–6.7 in.). **Chick.** Bill: black. Down: above pale buff or creamy-buff intermixed with light gray & spotted with brown & black, white band circles around neck, black stripe behind eye; below pure white. LF: gray to pinkish-gray. G&D: d0: 4.6–7.6 g (wing chord 10.5 mm); d28–33: 30 g, flies.

SANDPIPERS AND RELATIVES

(Family: Scolopacidae, Order: Charadriiformes)

The largest family in this order. Wading birds that occur in a wide variety of habitats, including grasslands and forests, but usually near water. **Adults.** Small to large shorebirds no longer than 66 cm (26 in.). Neck lengths vary per species. Diet: variability in bill morphology results in wide variety of food types. Many species probe in mud for invertebrates, some pluck amphibians or fish from water, others peck at seeds or fruits. Nest: usually a hollow in the ground; may be concealed in vegetation or in bare, open sites. Eggs: pyriform or oval, patterned and colored for concealment. Bill: varies from short to relatively long and slender; decurved, recurved, straight, or spatulate; tip somewhat depressed or may taper to point. Plumage: combination of grays and browns with white; some with reds or oranges during breeding season. Wings: long, tapered. Tail: short. Legs/feet: leg length varies per species, long toes, elevated hallux; tarsi scutellate in front; four toes (three in Sanderling, *Calidris alba*) with little to no webbing; phalaropes have lateral membranes developed into lobes. **Young.** Precocial, leave nest when down is dry, some capable of swimming, most self-feed (except woodcock and snipe that feed their young). Bill: slender, shorter than adult. Down: basic pattern is longitudinal stripes; some have darker stripes that may have white or buffy spots; others appear with broken patterns that look like mottling; crown pattern variable by species. Fledge 14–26 days (smaller sp) or 41–45 days (larger sp). Care: variable; female only in some, in others the role is reversed with male only, or duties may be divided between parents.

Upland Sandpiper (*Bartramia longicauda*)

Adult. 170 g; 28–32 cm (11–12.6 in.). **Chick.** d0: 15–22 g. Down: fully covered, above brown & tan with buffy or whitish edges (scaly appearance) & blackish-brown or buffy-brown mottling on dorsum & wings, ear coverts & nape with long black patch, white stripe through eye; below whitish.

Long-billed Curlew (*Numenius americanus*)

Large, long-legged, long decurved bill. **Adult.** (ID) M 531 g, F 642 g; 50–65 cm (20–25.6 in.); bill 11.3–22 cm (4.4–8.7 in.). Diet: carnivorous (insects, crustaceans, bottom invertebrates). **Chick.** Mobile 5 hr post-hatch. Bill: much shorter than Ad 1st 60 days. LF: semipalmate. Down: fully covered, thick, buff with brownish-black markings (broken, narrow stripe on sides of forehead that meet at occiput and extend down hindneck; bold spotting on back). G&D: d0: 44–56 g; d11: PinEm alar; d14: PinEm caudal. Fledges: flies 32–45 d.

Marbled Godwit (*Limosa fedoa*)

Adult. (AB, Canada) M 285–396 g, F 352–454 g, 42–48 cm (16.5–19 in.). **Chick.** Bill: egg teeth gone d1–2. Down above: pale buff-brown, dappled with dark spots on crown (darker), back & rump; temporal spots on sides of head, loral stripe almost to eye; middorsal stripe; diamond-shaped mark lower back; wings buff-brown; pinkish-buff. G&D: d0: 37 g (avg); d26–30: flies. **Juvenile.** Plumage like Ad but more heavily marked on back; wings with buff & irregular dark linear marks; tawny-cinnamon on ventral areas unmarked.

American Woodcock (*Scolopax minor*)   Plate 7

Crepuscular. Large dark eye near crown; very long bill for probing. **Adult.** M 176 g, F 219 g; 28 cm (11 in.). Diet: primarily earthworms. **Chick.** Down above: light brownish-buff back & sides blotched sepia & Vandyke brown; 3 stripes along back run longitudinal, central is broadest; broad crown patch tapers to bill, narrows to join larger neck patch, continues along spine to tail; stripe from commissure splits to one through eye & one below eye then joins again behind ear as dark patch. Below: pinkish-cinnamon, unmarked, may darken to light orange; chin & throat yellow-white. G&D: d0: 12–14 g, bill length 14 mm; d3: down full length; d30: indistinguishable from Ad. **Juvenile.** Bill & LF: grayish. Iris: light brown. Dark gray band on throat & upper breast.

Wilson's Snipe (*Gallinago delicata*)   Plate 7

Formerly Common Snipe (*G. gallinago*). **Adult.** (M > F) 79–146 g; 27–32 cm (10.6–12.6 in.). **Chick.** Leaves nest soon post-hatching (wanders or climbs on F back). Bill: dark gray, prom egg tooth. Down: thick, soft, pattern not well defined. Above: back & sides chestnut (brown) blotched blackish-brown; fine whitish tips on dorsum appear as if bird sprinkled with snow; dark streak from bill to (broken) under eye; dark (moustache) streak from bill (broken) through cheek; dark patch forehead to crown separated by transverse row of whitish specks; eye half encircled (under & behind) with white specks. Below: tawny-buff or rufous-tan; two indistinct dark stripes down front of neck. LF: long, deep olive-gray, big feet. Bill: black, short & thick. G&D: d0: 10.8 & 11 g (2 chicks); d6: PinEm back & scapulars; d7: mass doubles; d19–20: flies. **Juvenile.** d21: Ad mass, bill 45 mm. Plumage nearly complete (d42) with traces of down (until d56); like Ad except broader & buffier feather edgings on neck, back & scapulars; cinnamon rump. LF: gray.

Spotted Sandpiper (*Actitis macularius*)   Plate 7

Adult. (PA) 29.4–59.8 g; 18–20 cm (7–8 in.). Nest: depression in ground, cup lined with grasses. **Chick.** Mobile, leaves nest ~2 hr after last chick has hatched. Bill: dark upper, yellow-orange lower & tip. Skin: very dark. Down above: golden-brown & blackish-brown with long tips, sprinkled sepia; dark stripe down spine; dark crown stripe tapers to forehead & nape; dark streak runs from bill through eye to behind ear. Below: white except for drab-gray sides of neck & chest. Tail: long tufts, barred sepia & pale buff. G&D: d0: 6 g, tarsus 11–16 mm; (1 chick) d3: 8.2 g; d7: 15; d9: 18; d14: 26.3; d16–18: flies; d28: bill Ad length. **Juvenile.** Bill: darker upper, paler lower. LF: orangish-yellow. Iris: burnt umber. Plumage like Ad except very unorganized, down still attached; narrow subterminal bar & lateral specks on scapulars, upperwing covs & tertials; no spotting. Above: forehead & crown olive-gray, thin white eye ring.

Willet (*Tringa semipalmata*)

Adult. 215 g; 33–41 cm (13–16 in.). **Chick.** (NY) d0: 28 g (avg). Down: above dull grayish-white or pale brownish-gray, marbled dusky crown & dorsum, thin dark stripe from bill through eye to nape; below grayish-white, spotting on flanks.

Wilson's Phalarope (*Phalaropus tricolor*)

Adult. Largest of 3 N.A. phalaropes: 60 g; 22–24 cm (8.7–9.4 in.). **Chick.** Down: above buff or tawny-orange crown, back & wings, dark stripe nape to tail, dark patches on wings & rump. Crown: no large dark patch, "anchor-shaped" dark marking (narrow centerline with 2 lateral spots joined by a transverse line). Below: grayish-white or pale buff.

GULLS AND TERNS

(Family: Laridae, Order: Charadriiformes)

Very diverse family of birds mostly associated with water. **Adults.** Small to large birds in this order, 23–76 cm (9–30 in.). Diet: omnivorous and opportunistic; take food off ground, surface of water, diving, aerial, or even taking from humans. Nest: simple scrape to large mound or cup, usually on ground or on cliffs, roofs, and ledges. Eggs: subelliptical, buff, cream, or greenish, sometimes heavily marked. Bill: shape varies greatly; stout and blunt in bigger gulls, delicate and thinner in smaller species, or straight and acute in terns, some may be hooked; nostrils perforate. Plumage: mostly white, black, and gray. Wings: very long, somewhat narrow, pointed. Tail variable in length and shape: square, slightly rounded, forked (terns), or wedge shaped. Legs/feet: moderate size and length (gulls) or extremely small and short (terns); tarsus slightly longer than wings, scutellate in front and reticulate elsewhere; anterior toes are palmate; hallux if present (some are tiny or absent) small and elevated; leg color varies pink to red, yellow, or black. **Young.** Semi-precocial, can walk soon post-hatching, dependent on parents for food. Down: fully covered, mottled or spotted. Fledge: 19–56 days. Biparental care.

Western Gull (*Larus occidentalis*)

Adult. M 1050–1250 g, F 800–980 g; 56–66 cm (22–26 in.). **Chick.** Leaves nest d14. Bill: black, pink egg tooth. Down: grayish or buff with dark spots; head paler. Below paler, little to no spotting. LF: dark gray or black. G&D: d0: 60–75 g; d5: 100–150; d10: 150–300; d15–20: wing & tail PinEm; d20: 400–800 g; d30: 650–1100; d40–50: flies.

California Gull (*Larus californicus*)

Adult. M 490–1045 g, F 432–903 g; 53 cm (21 in.). **Chick.** Bill: black with cream tip. Down: whitish with dark spots darker on head. Below: buffy chest band (on some), throat may be spotted, otherwise unmarked. LF: variable brown or black. G&D: d0: 50 g; d5–6: wing PinEm; d10: weight triples. Leaves nest: d4, runs 9–20 d; flies 42–48 d.

Herring Gull (*Larus argentatus*)   Plate 7

Adult. (MA) M 973–1143 g, F 718–1385 g; 63.5 cm (25 in.). **Chick.** Bill: wedge-shaped & buffy brown or black with pinkish tip & egg tooth (gone d3). Down: background dense pale gray, dark blotches dorsally & on thighs (darkest on head & chin); buff-white to buff-gray below. Head fine buff & blackish-brown or gray-black blotches suffused with tawny. Eye area with fine markings & dark lines. LF: buffy-brown with pinkish-gray webs. G&D: d0: 52–68 g (4 incubator hatched), d5: 100–150 g.

Great Black-backed Gull (*Larus marinus*)

One of the largest gulls in the world. **Adult.** M 1380–2272 g, F 1033–2085 g; 71–79 cm (28–31 in.). **Chick.** Down: gray above, buffish-white below, paler on chin & throat. Spotting/speckling: sides of chin, sides, posterior abdomen, undertail covs; head with small, dark star-shaped or polygonal spotting.

Common Tern (*Sterna hirundo*)   Plate 7

Adult. (NY) 103–145 g; 31–35 cm (12–14 in.) (incl. tail). **Chick.** Bill: pink, black tip, lower mandible may turn orange by d12. Down: dense, fully covered, cinnamon-brown (or light tawny) to gray-brown (or black) above (except head), back lined with black spots or streaks; white below except blackish-brown chin & throat. LF: pinkish-tan, dull orange by d12. G&D: d0: 11–15 g (5 incubator hatched); d3: 25 g; d14: 100. **Juvenile.** MC: dull orange. Bill: upper mandible black, base of lower pale pinkish-tan to pale yellow-orange with black tip. Plumage above: light gray on back, scapulars, terts & upperwing covs, rump & tail (dark on outer 2–3 rect); crown light brown to blackish, white forehead & lores with brown to gray-brown streaking or tinge, blackish patch from eye through auriculars forms a collar through nape. Below: white, sides of breast are tinged buff. Iris: dark brown.

Forster's Tern (*Sterna forsteri*)   Plate 7

Adult. (OK) 127–173 g, 33–36 cm (13–14 in.). **Chick.** Down: light buffy-brown, black irregular spots; below paler, sooty throat. G&D: d5: 42 g; d10: 88; d25: 124. **Fledge.** Leaves nest ~4 d, flies 28–35 d.

LOONS

(Family: Gaviidae, Order: Gaviiformes)

Fresh or saltwater diving birds with dagger-like bills, short necks, long wings, and short legs set far back on a heavy body. Distinctive shape and powerful webbed feet set loons apart from all other birds. **Adults.** Large waterbirds. Diet: excellent divers, a wide variety of fish (up to 10 in. long), also crustaceans, amphibians, insects, other invertebrates, and aquatic vegetation. Nest: mound of vegetation, very near water's edge on a shore or small island. Eggs: subelliptical to oval; greenish-brown, dark speckles. Bill: straight, compressed, tapering, acute. Plumage: similar among species but distinctive from all other birds. Wings: relatively small but well-developed and somewhat pointed. Tail: short, stiff. Legs/feet: palmate, four toes (the first three fully webbed); tarsi reticulate and laterally compressed. **Young.** Semi-precocial, fed by adults. Bill: short, stubby. Down: uniformly gray above, white below; two downy coats, the second is paler. Fledge: 49–77 days. Biparental care: chicks are carried on back of adults while swimming, dependent on parents for food up to 10 weeks post-hatching.

Red-throated Loon (*Gavia stellata*)   Plate 3

Adult. 1000–2700 g; 53–69 cm (21–27 in.). **Young.** Bill: blackish, tip paler. Down: dense; two stages. *1st down*: varies, blackish-brown to blackish-gray & med grayish-brown with paler underparts. *2nd down*: med brownish-gray. G&D: d0: 55 g; d6–7: 200–250; d17–18: 550–850; d29–30: 850–1300. Fledge: 1st flight attempts 33 d, leave for ocean with parents then disperse.

Pacific Loon (*Gavia pacifica*)   Plate 3

Adult. 1000–2500 g; 53–74 cm (21–29 in.). **Young.** Bill: slender, dark, tip paler. *1st down*: short, thick, varies by individual & geography; grayer & paler than other species, pale grayish-brown to dusky-brown, undersides lighter. *2nd down* wk 3: paler than 1st & less dense. G&D: d0: 62–82 g (Victoria Is.), 86.2 g (Manitoba); wk 2: PrimUn; wk 3–4: ventral PinUn. Fledge: 50–55 d (fly from nest pond), stay with family until summer or fall.

Adult. 2200–7600 g; 66–91 cm (26–35.8 in.). **Young.** Leaves nest within 1st d of hatching. Bill: blackish-gray. *1st down*: black & dense, underside white. *2nd down (wk 3)*: long, brownish-gray (or sooty-brown) replaces natal down. Iris: cinnamon to walnut-brown. LF: blackish-gray, webs turn white by d56. Iris: walnut-brown, then red 1st winter/spring. G&D: d0: 71.4–108.5 g; d21: PrimUn; d56: FF; d84–91: fly. **Juvenile.** Bill: pale gray. Plumage above: thick gray with brown down remaining on nape & lower neck, loose down on legs, rump & areas of the head. Below: white. Chicks may become separated from family due to disturbance or rough weather; wild loons may adopt orphaned chicks (Timmermans et al. 2004). Independent by 1st fall.

CORMORANTS

(Family: Phalacrocoracidae, Order: Suliformes)

Large waterbirds with long necks, long hooked bills, and webbed feet for swimming deep under fresh or salt water for fish. The Double-crested Cormorant breeds along coasts as other cormorant species, but also inland. **Adults.** 56–92 cm (22–36 in.). Diet: mainly fish and invertebrates; some freshwater insects, amphibians, reptiles. Nest: simple heaps of seaweed (nests at coast) or large twig nests at inland sites; may contain other plant debris, sticks, and rubbish; on cliff ledges, trees, bushes, or the ground. Eggs: long subelliptical, pale blue to green. Bill: as long as head, straight, slender, sharp hook at tip; exterior nostrils absent; expandable bare gular sac on throat. Plumage: predominantly black, some with white patches. Wings: short, rounded. Tail: long, rounded. Legs/feet: webbed. **Young.** Altricial, young birds take food by reaching well into the parent's gullet or picking up disgorged food. Down: hatch naked, grow down by day 7. Fledge 35–70 days. Biparental care. Young birds, as early as 10 days, gather together in groups (called crèches).

Double-crested Cormorant (*Nannopterum auritum*)

Adult. 1200–2500 g; (FL) M 1808 g, F 1540 g; 70–90 cm (28–35.4 in.). **Nestling.** Skin: pink on head & neck with lavender & black markings; skin yellowish by wk 2. Down: naked at hatch (a few filaments on rump & lower wings), d7–14: grow black, dense, woolly down, fully covers body. LF: dark brown, turn dull black, tarsi relatively short. Iris: black, turns brownish (juv). G&D: d0: 27.6–34.7 g; d3–4: eyes open; d4–7: egg tooth gone; d6–7: down 1st appears; d13–14: prim PinVis; d14: fully covered in down; d16–19: prim PinUn, throat pouch turns yellowish; d21–23: wing feathers 2.5 cm long; d28: wing feathers 6 cm long; d35: FF except head & neck bare; d42: few feathers on head, neck still downy. **Fledgling.** Leaves ground nest 3–4 wks before able to fly, otherwise (if nest is in tree or on cliff) chicks fledge at 6–8 wks old.

PELICANS

(Family: Pelecanidae, Order: Pelecaniformes)

Two species raise young in N.A., the American White Pelican (AWPE) and Brown Pelican (BRPE). Breeding ranges do not overlap as AWPE nests inland on remote islands and BRPE nests along marine coasts in estuaries or offshore islands. Very large waterbirds with wingspans of 6 feet 7 inches to 9 feet, webbed feet, enormous bills, and large gular pouches that expand to trap prey. **Adults.** Diet: mainly fish caught while swimming and gathering, or plunge-diving. Nest: simple scrape or mound of debris on ground or may be a stick nest in tree. Eggs: long subelliptical, white. Bill: very long, straight, hooked, lacking nostrils. Plumage: white with black flight feathers (AWPE) or all brown (BRPE). Wings: very long, rounded. Tail: very short. Legs/feet: totipalmate (four toes united in one web), tarsi short, reticulate, and compressed. **Young.** Altricial, eyes open by end of hatch day, immobile.

American White Pelican (*Pelecanus erythrorhynchos*)

Adult. M 6329 g, F 4970 g; 127–165 cm (50–65 in.). **Nestling.** Bill & pouch: grayish-white. Skin: orange. Down: hatch naked, d1–2 white downy coat, d10–12 thicker & grayer. Iris white. G&D: d0: 110 g; d17–25: leaves nest; d25–28: scapulars & prim PinEm; d29–33: all PinEm. **Fledgling.** Leaves colony ~10–11 wks (1+ wks after 1st flight). Young form crèches (up to 100 birds) after d30. Plumage: dusky white, crown grayish (later fades to white), black wing pattern like Ad. LF: pale yellowish-green. Iris: hazel. Biparental care.

Brown Pelican (*Pelecanus occidentalis*)

Adult. 2000–5000 g, (FL) M 3702 g, F 3174 g; 100–137 cm (39–54 in.). **Nestling.** Bill: greenish-gray, wedge-shaped with slight hook & short (compared to Ad), becomes full Ad length after fledging. Skin: pink, purplish-pink d2–4, more purple by d5–6. Down: naked except distal portion of wing; white down on rump d10–12; d21–25 down fully covers. G&D: d0: 54.9–87 g (1st to hatch grows faster than siblings); unable to lift head, eyes open. Juv plumage complete by d70. LF: cream white (by d24). Voc: begging call shrill, rasping *squawk*.

HERONS AND RELATIVES

(Family: Ardeidae, Order: Pelecaniformes)

Largely aquatic, deep-water wading birds with long legs, necks, and bills. Many raise young in colonies. **Adults**. Medium to large birds. Diet: carnivorous, primarily fish, also amphibians, invertebrates including crustaceans and insects, and small mammals. Nest: platform of sticks; twigs in trees or bushes; reeds on ground. Eggs: elliptical to subelliptical, mostly bluish to pale greenish, or brownish in large bitterns, and unmarked. Bill: long, straight, acute. Legs/feet: tarsi and (four) toes long to very long, tarsi usually scutellate in front, middle nail is pectinate. Plumage: varies greatly in color, all have powder-down breast and rump patches used for water repellency, often with modified plumes, lores usually bare. Wings: long and rounded. Tail: short. **Young.** Semi-altricial, eyes open at hatch or soon after. Capable of climbing, become branchers (climb out of nest and perch on branches) at 2 weeks. Gape: very wide to accept large whole foods. Bill: flexible and soft. Down: hatch nearly naked, then sparsely covered. Fledge: 28–42 days (smaller sp), 42–56 days (larger). Biparental care.

American Bittern (*Botaurus lentiginosus*) Plate 8

Adult. (ON, Canada) 520–1072 g; 60–85 cm (23.6–33.5 in.), M slightly bigger. LF: bright yellow to green, tarsus 9.12 cm (M), 8.49 cm (F). Bill: length 70–82 mm (M), 63–72 mm (F). Plumage: sexes similar. Nest: most common over dense emergent vegetation, in upland grasslands on ground, well hidden. Eggs: elliptical to short elliptical, unmarked buffy-brown to deep olive-buff. **Nestling.** MC: light pink. Bill: pinkish-flesh, dark tip. Skin: yellowish, eye orbital bluish-gray to pale olive. Down: long, light buff, pale burnt sienna, "clay," or yellowish-olive, scant & lighter below. LF: pinkish to orangish. Iris: brownish or light olive. G&D: d0: unknown; d10: PinUn on back, scapulars & neck. **Fledgling.** Fledges 7–14 d, depends on Ad 14–28 d post-fledging. Plumage: duller than Ad; lack neck ruffs. Iris: more yellowish-green. LF: olive-green.

Least Bittern (*Ixobrychus exilis*)   Plate 8

Adult. 86.3 g, 28–36 cm (11–14 in.). Diet: small fish & insects; H fed regurgitated mudminnows. Nest: platform with canopy, usually over water. **Nestling.** Bill: H 10 mm, prom egg tooth (gone d16). Down: soft, long (10–12 mm), light to dark ochre above, whiter below. Face: H bluish-gray, N greenish-yellow. Iris: yellow. LF: front green, back & soles yellow. G&D: eyes open shortly post-hatching; d0: 10–20 g, 76 mm, tarsus 13 mm; d3–4: bittern stance; d4: prim PinEm; d5: climb out of nest; d15: 50–70 g, mostly FF except crown & underparts; d30: tarsus 40 mm. **Fledgling.** Fledges 13–15 d. 1st flight ~29 d. Plumage like Ad F but paler & browner crown; browner throat & breast heavily streaked.

Great Blue Heron (*Ardea herodias*)

Adult. (BC, Canada) M 2480 g, F 2110 g; 97–137 cm (38–54 in.). Diet: mostly fish, also invertebrates, reptiles, mammals & birds. Nest: single pairs or colonies; loose & bulky platform, constructed of sticks (high in trees) or vegetation (on ground). **Nestling.** Bill, skin & LF: pinkish-gray. Down: white or pale gray, denser on crown. Iris: bluish. G&D: d0: 49–54 g; d5: 200; d10: 300–450; d18: 860; d27: 1350; d40: 2100–2400. Voc: H *tik-tik-tik*. **Fledgling.** Near Ad size. Fledge 65–90 d. Plumage: solid gray crown, attain Ad plumage by 3rd fall.

Great Egret (*Ardea alba*)

Adult. M 935 g, F 812 g; 94–104 cm (37–41 in.). **Nestling.** Bill: H gray to black, turns yellow; dark streak from under eye to half of bill. Down: long, white, longest on crown. Face: H pale blue, orbital area yellow d7, gray or gray-yellow d14, yellow d21. Iris: H pale yellow. LF: pinkish (bluish cast), turn gray to green-gray d7, darker gray d21. G&D: d7: most PinEm; d14: most PinUn; head & body still downy; d 25: brancher; d28: FF, down remains on head, neck & body; d37: some down. **Fledgling.** Flies short distances d49–56.

Snowy Egret (*Egretta thula*)   Plate 8

Adult. 371 g, 56–66 cm (22–26 in.). **Nestling.** MC: pale pink. Bill: H pale pink-gray, darker distally; tip may be yellow, dark gray, or yellowish at base (upper); N (d7) bill variable (even within same nest): may be black, yellowish, or pinkish with darker base & tip, some begin to turn dark, some may have more yellow in lower mandible whereas others may be variegated on upper or lower or both. Skin: grayish with pinkish & greenish tint; lores greenish; orbital area dark blue with gray or pale blue-gray eye ring. Down: whitest on spinal tract, wings pale yellowish, shimmery silver elsewhere. LF: gray, pinkish or greenish above with gray toes (greenish to reddish pink above). Iris: pale gray to buffy-gray; d34 off-white. G&D: d0: 20 g (0.98 cm); d4: 42 g (15 cm) mm); d8: 96 g (19.7 cm), PinEm all tracts; d10: may climb in and out of nest; d13: 178 g (25 cm); d14: PinUn prim, sec, wing covs, rect, back & ventral tracts; d18: 226 g (23–32 cm), down only on forehead, crown & neck; d21: 268 g (27 & 33 cm), very little down remains; d27: 36 cm; d34: 41 cm, FF. Voc: soft buzzing, food begging 2–3 syllables. **Fledgling.** Fledges: d49+. Bill: pale yellow at base, pinkish (darker distally), black d49+. Skin: green. Plumage: all white, no plumes, long feathers on crown & base of neck. LF: greenish, dark scales on tarsus front, toes greenish-yellow; d34 yellow-green. Leaves colony 53–56 d post-hatch. *Sim sp: nestling Snowy and Cattle Egrets may be difficult to differentiate as colors of bare parts (bill, lores, legs & feet) are paler than adults and more similar to each other. Cattle Egret has shorter bill & tarsus.*

Cattle Egret (*Bubulcus ibis*)   Plate 8

Upper mandible of bill is curved, appears hunchbacked when standing, feathering on chin & throat has bulging appearance. **Adult.** (FL) M 296–460 g, F 270–512 g; 46–56 cm (18–22 in.). Diet: opportunistic, very diverse; from insects to reptiles depending on habitat, season & prey availability. Follows wild & domestic ungulates that stir up insects. Nest: colony of same or mixed species, saucer shape, mainly of sticks, in trees or other vegetation. **Nestling.** MC: bright pink, blackish on palate lining bill. Bill & lores: H yellow-ochre, horn, yellowish-green, or olive-green; some may show pink upper & lower, yellow tip, lower tip may be black; N (d5–10) turns almost black. Skin: various descriptions, plumbeous to olive-green, turns battleship gray; lores yellowish or olive; orbital area green, yellowish-green or olive-green, then turns gray. Down: whitest on spinal & femoral tracts, buffier elsewhere; longest on crown & back (21 mm), bare on nape & throat. LF: pale tan, horn, yellowish-green, or olive-green; also described as pale yellowish-green on tarsus back, front olive-green; just before fledging legs turn dark green to jet black. Iris: H various colors from straw-yellow or grayish to nearly white, N becomes paler yellow. G&D: d0: 19 g; d6: PinEm; d14: most down lost; d14–21: becomes brancher (most susceptible to falling); d56: flight feathers fully grown. Voc: continuous high-pitched *zit zit*. Food requirements: ~95 g/d in wild. **Fledgling.** Begins to fly d25, departs d30, independent d45. Bill: black, turns pale to yellow ochre (by d30). Skin: more greenish. LF: dark greenish, scales on tarsus front become slate-gray to black. Iris: yellow. Plumage: all white, light buff on crown. *Sim sp: see nestling "Snowy Egret."*

Green Heron (*Butorides virescens*)   Photo 5

Adult. (FL) 212 g, 41–46 cm (16–18 in.). Diet: carnivorous, mainly fish & some invertebrates; young fed regurgitated food. Nest: single, loose aggregations, or colonies, ground level to 10 m or more; platform of sticks. **Nestling.** MC: pink. Bill: yellow with gray or black tip; d3 upper mandible yellowish-pink; d11 upper & lower mandible greenish-orange; d24 upper & lower orange. Skin: pink or tawny, eye orbital green. Down: grayish-brown above, white below; crest on head. LF: greenish & pinkish-tan; d3 turn pale yellow; d11 pale green to yellow-green. Iris: d0–8 pale gray, d13–14 grayish-white. G&D: little info, d0: 11–16 g, eyes open; d1: 9.5 cm; d2: 16.5 g; d6: PinUn tips; d7: 80–90 g; d8: crown & rect PinUn; d9–11: 100–135 g; d13: most body PinUn; d14: 135–160 g; d15: brancher, can climb; d16–17: tarsus 4.7 cm, becomes brancher; d21: 173 g, short flights, can cling to branches; d31: FF, down still attached. Voc: *tik-tik-tik-tik* food begging, d5 *ha* call. **Fledgling.** Short flights d21–23. LF: d17 pale yellow to orange-yellow. Iris: d17 greenish to yellowish-white. Plumage above: greenish-gray with brownish wash; cov tips rounded (not pointed) with buff edgings & white spots at tip of rachis; greenish-black crown, buffy-white malar stripe extends down neck & bordered each side with black & brown stripes; neck with chestnut streaks. Below: buffy-white, streaked brownish. Care: fed by parents 14 d post-fledge.

Black-crowned Night-Heron (*Nycticorax nycticorax*)   Photos 6-8

The most widespread heron in the world. Some populations declining. **Adult.** Medium-size heron, 615–1014 g; 58–66 cm (23–26 in.). Diet: opportunistic, mainly fish & aquatic or terrestrial organisms. Nest: in colonies; platform of sticks in trees or cliff ledges. **Nestling.** MC & skin: pink shades. Bill: H upper light drab, gray tip, then yellowish d5. LF: tarsus bluish; toes lighter on top, yellowish on bottom; pinkish-ivory nails. Down above: long, grayish filaments, white on femoral. G&D: d0: 24 g, length 10 cm; d5–6: PinVis flanks & scapulars; d7: alar PinVis; d10: rect PinVis, PinUn flanks & scapulars; d12: dorsal PinEm, becomes brancher; d18: prim & sec PinEm; d21: rect PinUn; d35: FF, climbs to high branches waiting for food. Iris: H grayish-olive, d1–2 light yellow, brighter yellow d20–30. Voc: persistent begging calls. **Fledgling.** Some fledge 29–34 d or fall to ground but cannot fly; most vulnerable in urban settings, flies d45, follows Ad to feeding areas. Iris: bright yellow (20–30 d). Plumage above: variable brown to deep reddish-brown with large pale spots. Below: paler & brown striped. Care: biparental or another Ad in rookery.

Yellow-crowned Night-Heron (*Nyctanassa violacea*) Plate 8

Adult. (FL) M 716 g, F 649 g; 55–71 cm (21.7–28 in.). Diet: crustaceans (fresh & salt water). Nest: stick platform, in colonies or scattered pairs. **Nestling.** Bill: H upper mandible pinkish-yellow, pinkish-gray, or grayish-yellow; turns greenish-black d7, then black; lower mandible paler. Skin: pink shades on head & neck, pinkish-yellow belly. Down: long, white or pale to medium gray. Orbital ring & lores: H pale gray, d7 yellow to greenish-gray, d36 orange-yellow; eye orbital deep grayish-blue to pale blue. **Fledgling.** d36: brancher; flies d42. Bill: bulky. Plumage above: drab brown, buffy spotting on back & wing covs. Below: white, streaked brown.

IBISES

(Family: Threskiornithidae, Order: Pelecaniformes)

These waders, found in fresh- or saltwater wetlands, are set apart from all other long-legged and long-necked birds by their distinctive bills and striking plumages. Note: This guide does not cover individual spoonbill species, but spoonbills are described here for comparison. **Adults.** Medium to large waders: 48–96.5 cm (19–38 in.). Diet: variety of terrestrial or aquatic invertebrates and vertebrates, occasionally vegetation; ibises probe in mud or water, spoonbills sweep bills through water. Nest: most in colonies, pile of sticks and vegetation, on ground, bushes, or trees near water. Eggs: most are subelliptical, dark bluish-green or greenish-blue, fading to paler blues and greens with incubation; some with light speckling. Bills are very long: ibis bill is decurved, slender, and somewhat terete; spoonbill is straight, broad and spatulate. Plumage: Dark for ibis (except White Ibis, *Eudocimus albus*), with metallic green and bronze, some with purplish overtones; spoonbills are white or pink with carmine accents (breeding season). Wingspans: long, broad, 36–45 in. (ibis), 50 in. (spoonbill). Tail: short. Legs/feet: medium-long, tarsi usually reticulate with slightly elevated hallux. **Young.** Semi-altricial, immobile, hatch with eyes open. Bill: nestling bills of ibis and spoonbill are straight and tubular; d9 spoonbill

bill tip begins broadening and flattening; d16 becomes spatulate. Down: fully covered but very sparse (skin visible underneath). Fledge: 23–56 days depending on species. Biparental care.

White-faced Ibis (*Plegadis chihi*)

Adult. (UT) M 563–807 g, F 433–677 g; 46–58 cm (18–23 in.). **Nestling.** Bill at hatch: short (15 mm from commissural point), straight, pink with 3 bands (1 around base, 1 at middle, 1 at tip); by d14 band turns darker & more distinct, by d12 bill is slightly decurved. Skin: pink, reddish-pink bald spot on the crown (d2). Down: sparse, brownish-black, darker/denser (d7) on dorsum; naked below. LF: olive-gray, dark spots on back of tarsi (all black d7). Iris: brownish. G&D: eyes partly open at hatch (open d2); d0: 30.7 g; d4: PinEm alar, capital & caudal; d4: prim PinEm; d4–7: patch of small white feathers may appear (some birds); d5–6: egg tooth gone; d7: prim PinUn; d9: climb out of nest, walk around; d12: hide in vegetation; by d14: skin darker, black PinUn all over body; by d21: very mobile, natal down mostly gone, bald spot on crown turns light orange in some birds; by d28: FF, remiges well developed. **Fledgling.** 1st bird to fledge leaves nest by d9, short explorations near nest; short flights by d28, better flights by d35. Plumage above: mostly dark greenish-olive with oily appearance. Below: lighter than upperparts, grayish hair-brown on chin & throat, crissum blackish-green. Leaves colony 6–7 wks after hatching.

VULTURES, NEW WORLD

(Family: Cathartidae, Order: Cathartiformes)

Adults. Large to massive birds: California Condor (*Gymnogyps californianus*) 7000–9900 g, 117–134 cm (46–52.8 in.); smaller vultures ~2 kg, 60–81 cm (24–32 in.). Diet: primarily carrion, some species occasionally kill live prey. Nest: scrape on bare ground on cliff ledge, or cavities in trees or rocks. Eggs: subelliptical, white (vultures) with brown spots and blotches; blue-green (condor). Bill: moderately hooked with large, oval, perforate nostrils. Plumage: mostly dark, small bare heads may be brightly colored; lores have bristle-like feathers. Flight feathers: 10 primaries, 22 secondaries, 12 rectrices. Wings: long, broad, rounded. Tail: long, rounded. Legs/feet: weak, anisodactyl, tarsi reticulate; short nails. **Young.** Semi-altricial. Down: fully covered; two downy stages in condor; one stage in Turkey Vulture; second stage not described in Black Vulture. Fledge: 8–13 weeks (vultures), 6 months (condors). Care: biparental, young remain dependent many months after fledging.

Black Vulture (*Coragyps atratus*)   Plate 9

Strong social bonds maintained for life, roost communally. Sympatric with Turkey Vulture but (up close) slightly smaller, short tail, head gray, bill long & straight, blacker plumage. Lacks sense of smell. **Adult.** (TX) 2159 g; 60–69 cm (23.6–27 in.). Nest: bare ground in cave, tree cavity, or abandoned building. **Nestling.** Bill: black, sharply decurved & hooked at tip. Skin: dark gray to black; face appears bare. Down: long, thick, pale pinkish-buff. G&D: relatively slow growing compared with other vultures & raptors of similar mass; d0: 70 g; d17–23: wing PinEm; d35: prim PinF (only 1–2 cm) unsheathe (tips); d39: N almost full size, tail PinF 5–7 cm; d52: most wing cov PinUn & breast PinEm; d65: wings FF; d79: FF, some down on breast & belly. **Fledgling.** 1st flight: 75–80 d. Bill: black, long, hooked; nostrils perforate, narrow & small. LF: blackish-brown. Iris: dark brown. Plumage like Ad except head is black, smooth textured, becomes lighter & wrinkled with age, short & sparse bristle-like feathers extending to upper neck. Parental care: may feed young several months post-fledging.

Turkey Vulture (*Cathartes aura*)   Plate 9

Highly social, communal roosts, flocks generally accept juveniles as new members into the group. **Adult.** 2000 g; 64–81 cm (25–32 in.). **Nestling.** Bill: dark, short, hooked, large nares are perforate. Down: long, white, fluffy, shorter on head, thickens & turns "dingy white" (by 25 d), d60: only traces of down. Face, throat & crop areas: black & bare. LF: gray at first, then pinkish. G&D: d0: 60 g, eyes *may* be open; d14: egg tooth still present; d14–18: distal prim PinEm; d18: able to stand; d21: prim PinF longer, face becoming grayer, able to stand; d26: back PinUn, prim feathers half in sheaths & half exposed feather; d33: back feathers more apparent, tail feathers poke through down; d34: wing covs evident on wrists, down thinning on crown; d37: tail ~3 in.; d42: wing feathers longer; d43–44 (or earlier): breast PinUn nearest wing, neck ruff evident; appears very dark above; d53: 1st feathers appear on legs below down; d54 to flight: remaining white areas of down fill in with dark feathers; amount of white remaining varies by nestlings' willingness to self-groom; may have blue speckles/flecks on shoulders. **Fledgling.** Leaves nest: ~60 d post-hatch with short flights at first, extended flight 70–80 d. Bill: tip black for several months, by fall turning white. Plumage: like Ad but blacker with more iridescence, remnants of down, gray legs & head (turn pinkish by 1st fall).

OSPREY

(Family: Pandionidae, Order: Accipitriformes)

Osprey (*Pandion haliaetus*) Plate 9

Distinctive features (evident in fledglings) separating osprey from other raptorial birds are wing shape and dark eyeline. Wing & culmen length are used to age young. **Adult.** Large raptors, M 1220–1600 g, F 1250–1900 g; 58.4 cm (23 in.). Diet: dive for fish, usually in shallow waters. Nests: bulky with large sticks & other debris, near fresh or salt water. Eggs: subelliptical, white to yellowish or pinkish usually heavily marked with reddish-brown spots. Bill: hooked for tearing flesh of fish, pale blue-gray cere. Flight feathers: 10 primaries, 18–19 secondaries, 12 rectrices. Wingspan 150–180 cm. Wings: long, narrow, pointed. Tail: medium, rounded. Legs/feet: may be greenish or light bluish-gray, zygodactyl toe arrangement (outer toe reversible), tarsi reticulate, feathered to toes, soles of feet have spiny scales. Talons: black, long, strong & acutely curved for grasping, holding, or killing prey. **Nestling.** d0: ~40–60 g. Semi-altricial; eyes open at hatch, confined to nest, limited movement, cannot self-feed. Bill: dark gray. *1st down*: short, thick, completely covers bird but sparse on head, nape & sides behind wings, brownish above, whitish below, mottled appearance. Head creamy white with brown markings, dark brown "mask" & auricular patch, wide buffy stripe runs from head down central back. *2nd down* d11: thicker & darker. d21: 1st juv feathers appear. **Fledgling.** Fledges: 1st flight, 50–59 d. Plumage: similar to Ad DefB except buffy breast & dark feathers on back & wings with buffy to white edges or scaling. Carpal patch mottled with white, not as conspicuous as Ad. Tail: dull white with more dark brown bars & broader white tips. Iris: generally darker than Ad iris but variable, from brownish-yellow to orange-red, turn yellow by 1st fall. Behavior: apparently fishing is innate (do not learn to hunt by observing parents), may dive to recover dropped food by parents. Biparental care. Independent 10–20 d post-fledging.

HAWKS, EAGLES, AND KITES

(Family: Accipitridae, Order: Accipitriformes)

Accipitrids hunt diurnally with hooked beaks for tearing flesh and sharp talons for grasping, holding, or killing prey. They have a soft, fleshy, colored cere at the base of the beak. Females are larger than males. Young are considered semi-altricial in most accounts as eyes are open or partly open at hatch, and they have down in all feather tracts. Near-fledglings may leave the nest and become branchers (perch on nearby branches) and may climb in and out of their nest. Biparental care.

HAWKS

Adult. Medium to large raptors. Diet: most species are opportunistic, hunt live prey or scavenge; hunting method helpful for identification. Nests: large cups with twigs and other debris, on trees, bushes, ledges (rock or building), or the ground. Eggs: rounded, white to cream, sometimes with dark blotches. Bill: strongly hooked. Plumage above: most predominantly brown, gray, or black; many species with cream-colored breast at first; below often paler and barred or streaked with brown. Wings: broad and rounded in buteos and more sharply pointed in accipiters. Tail: medium relative to body. Legs/feet: Accipiters have long toes (especially the middle front and hallux) for grasping birds and scales are scutellate on tarsus front. In buteos, tarsus is scutellate on front and (except Rough-legged and Ferruginous Hawks) unfeathered. **Young.** Eyes open at hatch or soon after. Size: varies with species. Down: two stages, white at hatch (except cryptically colored Harris's Hawks). Juvenile feathers appear after several wks. Newly hatched chicks are able to hold up heads & feed by sight. Fledge: varies by species.

EAGLES

Adult. Very large raptors. Wingspan: 79 in. (Golden Eagle); 80 in. (Bald Eagle). Diet: hunt live prey or scavenge; most species are opportunistic, taking fish, rabbits, ground squirrels, ducks, and occasionally larger mammals. Nest: large cup with twigs and other debris; in trees, bushes, and ledges (rock or building) and on ground in regions where cliffs and trees are scarce. Eggs: rounded, white to cream, sometimes with dark blotches. Bill: large, strongly hooked. Wings: broad for flapping and soaring. Tail: medium relative to body. Legs/feet: feathering on lower tarsi varies by species. **Young.** Eyes partly open at hatch or shortly after, limited movement, able to hold up head. Size: varies per species. Down: two stages, white at hatch; mostly fully covered. Juvenile feathers appear after several weeks. Fledge: 8–14 wks (Bald Eagle), 45–81 days (Golden Eagle).

KITES

Adult. Medium-size raptors, 35.6–55.8 cm (14–22 in.). Diet: some catch insects in air, pluck reptiles and insects from treetops; a few species feed on snails by plucking from marsh grass or treetop; another species feeds on small rodents by hovering then dropping to the ground. Nest: sticks and twigs with shallow cup. Eggs: elliptical or subelliptical, white marked with brown to reddish-brown. Bill: short, hooked tip. Plumage: variable by species, in whites, grays, and browns. Wings: long, broad or narrow, pointed. Tail: rounded, squared, or forked (Swallow-tailed Kite, *Elanoides forficatus*). Legs/feet: bare, nails acute. **Young.** Down: two successive coats. Fledge: 23–42 d, depending on species. Care: in some sp M brings food but F feeds.

White-tailed Kite (*Elanus leucurus*)

Adult. (CA) 346 g (avg); 32–38 cm (12.6–15 in.). **Nestling.** MC: pink. G&D: d0: 17.4 g, mass gain ~12 g/d to d28; d10: tail 10–12 mm; d20: tarsus near Ad length of 34.4 mm (mean); d28: equal to or greater than Ad mass. *1st down*: pale vinaceous buff, lighter on lower back & sides. *2nd down* d7: gray & longer than 1st; d40: FF. **Fledgling.** 1st flight 28–35 d post-hatch. MC: red. Bill: black. LF: yellow. Iris: brown. Plumage above: head mostly white, buffy-tan elsewhere, nape dark gray, back light brownish or sepia. Lesser & median covs blackish (some with white tips), greater covs dark gray with white tips, prim covs blue-gray tipped white; dark spot (distal) under wing. Below: mostly white, tawny band on upper breast. Tail: med gray, dark gray subterminal band (~5 mm wide). May learn to hunt by joining parent while hovering & dropping to ground. No record of siblicide.

Golden Eagle (*Aquila chrysaetos*)

Adult. (ID) M 3000–4475 g, F 4075–5280 g; 70–84 cm (27.6–33 in.). **Nestling.** Bill: black with prom egg tooth. Cere: fleshy, yellowish-white. LF: light yellow, talons white to flesh. *1st down*: short grayish-white or white with dark (or white) tips, gone by ~d15. *2nd down* d6: long, white; becomes dense & water resistant; grows until ~d30. G&D: d0: 100 g; d10: 500; d15: prim PinEm; d18: sec, scapulars & rect PinEm; d21: prim PinUn; d22–25: upperwing covs PinEm (PinUn d27); d22–28: dorsal & ventral PinEm (PinUn d29–35); d29–35: PinEm head (PinUn d36–49) & thighs (PinUn d36–42); d50–60: reach max body mass; d56: juv feathers replace down (complete ~60 d); d65: 7th prim 269–316 mm & left center rect 191–253 mm; d80–105: feather growth complete. *Young birds prone to heat stress & death in extreme temp. Parents will feed grounded nestling.* **Fledgling.** Fledges (highly variable) 45–81 d. May fall, jump, walk, or fly from nest. Sustained flight > 64 d. Cere yellow. LF: legs feathered to toes. Plumage (retained 9 months or so): darker brown than Ad with white (amount varies) at base of sec & inner prim; 2/3 of tail white at base (may have some dark flecks) with black band on tip. Rect: mostly white, dark terminal band varies in width. Iris: brown.

Northern Harrier (*Circus hudsonius*)   Plate 10

Adult. F avg about 50% heavier & 12.5% larger than M. (NJ) M 283–472 g, F 375–661 g; 46–50 cm (18–19.7 in.). **Nestling.** Bill: blackish, base becomes bluish in 1st yr. Cere pale pinkish-tan, becomes yellow by d14. LF: pale to orange-yellow; yellow (F thicker & shorter tarsi than M) by d21. Iris: brown; d11–14 can distinguish sexes (M has grayish cast, F has brownish cast). *1st down*: short, white, tinge of pinkish-buff, densest on dorsum. *2nd down*: longer, pale smoke-gray, lighter below. G&D: d0: 23.8 g, eyes open (or very soon after); d2: can walk around nest; d7: PinUn wing tips, PinEm back, shoulders, breast & tail; d9–10: tail feathers emerge; d15–20: brancher. **Fledgling.** Fledges: 1st flights 27–35 d. Plumage somewhat similar to DefB of Ad F but darker & warmer: blackish-brown, pinkish-tan streaks on head, upperwing covs with buff edges, prim dark & white (narrow) barring, barring on tail more defined than Ad. Below: upper breast streaking heavy & dark, wider on flanks, finer elsewhere. Siblicide: very rare. *Apparently highly susceptible to imprinting or habituating on caregiver.*

Sharp-shinned Hawk (*Accipiter striatus*)   Plate 10

Smallest North American accipiter. **Adult.** (WI) M 82–125 g, F 144–208 g; M 24–28 cm (9.4–11 in.), F 29–34 cm (11–13.4 in.); wingspan: M 53–56 cm, F 58–65 cm. **Nestling.** MC: light cobalt blue. Gape (surrounding tissues of beak): yellow or greenish-yellow. Cere whitish to yellow & greenish-yellow. *1st down*: short, white tinged creamy. *2nd down* d7: longer, woolly, white, pinkish-buff belly. G&D: M develops several days sooner than F; d0: eyes open; d1–2: leaves nest, climbs around on branches; d7: M PinEm contour & prim; d9–10: M PinEm tail; d14: Prim PinUn; d28: FF. Voc: *peep* calls d1. **Fledgling.** Fledge dates vary by geography & by sex: M leaves 21 d, F 27 d. LF: yellow, black talons. Iris: variable;

H grayish, then yellow, then orangish to ruby-red. Plumage: still downy, white background with dark sepia streaking on forehead, crown, nape, throat (white) & sides of neck. Supercilium: pale whitish, narrow dark streaks. Tail: square, 3–5 broad, blackish bands tipped white. Remiges: inconspicuous barring above, large white patches on some scapulars & inner sec. Below: white or whitish-cream, streaked with tan, brown, or dark cinnamon (largest on breast).

Cooper's Hawk (*Accipiter cooperii*)   Plate 10, Photo 10

Medium-size hawk. Short, rounded wings & long, rounded tail. **Adult.** Size varies greatly: M < F, western birds smaller than eastern; avg mass (breeding) w. US M 280 g, F 473 g; avg mass breeding e. US M 338 g, F 566 g; M ~39 cm (15 in.), F ~45 cm (18 in.). **Nestling.** Altricial. MC: purplish. Bill & cere: pale pinkish-tan, lower mandible darker. LF (& talons): nearly white, turn lemon-yellow with black talons by d21. Iris: blue-gray, turns yellow by 1st yr. G&D*: (1 captive): d0: ~28 g, 9 cm; d9–12: 147 g; d10: egg tooth gone; d11: PinUn; d11–13: PinEm all over; d14–18: 238 g; d20: ~282; d27–28: ~312, nearly FF. *1st down*: white, fully covered. *2nd down*: d5–9 woolly, white to buffy. Voc: *cheep* calls (hatchling). **Fledgling.** Fledge date variable geographically: (NY) M leaves 30 d, F leaves 34 d; (AZ) leaves 31 d. Plumage overall: clove to dark brown feathers (edged tawny) on white bases. White supercilium. Below: pale buff to nearly white or pale gray, breast & flanks with dark teardrop-shaped streaks. Tail: rounded, 6 broad dusky bars narrowly edged white. *d9–20 weights from one bird in rehab (Carrie Laxon).*

Northern Goshawk (*Accipiter gentilis*)

Adult. (WI) M 735–1099 g, F 845–1364 g; (AZ) M 631–774 g, F 907–1100 g; 53–64 cm (20.9–25.2 in.). **Nestling.** Bill: gray. Cere, gape & LF: greenish-gray. *1st down*: short, silky, white (may have grayish tinge on head & back); *2nd down* d7: thicker, white (may be grayish on dorsum). G&D: d4–7: 13 cm long; d9–12: 15–18 cm; d14–17: ~20–23 cm, PinEm on prim, sec & tail; d19–22: walks on feet, PinUn prim, sec & tail; d24–26: auriculars feathered; d36–38: body ~90% feathered with down present on side of neck & thighs, tail ~3/4 full length. **Fledgling.** Becomes brancher 34–35 d; 1st flights M 35–36 d, F 40–42 d. Cere & gape: pale yellow. LF: pale yellow (juv), half of tarsus feathered in front. Plumage above: dark brown to brown-black, streaked buff white & cinnamon, zigzag pattern on tail (wavy dark brown bands with thin whitish edges), supercilium (pale white) on brown head. Below: buff-white, throat broadly streaked cinnamon to blackish-brown.

Bald Eagle (*Haliaeetus leucocephalus*)

Adult. (AK) M 3637–4819 g, F 4631–6400 g; 71–97 cm (28–38 in.). **Nestling.** Bill: blackish-gray. Cere: pale gray, yellow by d4–8, pale olive by d9–12, becomes darker with age. Gape, LF, skin: pink. Iris: brown. *1st down*: light gray. *2nd down* d9–11: thick & dark gray. G&D: M gains 102 g/d, F gains 130 g/d; max growth 3–4 wks post-hatch, M more rapid than F. Asynchronous hatching results in chicks of greatly different sizes & ages in same nest. d14–21: PinEm wings, tail & humeral tract; d24–31: PinEm body (contour); d28–35: PinEm head & back; d42–56: feathers appear on tarsi; 11–14 wks: FF. **Fledgling.** Fledges (highly variable) 56–119 d; some may be grounded for wks before gaining flight ability, parents continue to feed. Bill & cere: blackish-gray. L/F: lower tarsi unfeathered, yellow by d9–12, talons black. Plumage variable (some darker than others): dark brown contour feathers over white bases all over. Below: amount of white & mottling varies. Tail: mottled with white, gray & dark brown; some with indistinct brownish-black terminal tail band.

Mississippi Kite (*Ictinia mississippiensis*)   Photo 9

Adult. (OK) M 245 g (avg), F 311 g (avg); 34–37 cm (13.4–14.5 in.). **Nestling.** MC: pale yellowish-gray. Cere & rictus: pale to bright yellow. Eye ring & lores: dark gray to black. Bill: black. LF: pale yellow to orange-yellow. G&D: d0: 13–21 g, gains 7.7 g/day to 10 d; d10: 75–120. Down: pure white. **Fledgling.** d25–30: moves to branches, d30–35: flies, Ad mass.

Red-shouldered Hawk (*Buteo lineatus*)   Photo 11

Adult. (*B. l. lineatus*) M 486–582 g, F 593–774 g; 43–61 cm (17–24 in.). **Nestling.** Bill: black. Cere: H light, turns greenish-yellow. LF: H talons gray. *1st down*: yellowish-white with darker back & wings; *2nd down:* woolly & thicker, pure white belly, grayish-white above. Iris: pale to med gray-brown. G&D: weight increases most rapidly during 1st 21 d, levels off ~d34, asynchronous hatching results in chicks of different sizes & ages in same nest. One observation found 5–7 d span between the 1st & 4th egg. d0: 35 g; d5: 55, can back up to defecate over nest edge; d10: 165 g; d14: prim PinEm; d15: 320 g; d20: 460; d21: rem PinUn; d23: PinUn on body & rect, stands frequently & walks; d26: head PinUn & become FF (back 1st, then sides of breast, head is last); d30: 534 g. **Fledgling.** 470–690 g (varies by sex, F > M). Fledges 33–45 d, varies considerably. After young leave nest, wings grow 3–8 cm longer & tail grows 5–12 cm. Cere: greenish-yellow. LF: pale yellow or greenish, black talons. Plumage: above dark & light browns with tawny edges, below whitish to buffy with elongated or oval spots, flanks streaked. Tail dark brown dorsally, thin light brown bands with thin white band on tip.

Broad-winged Hawk (*Buteo platypterus*)

Adult. 265–560 g; 34–44 cm (13.4–17 in.). Two color morphs, light and dark. **Young.** Bill: blackish, cere pale yellow. Gape: pinkish-yellow edge. LF: H talons gray. Down: thick, dirty white, then white with grayish undercoat. G&D: d0: 26–30 g; d9: rem PinUn; d11–13: rect PinUn. **Fledge.** Fledges 30–36 d, 370 g. LF: yellow, black talons. Plumage: *light morph* similar to Ad but with brown streaking (amount varies) on breast, sides & belly. *Dark morph* similar to Ad but breast more rufous & with more tawny streaking.

Swainson's Hawk (*Buteo swainsoni*)

Adult. (AB, Canada) M 693–936 g, F 937–1367 g; M 48–56 cm (19–22 in.), F 51–56 cm (20–22 in.). **Nestling.** *1st down*: thick, pale white (with yellowish tinge) or yellowish to grayish-white (with pale pinkish-buff tinge). *2nd down* d6: white (may have gray cast). G&D: d0: 39.4 g (avg); d9–10: prim PinEm; d13–17: can stand; d14–15: tail PinEm, dark patches appear as rect, scapulars & remiges show; d16: breast patches darken; d28: PinEm head (down still present); d29–33: exercises wings; d34: appears FF. **Fledgling.** 1st flight: 38–46 d. Bill: slaty to blackish. LF: paler than Ad. Iris: gray or blue-gray (becomes yellowish in 1st yr). Plumage of *light morph*: crown streaked dark brown, forehead whitish, buffy supercilium, narrow eyeline, malar streak broad & dark brown; upper back buffy & appears black streaked, browner on back. Below: creamy-white (some may have cinnamon-buff tones), whitish chin & throat with dark brown streaking. Wings: fuscous. Tail: fuscous, dark brown bands (increase in width toward tips) & broad subterminal band (black with whitish tip). *Dark morph*: similar to light morph but markings are heavy & browner with rufous spotting below, brown band or "bib" across throat & upper breast, belly darker.

Red-tailed Hawk (*Buteo jamaicensis*)   Plate 10

Adult. M 1000 g, F 1200 g. (*B. j. calurus*, w. US): M 900 g, F 1300 g; M 45–56 cm (17.7–22 in.), F 50–65 cm (20–25.6 in.). **Nestling.** MC: cream to buff. Bill: black. LF: yellow (more yellow in juv), tarsus unfeathered throughout life. Down: short, white, no down on femur, white occipital spot may be visible; becomes longer & lighter on head by d9. Iris: dark, then grayish. G&D: d0: 58 g; d7: begins to pick at prey in nest, captive birds may peck & fight with nestmates but by d14 aggression with siblings disappears; d9: 7th prim PinEm (best indicator of age after d24), white occipital spot on head becomes longer & lighter; d14–15: sits up on tarsus (captive birds head-bob, stand erect, flap & stretch wings); d21: aggressive behavior (strike with talons & wings) toward intruders; captive birds may self-feed; d29–30: stretches & flaps wings, head 90% downy, occipital spot still evident, upperwing 90% feathered, breast 50% feathered, feathers appear on legs, uppertail covs well grown, ear openings completely covered; d33–35: head 50% feathered, dorsum 95% feathered, breast 90% complete. **Fledgling.** Fledges 42–46 d, sustained flight 2.5 wks post-fledging. Plumage (3 morphs): *Light morph*: above dark fuscous on white base, breast creamy white; tail brown with 8–12 fuscous bands tipped whitish; prim with indistinct sepia bars & tips. Below: white, usually unmarked except throat & thighs sometimes spotted fuscous, abdominal band with sepia markings. *Dark morph*: above fuscous, base of uppertail covs cinnamon-rufous; undertail covs pale cinnamon marked with fuscous, edged cinnamon-rufous. Wings & tail as light morph. *White morph*: similar to light morph but more heavily mottled white above; below whiter with paler brownish (thinner) streaking; tail white with curved paler brown bands. Iris: light yellow, then brighter yellow (1st winter) to brown (Ad). Voc: soft *peeping* (d1), high whistling sounds especially when Ad overhead (d10). Nearly independent 6–7 wks post-fledging.

Ferruginous Hawk (*Buteo regalis*)

Adult. (w. US) M 977–1347 g, F 1501–2047 g; 56–69 cm (22–27 in.). **Nestling.** *1st down*: white or pale gray (5–7 d). *2nd down*: long, thick, mostly white. LF: yellowish or olive. G&D: d0: 51.6 g; d15: PinF 1st appear; d18–20: able to stand & walk; d32: M 1000 g & F 1500 g; d33–34: flaps wings. **Fledgling.** Fledges 38–50 d (leave ground nests earlier than tree nests). Heat stress may cause early departure. LF: orange-yellow. Plumage: *Light morph*: above clove-brown to yellowish-brown; head &

back tipped ochreous-buff, upperwing covs tipped russet; tail white at base, brownish & grayish, whitish on inner webs, 1–5 diffuse & indistinct dusky bars. Below: white to pale buff with buff on upper breast & sides & black spotting. *Dark morph*: dusky brown all over, breast may be tinged rufous. Tail: bands may be darker. Siblicide rare.

BARN OWLS

(Family: Tytonidae, Order: Strigiformes)

One subspecies in this family raises young (sometimes year-round, climate permitting) in mainland N.A. (and south to Nicaragua and Hispaniola).

Barn Owl (*Tyto alba*)   Plate 10

Barn Owls are widely distributed and the most widespread of all owls with some populations declining. They are nocturnal predators with directional hearing for accurately locating prey. Two external features involved in their remarkable hearing are a triangular or "heart-shaped" facial disc that guides sound into the ears and large, external, asymmetrically placed ear openings. Toes: second and third toes equal in length with pectinate claw on nail of third toe. Asynchronous hatching results in chicks of widely varying development. **Adult.** Medium-size owl, size varies by ssp and sex: (*T. a. alba*) M 330 g, F 370 g; (*T. a. guttata*) M 306 g, F 357 g; (*T. a. pratincola*) M 474 g, F 566 g; M 33–41 cm (13–16 in.), F 33–40 cm (13–16 in.). Diet: mainly mammals; some birds, reptiles, amphibians, insects. Nest: wide variety of existing cavities in trees, cliffs, boxes, buildings, haystacks; sometimes excavates burrows in banks or arroyos. Eggs: white, subelliptical to elliptical. Bill: hooked; cere present; nares open at edge of cere. Plumage: very soft, pale, variable in coloration and pattern. Wings: long, rounded. Tail: short, square. LF: zygodactyl, long, tarsi twice as long as middle toe without nail, sparsely feathered, feathers on back of tarsi pointed upward; middle nail pectinate (serrated). **Nestling.** Semi-altricial for the most part, hatches with eyes closed but has down in all tracts. Distinctive facial and bill features evident. Bill: ivory. Skin: pink. Down: fully covered. *1st down*: sparse, white to grayish-white on upperparts. *2nd down* d14: long, dense, white to pale gray or buff. G&D: d0: 12.3–21.2 g; d40: maximum mass attained. **Fledgling.** 1st flight 50–55 d, clumsy at first. Biparental care. Siblicide rare; may eat dead chicks.

OWLS

(Family: Strigidae, Order: Strigiformes)

Adult. Small to large raptors, F > M (the difference is greatest in the larger species). Diet: carnivorous; wide variety of prey. Nest: cavity or platform, man-made boxes, one species excavates a burrow. Eggs: white, elliptical to rounded. Bill: short, powerful, culmen strongly decurved, hooked; cere present; nostrils open at edge of cere except in genus *Athene*. Plumage: soft, fly silently; facial disc circular; many have "ear-tufts." Wings: form varies, one to six of outer primaries with notched edges. Tail: length varies; usually somewhat rounded, some square. LF: zygo-dactyl, fourth toe reversible, strong, very sharp talons; feathers on back of tarsi point downward; middle toe shorter than inner toe; some species feathered to toes. **Young.** Semi-altricial for the most part, hatch with eyes closed but have down in all tracts. Down: fully covered; two downy stages reported for most species. Juvenile feathers of body very loose in struc-ture especially on undersides. Fledge: nest period dependent on species, however some leave before they can fly. Some become branchers whereas others (usually ground nesters) leave on foot. Fledgling males generally darker than females. Biparental care; length of dependency post-fledging variable by species.

Flammulated Owl (*Psiloscops flammeolus*)

Second smallest owl in N.A. Strictly nocturnal, highly migratory. **Adult.** (CO) M 44.5–62.5 g, F 44–81.5 g; 15–17.8 cm (6–7 in.). Diet: mostly insects; moths & beetles mainly. Nest: prefers large abandoned wood-pecker cavity. **Nestling.** Down: fully covered, thick, snow-white. Bill & LF: pinkish-tan at hatch, then darker. Cere & skin surrounding eye or-bit: pinkish-buff. G&D: during rapid growth phase gains 4g/d. (CO) d0: 6–9.5 g, (NM) d0: 6–8 g; d3–5: egg tooth lost; d6–7: prim PinEm; d10: fluffy gray feathers emerge; d11: 30–40 g (last hatched = smallest); d21–28: little down remains. Voc: *trill* begging call. **Fledgling.** (CO) 45.5–73.5 g, (NM) 48.2–71 g, rem 60% & rect 50% of full length (fully grown 20–25 d). Fledges: d25 after hatch, does not fly well, either falls or 1st flight is to ground, climbs back up and repeats flight attempts, attended to by parents. Runts usually catch up to size of siblings by fledg-ing. Plumage: fluffy gray feathers with transverse barring of narrow dark & pale lines, most distinct on crown. Facial disc edged rufous to buff, facial feathers not fully developed at fledge. Rem & rect gray-brown with lighter barring, paler on outer webs. Independent 25–30 d post-fledging.

Western Screech-Owl (*Megascops kennicottii*)   Plate 11

Small owl with feathered ear tufts & legs feathered to toes. Plumage variable: brown to gray-brown in Northwest (gray in s. deserts), ~7% reddish-brown in coastal populations in Pacific Northwest. **Adult.** F > M, M 100–230 g, F 120–305 g; 19–25.5 cm (7.5–10 in.). Nest: in cavity (natural or box). Diet: invertebrates (insects primarily), also wide variety of small animals. **Nestling.** Down: covered, white, then grayish barred with sepia, narrower barring below. G&D: d0: 13.8–14.3 g, ~64 mm long; d7: 51.2–58.8 g (3 nestlings), eyes open, egg tooth gone. **Fledgling.** 114–137 g (3 birds). Fledges d28. Iris: pale to lemon-yellow. Plumage similar to Ad DefB, dusky & white barring, may still have remnants of down at fledge; prim with more distinct barring & heavily spotted with buff. Independent ~5 wks post-fledging.

Eastern Screech-Owl (*Megascops asio*)   Plate 11

Adult. 121–244 g; 16–25 cm (6.3–9.8 in.) (smallest M) to 25 cm (9.8 in.) (largest F). Nest: in cavity (natural or box). Diet: invertebrates (insects primarily), also earthworms & crayfish, small songbirds & rodents. **Nestling.** Skin: pink. *1st down*: fully covered, creamy white; *2nd down*: tawny-olive (tipped white) & barred sepia. G&D (TX): d0: 13 g; d7–10: juv feathers appear in following order: scapulars to wings, tail, back, underparts & head; d14: prim PinEm (2–3 mm); d14–16: 72–84 g, can thermoregulate. **Fledgling.** (TX) d28: 125 g (less in rural birds than suburban possibly because of less food availability in rural areas). Jumps or flies limb to limb; if chick falls to ground will walk-hop to next tree & climb up. By d20 morphs distinguishable: plumage above similar to Ad DefB except sides of head lack dark ring, ear tufts one color, body with narrow dark cross-bands; down mostly gone (except few on head & back). Below: *gray morph* is grayish-white; *rufous morph* is grayish-white but barring is narrower with pinkish-cinnamon tinge. Wings & tail ½ full length.

Great Horned Owl (*Bubo virginianus*)   Plate 11, Photos 12-13

Found in wide variety of habitats year-round from northernmost to southernmost regions in N.A. & parts of S.A. Hunts primarily from perch at night; hunts during daylight especially when feeding young. **Adult.** (*B. v. pallescens*) M 724–1257 g, F 801–1550 g; (*B. v. virginianus*) M 985–1588 g, F 1417–2503 g; 46–63 cm (18–25 in.). Nest: wide range of sites from cavities, platforms & ground; commonly takes over hawk nests.

Diet: wide variety of prey (90% mammals especially rabbits & hares, 10% birds); occasionally some invertebrates. **Nestling.** Skin & LF: pink. Bill: slate gray. Iris: yellowish-hazel (2–3 wks), then slight traces yellow appear; bright yellow by 4–5 wks. *1st down*: pure white; *2nd down* d7–21: longer, soft, grayish-buff with dusky mottling on back. G&D: d0: 49–54 g (gain 34.7 g/d); wk 1: egg tooth disappears, rem & rect PinUn; wk 2: eyes open, prim & sec PinUn; wk 3: ear tufts more evident; wk 3–4: F 1000 g, M 800. **Fledgling.** Leaves nest ~75% of Ad mass. 6 wks: move around on branches; 1st flight (short distance) at 7 wks. Long dependency; remains near (& fed by) Ad through early to midfall. LF: completely feathered. Plumage variable: soft, loose & long on rump, thighs & under tail; (wk 8–9) down still present. Barring dusky & heavier on back than breast. Head: short feathers, cinnamon-buff with paler tips; ear tufts becoming evident; (wk 11) white bib & facial disc well defined; (wk 26) ear tufts fully grown.

Snowy Owl (*Bubo scandiacus*)

Raise young in northernmost areas on Arctic tundra. **Adult.** (AB, Canada) M 1606–2043 g, F 1838–2951 g; 52–71 cm (20.5–27.9 in.). Nest: depression in elevated mound on ground. **Nestling.** Semi-altricial. Skin pink. *1st down*: white or light cream; *2nd down* d5–10: gray colored, considered part of juv plumage. Iris: grayish-yellow, or various other descriptions of pale gray to greenish-yellow. G&D: d0: 35–55 g (by d7 mass triples); d9–11: eyes open. **Fledgling.** 21 d, leaves nest on foot but cannot fly, strong flight 45–50 d, hunts independently at 60 d. Bill: black with large nostrils covered by long rictal bristles; wide gape. LF: pink metatarsal pad, white claws; as chick grows long feathers cover whole leg & over toes. Iris: yellow (22–28 d). Plumage above: dark mouse-brown, mantle & scapulars with faint whitish barring, white facial disc & chin; prim & covs, sec & rect white with dark brown markings. Below: dark mouse-brown with grayish-white tips & speckles. 5–6 wks: F may show more extensive barring than M on prim, tertials & sec.

Northern Hawk Owl (*Surnia ulula*)

Adult. (MN & Canada) M 242–375 g, F 250–454 g; 36–45 cm (14–17.7 in.). Nest: typically in broken snag. **Nestling.** Down: short, dense, white or white with buff-yellow. G&D: d0: 18 g (gains ~9 g/d); d14–21: facial disc evident. **Fledgling.** Fledges 25–35 d. Plumage: shades of gray & white, down still evident on head; facial disc mostly black (more contrast at 28 d), large white V mark between & above eyes, variable amount of white in disc surrounding eye & at cheek. Juv similar to Ad DefB except gray-brown & fewer white markings, crown tipped gray, white, or buff; below dull grayish-white with brown barring (narrower on upper breast & legs).

Northern Pygmy-Owl (*Glaucidium gnoma*) — Photos 14-15

One of the smallest owls of N.A. The similar Ferruginous Pygmy-Owl (*Glaucidium brasilianum*) occurs in extreme s. TX & s.c. AZ, but range & habitat do not normally overlap. **Adult.** (CA): M 54–74 g, F 64–87 g; (AB, Canada): 54–64 g; 16–18 cm (6.3–7 in.). Diet: variety of small vertebrates & insects. **Nestling.** Semi-altricial. *1st down*: white, short & dense, fully covered; *2nd down*: may begin few d after hatching. G&D: little known. **Fledgling.** Fledges ~23 d, fly weakly at first. Bill: gray or gray-green. Nostrils oval. LF: legs feathered to feet. Plumage: *Gray morph* similar to Ad except crown & nape grayish & unspotted. Above: brown & largely unspotted, may have white flecks on forehead, sootier eyespots on nape. Tail: whitish to pale buff-cinnamon bars.

Burrowing Owl (*Athene cunicularia*) — Plate 11

Populations declining in many areas, listed as endangered in Canada, under special protection in Mexico, and endangered or threatened in 9 w. US states. **Adult.** (CO) M 146 g, F 156 g, (FL) 150 g; 19–25 cm (7.5–9.8 in.). Nest: burrow in ground excavated by other animals. Diet: primarily insects & other invertebrates, but also wide variety of small vertebrates. **Nestling.** Altricial. Skin: pink. Down: scantily covered (bare areas) with white or grayish-white down, by 5–6 d appears darker gray. G&D (w. birds): d0: 6–12 g, tarsus 9.7 mm; d5: eyes open; d6: PinEm alar, femoral, crural & spinal tracts; d9: egg tooth gone; d14: PinEm all tracts (complete 30–35 d); d15: prim PinEm. **Fledgling.** d14: leaves nest burrow; d21: runs, hops, flaps wings; d28: short flights; flies well (44–53 d) but stays near burrow. Bill: cream, yellowish-white, or greenish-yellow; cere gray to grayish-green. Gape: pinkish. LF: long with short white to beige feathers. Iris: lemon (or in some populations a few may be brown or olive). Plumage similar to Ad except chest not mottled but solid buff (or darker), light buff lower breast & belly (some plain, some appear mottled), buffy patch across dorsal surface of wings. Above: base color brown; head round with distinct facial disc & white eyebrow; crown & forehead dark with buff-white spots or streaks; buff-white barring (some incomplete, appear as spots) on nape, back & upperwing covs. Tail: dark brown dorsally with buff-white bars (3–4, may be incomplete, may appear as spots).

Spotted Owl (*Strix occidentalis*)   Plate 11

Loss of habitat has placed this owl on threatened to endangered listings depending on state or territory. Very tame & curious. Young are raised in forested habitats of their range from elevations of 1200 m (n. part of range) to 2700 m (s.w. US). **Adult.** Northern Spotted Owl (*S. o. caurina*): M 490–690 g, F 540–850 g; California Spotted Owl (*S. o. occidentalis*): M 470–685 g, F 535–775 g; Mexican Spotted Owl (*S. o. lucida*): M 449–625 g, F 480–680 g; length 47–48 cm (18.5–19 in.). Nest: platforms & cavities in trees or on cliffs, natural or previously used by other animals. **Nestling.** Down: white, sparsely covered. G&D: d0: unknown; d5–9: eyes open; d10–20: juv plumage begins to replace down; d34–36: juv plumage complete with downy tips remaining; wings & tail by 65 & 75 d. **Fledgling.** (CA) 400–450 g. Fledges 34–36 d (or prematurely 15–25 d), parental care 60–90 d post-fledging. Bill: gray, then yellow-green. Skin around eyes black & bare until facial feathers grow in. LF: fully feathered. Iris: dark brown. Plumage above: light brown with dark transverse barring. Rem & rect: dark brown with light brown transverse barring (variable width).

Barred Owl (*Strix varia*)   Plate 11, Photo 16

Adult. M 468–774 g, F 610–1051 g; 43–53 cm (17–21 in.). **Nestling.** d0: 46 g. Bill: buff-yellow. *1st down*: fully covered, thick, soft, white. *2nd down* d14–21: gray-buff with white tips, dark grayish-brown barring on back & wings. Below: breast similar but paler than back, elsewhere paler & longer with yellowish-white to white tips. LF: fully feathered (most populations). Iris: dark, brown to black. G&D (Nova Scotia): young gained ~15 g/d 1st 30 d. **Fledgling.** Fledges 28–35 d, flightless at first, climbs on branch, drops to ground, climbs back up to branch, short flights at 10 wks. 50–75% of Ad mass at fledging. Down remains on head at 4 months. Cannibalism may occur in broods of 4 or more.

Great Gray Owl (*Strix nebulosa*)

Adult. M 825–1050 g, F 1025–1700 g; 61–84 cm (24–33 in.). Nest: old nests built by ospreys, ravens, or hawks. **Nestling.** Semi-precocial. Down:

gray-white above, white below. G&D: little info. d0 (mass of 4): 36–39 g (gains ~20 g/d); d7: juv feather PinEm; d14: facial disc pale gray with black concentric rings. **Fledgling.** 360–755 g. Fledges 21–28 d, does not fly but jumps from nest & climbs up leaning trees, flies at 7–14 d post-fledging. Care: both until 3–6 wks after fledging, then M may provide care. Plumage above: dark barring with white spots, below: olive-brown & barred. Wings & tail (d11): gray tipped, fully grown 6–7 wks.

Long-eared Owl (*Asio otus*)

Adult. (MT) 223–304 g, F 289–409 g; 35–40 cm (13.8–15.7 in.). **Nestling.** Semi-altricial, hatch with eyes closed. *1st down*: short, dense, white. *2nd down*: buffy or cream-white, grayish-black barring. **Fledgling.** 219 g, fledges: d21, does not fly at first, stays in vegetation, then short flights d35. Bill & cere black. LF: densely feathered legs & toes. Iris: yellow. Plumage above: brown feathers with creamy white tips & barring, below dull buff-white, barring pale brown on breast, paler on belly & flanks.

Short-eared Owl (*Asio flammeus*)

Adult. M 206–368 g, F 284–475 g; 34–43 cm (13.4–16.9 in.). Nest: on ground. **Nestling.** Semi-altricial. *1st down*: short, dense & fully covers, buff above, white below. *2nd down* d5: longer, light buff with dark bands, turns darker to dark rust by d11. G&D: d0: 16–18 g, gains 8 g/d 1st 5 d; d6–10: gains 19 g/d; d16–20: gains 12 g/d; d5–9: eyes open; d9: contour PinUn, facial disc PinUn (black around eye orbits); d11: prim & tail feathers evident; d15: ear tufts emerge. **Fledgling.** d12–18 leaves nest on foot, 1st flight 27–35 d. Bill: black. Iris: brown. Plumage: like Ad DefB but head & upperparts duskier with buffy tips, facial pattern (unlike Ad) brown-black tipped buff.

Boreal Owl (*Aegolius funereus*)

Adult. (ID) M 93–139 g, F 132–215 g; 21–28 cm (8.3–11 in.). **Nestling.** Skin: pink to bright red, turns dark violet-red by fledging. *1st down*: cream-white; *2nd down*: resembles juv feathers but looser & soft. Iris: d11 grayish-yellow, d18 bluish, d20 yellow. G&D: d0: 9 g (gains ~5.2 g/d); d7: PinEm; d8–11: eyes open; d9: PinUn; d14: prim PinUn, white feathers around gape evident; d14–17: Ad mass; d18: facial disc & spots on wings evident; d21: white spots crown & tail. **Fledgling.** Fledges 28–36 d, weigh

16 g (avg) at 30 d. Bill: cream to bluish-yellow, gray cere. Plumage above dull chocolate-brown (21 d), facial disc brown bordered black with broad white X between eyes; below brown, mottled cream & brown at vent, legs & feet.

Northern Saw-whet Owl (*Aegolius acadicus*)   Plate 11

Adult. (BC, Canada) M 77.4 g, F 131 g; 18–21.5 cm (7–8.3 in.). F mass 50% more during egg laying. **Nestling.** Semi-altricial. Down: fully covered, white. G&D: d0: 7.5 g; d4–14: gains 7 g/d; d5: PinEm; d7–10: eyes open; d13–14: prim PinUn. **Fledgling.** ~96 g (may weigh 10 g more before fledging). Fledges d33. Bill, eye ring & claws: black. LF: buff. Iris: N dull olive, Fl bright yellow. Plumage: may still have downy tips. Above: dark reddish-brown or sooty & unmarked; white streaking on head, facial disc dark brown, conspicuous (Y-shaped) white patch between & above eyes; wings & tail with white markings (as Ad). Below: upper breast brown or sooty-gray, burnt umber to buffy breast, belly & undertail covs.

KINGFISHERS

(Family: Alcedinidae, Order: Coraciiformes)

These birds are very distinctive with large heads, short necks and tails, small feet, long prominent bills, most with "shaggy" crests, and bold markings of blues and greens with white or rust accents. **Adults.** Small to medium-size birds. Diet: mainly fish, also crustaceans, insects, amphibians, reptiles, sometimes small birds or mammals. Nest: in burrows on banks excavated by digging or sometimes in tree cavities. Eggs: elliptical to subelliptical and white. Bill: long, sturdy, straight, compressed, acute; has linear nares. Plumage: usually colorful, head crested. Wings: moderately long, pointed. Tail: moderately long, slightly rounded. Legs/feet: syndactyl, hallux much shorter and partly connected to inner toe, two outer toes on both feet fused together for half of their length; tarsi very short and partly bare, irregularly scutellate in front. Nails: very acute, middle nail somewhat flattened. **Young.** Altricial. Mouth: pink interior. Bill: heavy, longer than bird's head. Down: hatch naked. Once feathers develop, nestlings have blue and white coloring. Fledge: 30–38 days. Biparental care. *Note: Nests of kingfishers are sometimes unearthed after excavation around the nest burrow. Fledglings usually fly fairly well out of the burrow. Any young on the ground should be investigated.*

Ringed Kingfisher (*Megaceryle torquata*)

Large kingfisher; has expanded range into TX along Rio Grande & nearby areas. **Adult.** 305–341 g; 28–35 cm (11–13.8 in.). **Nestling.** Skin pink. Down: none. LF: calloused heel pads. G&D: d10: eyes open, PinEm; d24: FF. **Fledgling.** Fledges ~35–38 d. Plumage similar to Ad F but crest darker, paler below.

Belted Kingfisher (*Megaceryle alcyon*)   Plate 12

Adult. 140–170 g; (PA) 125–215 g; 28–35 cm (11–13.8 in.). Nest: in earthen burrow on vertical bank near or over water (sometimes ditch, landfill, sand, or gravel pit far from water). Eggs: pure white, smooth, glossy (about 1 x 1.5 in.). **Nestling.** d0: 9–13 g (8.46–10.36 g, 8 incubator hatched; Wetherbee and Wetherbee 1961). MC: pinkish. GF: none. Bill: blackish, heavy, longer than head, compressed, straight, lower mandible extends beyond upper, egg tooth on upper & lower. Skin: bright pink. Down: none. LF: N have "heel pad," gray nails. G&D: growth greatest 1st 10 d; d6: humeral PinEm; d12–13*: 155.8 g (22.2 cm), 161.8 g (22.3 cm), dorsal tracts mostly PinF, many tips Un, wing PinUn tips only, F shows chestnut pinF on chest; d13: FF on dorsum, humeral PinUn; d16–18: appears FF, attains Ad mass, N moves into burrow from nest chamber. Wing & tail PinUn last 10 d before fledging. **Fledgling.** Leaves burrow: 27–29 d (or longer) post-hatching, capable of limited sustained flight. Remains with parents ~3 wks, when young have mastered ability to catch fish. Plumage similar to Ad but bluish-gray with darker crest; more white in wing covs & white spots on central two tail feathers, breast band has more cinnamon or brown. Diet of young: N stomach acid helps digest whole fishes & arthropods (with shell); as fledgling, stomach changes occur so must regurgitate (cast) pellet of bones, scales & shells of above foods. N requires ~8.3 fish/d for 28 d (11.2 fish/d at peak growth). *d12–13 data from 2 carcasses, Nevada County, CA.*

Green Kingfisher (*Chloroceryle americana*)

Smaller than Ringed and Belted Kingfishers. **Adult.** 35–40 g; 30 cm (11.8 in.). **Nestling.** Skin: pinkish. Down: none. **Fledgling.** Flies ~25 d. Plumage similar to Ad F but duller buff spots on crown & covs & belly band incomplete.

WOODPECKERS

(Family: Picidae, Order: Piciformes)

Members of this family have smaller eggs, remarkably shorter incubation, and longer nestling periods compared to some open-nesting altricial birds in comparable range of body and egg masses (Hadow 1976; Yom-Tov and Ar 1993). **Adults.** Small to medium-size relative to the size of passerine birds. Diet: insects, fruits, nuts, sap of trees. Nest: excavated cavities in trees. Eggs: glossy, white. Bill: strong, usually straight, some slightly curved, some more acute; chisel-shaped for excavating nests and food from wood; bristle-like feathers completely conceal nostrils. Tongue: barbed, sticky, can be up to 4 in. long; used for removing ants and other bugs; length varies by sex in some species (more study needed). Sapsucker tongues are fringed and spoonlike on the tip for lapping up sap. Plumage: most are black and white or brown (flickers) with some to no red or yellow on heads; may be barred, spotted, or streaked; variations of red or yellow influenced by carotenoid pigments. Wings: moderately long, more or less pointed, outermost primary very short or rudimentary. Tail: 12 feathers, pointed, rounded, or graduated; rectrices acuminate with stiffened tips to brace against tree surface to support weight while pecking. Legs/feet: strong, zygodactyl arrangement (except in 3-toed species); fourth toe permanently reversed; tarsi scutellate in front, reticulate in back. Nails: strong, decurved, very acute. **Young.** Altricial. In most species studied, asynchronous hatching results in a runt that remains smaller and behind in development compared with siblings. Mouth: generally pink but not brightly colored. Gape: young nestlings have white or whitish-pink, highly sensitive swollen lumps or "knobs" at the flanges; gaping elicited by touching these areas; knob usually gone shortly post-fledging. Bill: heavy. Skin: pinkish. Down: none at hatch. Plumage: first juvenile feathers typically dark with lighter spots on back; similar to adults though head patterns may differ. Many species have white undersides. Crown patch: red (to pinkish); both sexes may show color, amount of color usually greater in M than F. Spinal tract: well-defined on neck, forked on the lower back. First prebasic molt (flight-feather replacement): in some species, occurs before young leave nest beginning with innermost primaries (in most species are 1/10th the length of fully grown feathers) (Pyle 1997b). Preen gland: tufted. Legs/feet: nestlings have a callused pad on the heel (back of the leg joint) and rest or move on the whole tarsus when in the nest. Begging sounds: loud harsh noises; captive nestlings beg incessantly.

Lewis's Woodpecker (*Melanerpes lewis*)

Adult. M 105–122 g, F 88.3–106 g; 25–28 cm (10.2–11 in.). **Nestling.** Bill: dusky to black. Down: H naked, then fuzzy down with no filaments on skin. G&D: little info; d7: PinEm, red on ventral tracts at belly (7–14 d); d14: PinUn. **Fledgling.** Fledges 28–34 d; do not return to nest. Bill: black or dusky. LF: bluish appearance due to white scales on black, turn bluish-gray. Iris: brown. Plumage similar to Ad DefB but duller dusky-gray & black, white spots on hindneck, variable amount of red on facial area, pinkish-red on breast, silver-gray collar.

Red-headed Woodpecker
(*Melanerpes erythrocephalus*)   Plate 12, Photo 17

Live in tightly knit social family units, extremely territorial, 1–2 broods/yr, sometimes into Sept. **Adult.** (ON, Canada) 56–90.5 g, M 8–9% > F; 23.5 cm (9.25 in.). **Nestling.** Bill: white egg tooth upper & lower mandibles. G&D: d0: 8 g, d6–15: eyes open. N fed small insects; larger whole insects & fruits added with growth. **Fledgling.** Fledges 24–27 d. LF: dark gray; nails black. Plumage: mottled brown head & neck; white breast, belly & rump variably marked with brown streaking. Dark brown back & tail, upper wings with paler edgings.

Acorn Woodpecker (*Melanerpes formicivorus*)   Plate 12, Photo 18

Most commonly found in close association with oaks in pine-oak woodlands of w. US & areas of s.w. US (& s. through Mexico). Best known for acorn "granaries" and highly social family system. **Adult.** (CA) M 81.8 g (avg), F 77.5 g (avg); 23 cm (9 in.). **Nestling.** Knobs: lower mand only, barely evident by 55–60 g. G&D: d0: 3.9–5.5 g, doubles in 24–36 hrs, sits upright within hrs of hatching, balances on heel pads; d3: 14 g; d6: 28; d10: PinEm on dorsum, eyes open; d12: egg tooth gone; d13: 46 g; d14: PinEm all tracts; d18: 64 g; d21: 75, PinUn; d26: 81 g. Behavior: waves head back & forth with mouth open, attempts to swallow anything it touches, begs when "shadows" block off light entering nest cavity, locomotion backward until fledging. Voc: *tse* and *rasp* begging calls, then *squee* and *trtrtr* (21–28 d). **Fledgling.** Fledges 30–32 d (runts much longer), weight drops just before fledging to ~80 g (runt mass much less); usually flies from nest, some climb out and branch before 1st flight. Iris: dark brown. Plumage: similar to Ad but duller. Back, tail, wings, sides of head & chin dull black; white eye ring; throat with bare areas under faint yellowish feathers still unsheathing, breast thinly streaked black on white; lower belly white with fine grayish streaking (more gray on flanks & vent area), feathers on flanks & rump "fluffy"; crown: dull reddish-orange patch; remnant of gape knob; wings barely extend beyond rump. Juvenile M & F are easily distinguished in late fall after preformative molt: M has solid red crown that directly adjoins white forehead; F has black band separating white forehead from red crown. Also, M has slightly longer bill. Biparental care plus nest helpers: 3–4 months post-fledging. *Reuniting: a healthy near-fledgling can be returned within 30 days if nest tree can be determined and can be closely monitored.*

Gila Woodpecker (*Melanerpes uropygialis*)

Adult. 51–79 g; (AZ) M 54.6–80.6 g, F 53.8–67 g; 22–24 cm (8.7–9.4 in.). **Nestling.** G&D: unknown. **Fledgling.** Fledges ~28 d. Bill: dull black. Iris: brown. Plumage: M similar to Ad M DefB, but paler, grayer, barring less distinct; crown & breast with fine black streaking; crown patch darker red & smaller. F similar to juv M but no red (or some with 1–2 red feathers) on crown.

Golden-fronted Woodpecker (*Melanerpes aurifrons*)

Adult. M 73–99 g, F 66–90 g; 22–26 cm (8.7–10.2 in.). **Nestling.** Bill: upper beak shorter than lower, white egg tooth. Skin: pink. G&D: little info; d2: PinVis; d8: contour PinVis, rem & rect PinEm, lower & upper mandible same length; d9: eyes begin to open; d16–17: PinUn; d21: FF, climb on cavity walls. Voc: *peep* & low *buzzing* sounds. **Fledgling.** Fledge 30 d, do not return to nest. LF: grayish-green. Iris: brown. Plumage: similar to Ad DefB, but duller; crown & breast with fine dusky streaks, no yellow or orange on nape or nasal tufts, barring on back indistinct black & white. Crown patch: M small patch red, F with no red or 1–5 red feathers.

Red-bellied Woodpecker (*Melanerpes carolinus*)   Plate 12

Adult. 56–91 g; (FL) M 73.2 g, F 66 g; 24 cm (9.4 in.). Nest: cavity. **Nestling.** Knobs evident, reduced d15. Bill: lighter than Ad. LF: dark gray with black nails. G&D: d0: 8 g; d6–15: eyes open; d8: PinVis on dorsum; d10–11: 12.5 cm (avg), some PinEm, egg teeth still present, knobs smaller, hock-sits; d15: 70.5 g (avg), 15.8 cm (avg), knobs nearly gone, M distinct from F by faint red on nape & crown, cling to vertical cavity walls; d21:

70.5 g (avg), 17.4 cm (avg), FF, egg teeth & knobs gone. Diet: small insects (larger whole insects & fruits gradually added). **Fledgling.** Fledges d24–27. Dependency may be several wks post-fledging. Voc: raspy *kwirr* begging, *wee-urp* following feeding.

Williamson's Sapsucker (*Sphyrapicus thyroideus*)

Adult. (NV) 44–55 g; 23 cm (9 in.). **Nestling.** Knobs: flesh pink, retained up to 7 d post-fledging. Skin: bright flesh-pink. G&D: little info, egg mass 3.8 g (avg); d7: PinEm; d7–14: eyes open; d14: FF (sex evident by plumage patterns). **Fledgling.** Fledges d31–32. Plumage: resembles Ad. Juv M not as glossy, white instead of red throat, more white on nape, bases of black back feathers have "white wedge marks," more barring above, paler wings & tail, belly pale yellow. Juv F similar to Ad F but duller & browner, more barring throughout, no black on breast.

Yellow-bellied Sapsucker (*Sphyrapicus varius*)   Plate 13

Adult. (PA) 40.7–62.2 g; 21–22 cm (8.3–8.7 in.). Diet of nestlings: small insects for H, insects coated in sap before feeding; & fruit of serviceberry. **Nestling.** Bill: light gray. Skin: pink. G&D: d0: 2.5 g; d8: eyes open; d9: 24.9 g. **Fledgling.** Fledges 26.9 d, flies weakly. Plumage similar to Ad but dark olive-brown overall: head & breast brownish, red & white head markings subdued, white nape, buffy-white scalloping on crown & throat. Black back with black & white mottling, rump with more barring on white areas. Belly unmarked. Juv M may have slight red tint on throat. *Hybridize with Red-naped & Red-breasted Sapsuckers where sympatric.*

Red-naped Sapsucker (*Sphyrapicus nuchalis*)   Plate 13

Adult. (AZ) 36–55 g; 19–21 cm (7.4–8.3 in.). **Nestling.** Skin: pink. G&D: d7: nearly all PinEm; d12–17: 41.9 g; d14: FF; d20–24: 46.8 g. **Fledgling.** Fledges 23–28 d. LF: greenish-gray. Iris: grayish-brown. Plumage above: back blackish-brown with black & white mottling, rump barring on white areas, crown & auriculars dark brown to slate, thin buff white lines above & below eye, red nape. Below mostly dark brown or blackish, paler on lower throat, neck & breast with scallop-shaped dusky bars. Lower belly pale yellowish or whitish. Sides/flanks dull brownish with olive-brown barring. Juv M (only) with few red feathers in chin, forehead & crown. *Hybrids: see "Yellow-bellied Sapsucker" & "Red-breasted Sapsucker."*

Red-breasted Sapsucker (*Sphyrapicus ruber*)   Photo 19

Adult. (BC, Canada) M 58.3 g, F 57.7 g, (CA) 48.9 g (avg); 20–22 cm (8–8.7 in.). **Nestling.** Skin: pink. G&D: little info, d7: PinEm; d10: eyes half open, d14: FF. **Fledgling.** Fledges 23–28 d, mass near AD, flies poorly. LF: grayish. Iris: grayish-brown. Plumage above: back blackish-brown with black & white mottling, rump barring on white areas, crown dark brown. Below: dark brown or blackish, no white on throat, amount of red on throat & breast variable, sides/flanks more extensively marked with dark or olive-brown. *Hybrids: see "Yellow-bellied Sapsucker" & "Red-naped Sapsucker."*

American Three-toed Woodpecker (*Picoides dorsalis*)

Three toes, no red, heavy bill. **Adult.** (BC, Canada) M 50–64.5 g, F 47–59 g; 20–22 cm (8–8.75 in.). **Nestling.** G&D: unknown, egg mass 4.4 g. **Fledgling.** Fledges 24 d. LF: dark grayish-horn. Iris: gray brown. Plumage M: similar to Ad M DefB except duller; black crown with white flecking & yellow center (may extend in front of eye). Below: muted buff, brownish spots on flanks, rect pointed with black & white pattern (less distinct than Ad). Juv F similar to juv M but no yellow (or few feathers tipped yellow) in crown.

Black-backed Woodpecker (*Picoides arcticus*)   Plate 13

Adult. (CA) M 61.3–88 g, F 63.4–71.3 g; 23.5 cm (9.25 in.). **Nestling.** MC: whitish. Skin: whitish & translucent, becomes more pinkish. LF: whitish, then gray by d12. G&D: d0: 2.7 g; d9–12: eyes open. **Fledgling.** Fledges 22–26 d. LF: slate, dark or bluish-gray. Plumage similar to AD but duller, crown dull black, M may have yellow crown patch, F patch reduced or absent; below washed buff, brownish spotting on flanks.

Downy Woodpecker (*Dryobates pubescens*)   Plate 13

Adult. (CA, avg 9 birds Nevada County, June–July) 23.7 g; (PA) M 26–29 g, F 22.2–28.5 g; 14–17 cm (5.5–6.5 in.). Resident, although some evidence suggests some seasonal dispersal. Diet: 76% animal, 24% vegetable; young fed soft-bodied insects 1st 2–3 d. Nest: cavity (primarily). Iris: brown or red. M has red patch on nape. Outer rectrices usually barred. Bill: much smaller than length of head. **Nestling.** Knobs: light pink or white. Tongue tip white. Bill: pinkish-white, upper mandible (with egg tooth) shorter than lower. Skin: pink or pinkish-red, faint dots on wings (except last hatched). Down: none. LF: H pinkish, white nails; N gray turning greenish-gray, gray nails. G&D: d0: 2–3 g; see Table C-3, Appendix C. Voc: *pip-pip-pip* and rasping notes; d16: *whinney*. **Fledgling.** Fledges 18–21 d (20–25 d from another source), flies well. Bill: more convex above (straight in Ad). Iris: brown, yellowish halo. Plumage similar to Ad except black areas more dull or brown, narrower rect. Below (varies geographically): more grayish or white, unmarked except breast sides & flanks finely streaked. Crown: red (M), F no red. Independent 3 wks post-fledging. *Sim sp: Ad bill of Hairy Woodpecker larger & about twice as long, calls louder. Hybridizes with Nuttall's Woodpecker.*

Nuttall's Woodpecker (*Dryobates nuttallii*)

Adult. (w. US) M 32.4–46.9 g, F 32.8–40 g; 19 cm (7.5 in.). **Nestling.** Unknown. **Fledgling.** ~32 g. Fledges 15 d. Iris: grayish-brown to brown through 1st fall or winter. Plumage: like Ad but slightly more grayish to buffy below. Above: whiter, both sexes show red in crown (unlike Ad), usually small patch in center of crown in M, F has fewer & more scattered, red-tipped feathers. *Hybridizes with Downy Woodpecker.*

Ladder-backed Woodpecker (*Dryobates scalaris*)

Adult. (w. US) 25–41.4 g; 16–18 cm (6.3–7 in.). **Nestling.** Unknown. **Fledgling.** LF: olive-greenish or dull, grayish olive-green. Plumage M: similar to Ad M DefB except duller with brownish wash, dull black crown with some white flecks & mottled red center. F similar to juv M, but little to no red crown.

Red-cockaded Woodpecker (*Dryobates borealis*)

Endangered status. **Adult.** (NC) M 48.6 g, F 47.4 g; 20–23 cm (8–9 in.). **Nestling.** Knobs: lower mandible, curve upward. Bill: upper mandible pinkish-white, shorter than lower. Skin: bright pink. Iris: dull brown with buff-yellow ring around. G&D: d0: 3.3 g; d4–5: PinVis; d7–8: red spot on forehead of M evident. **Fledgling.** 42–45 g, fledges d26–29. Plumage: white cheek patches mixed with gray, forehead has white flecks (near base of bill) and red spot may have turned black.

Hairy Woodpecker (*Dryobates villosus*)   Plate 13

Adult. (MT) M 79 g, F 67.5 g; (PA) M 60.8–79.6 g, F 59.3–65.9 g; 23.5 cm (9.25 in.). **Nestling.** Knobs: white to pinkish. Bill: pinkish-white, upper mandible shorter than lower. G&D: d4: PinVis; d6–7: PinEm; d15: well feathered, some still in sheaths at bases. **Fledgling.** Fledges d28–30. Bill: culmen shorter, wider & more convex above than Ad. Iris: brownish to grayish-brown. Plumage: M similar to Ad M DefB except duller, black is tinged grayish or brownish, white flecks on forehead nearest bill, no eye ring, white supercilium ends at back of eye (sometimes tinged yellow), red rarely on nape, crown with some red (or orange or pink) tips. F similar to juv M except very little to no red on crown, flanks may be lightly barred black with buff wash. *Sim sp: see "Downy Woodpecker."*

White-headed Woodpecker (*Dryobates albolarvatus*)

Adult. (OR) M 55.6–68 g, F 52.6–66.4 g; 21–23 cm (8.2–9 in.). **Nestling.** G&D: d5: PinVis; d12: PinEm; d15: PinUn; d19: FF. **Fledgling.** Fledges ~26 d. Plumage similar to Ad DefB but duller dusky black, may have whitish tips on ventral feathers sometimes appearing as barring on abdomen; M crown patch pale scarlet, F has little to no crown color.

Northern Flicker (*Colaptes auratus*)   Plate 13, Photo 20

Adult. (OR): 121–167 g; (PA): M 106–143 g, F 104–137 g; 28–31 cm (11–12 in.). **Nestling.** MC, bill & LF: beige to pink. Knobs: white. Skin: light pink. G&D: d0: 5.5 g; d6: PinVis; d7–8: wing & tail PinEm; d10–11: PinUn; d12–15 more contour PinUn. Voc: buzzing (like a beehive) starts shortly after hatch. **Fledgling.** Fledges 24–27 d, rem & rect full length 5–10 d post-fledging. *Yellow-shafted*: plumage similar to Ad DefB except duller, M has some red in crown, less distinct barring on upperparts, more grayish on head & neck, smaller black breast crescent, larger spots on underparts, malar stripe less distinct or lacking in F. *Red-shafted*: plumage similar to Ad DefB except similar to those of Yellow-shafted; crown has less red (M little or none, F usually no color) than Yellow-shafted, malar stripe reddish to salmon pink, may be darker in M than F.

Gilded Flicker (*Colaptes chrysoides*)

Adult. (AZ) 92.2–129 g; 28 cm (11 in.). **Nestling.** Bill & LF: beige to pink. Skin: pink. G&D: as Northern Flicker; d2: 11.8 g (avg). **Fledgling.** Fledges 24–27 d. Plumage similar to Ad but duller & softer.

Pileated Woodpecker (*Dryocopus pileatus*)

Distinguish by size, mass, habitat. **Adult.** (PA) M 308 g (avg), F 266 g; 40–50 cm (15.7–19.7 in.). **Nestling.** Skin: pink. G&D (NY): d0: < 15 g; d1: 29; d2: 41; d3: 51, PinVis; d4: 67 g; d5: 79, rem & rect PinEm; d6: 99 g; d7: most body PinEm; d8: 165 g; d9: eyes open; d10–12: 185.5 g; d14: red malar stripe appears (M only); d16: 209 g; d21: 226; d26: wing cov PinUn; d29: PinUn head & dorsum. **Fledgling.** Fledges: 24–28 d, some can fly when they leave nest, others take longer. Plumage: similar to Ad DefB but duller, looser. Sex evident 9–10 d: crown PinVis through skin & red malar stripe M appears d14.

FALCONS

(Family: Falconidae, Order: Falconiformes)

   Adults. Small to large raptors. Diet: mainly birds; some mammals and insects. Nest: sticks, in tree cavities, caves, or on cliff ledges. Eggs: sub-elliptical to elliptical; pale, some with reddish-brown spots. Bill: strongly hooked; upper mandible toothed near tip; brightly colored cere in which the circular nostrils open centrally. Plumage: compact, streamline appearance; lores have bristle-like feathers; distinctive "moustachial" stripes. Wings: long, narrow, and pointed; forest falcons are broad and rounded. Tail: long and somewhat rounded. Legs/feet: anisodactyl and strongly curved talons; hallux same length or slightly longer than shortest front toe; underside of talons grooved; tarsi front with small irregular reticulate scales. **Young.** Altricial. Eyes open a few days after hatch. Fleshy gape flanges. Fledge: 25–56 days. Biparental care.

American Kestrel (*Falco sparverius*)　　　　　Plate 9, Photo 21

Smallest North American falcon, distinct sexual dimorphism. **Adult.** (CA) M 112 g, F 122 g; (KY) M 113 g, F 132 g; (FL) M 90.8 g, F 103 g; M 22–27 cm (8.7–10.6 in.), F 23–31 cm (9–12 in.). Nest: natural cavity (excavated by other sp), box, sometimes buildings. **Nestling.** Bill (& cere): H whitish-pink, notch & tooth evident, not hooked, egg tooth 1–2 mm (gone 7–14 d). Skin: pink, eye orbital light bluish-green (d21). *1st down*: fully covered, sparse, silvery white. *2nd down:* longer, white. LF: H pink, then yellow (sometimes greenish). G&D: d0 (CA): 10–12 g, (FL) 8–9 g; d1–2: eyes partly to fully open; d7: PinEm wings, shoulders, back & crown; eyes fully open; talons darker; d8: prim PinUn; d9: rect PinUn; d14: sex may be evident (F have barring on mantle); d21: FF, down still attached on crown & upperwing covs. Voc: *peeps*, then *klee* by d3; *whine* (when begging), *klee* & *killy* (d14). **Fledgling.** Fledges 28–31 d, may occur over several days. Bill: pale bluish-gray. Biparental care, fed 12–14 d post-fledging.

Merlin (*Falco columbarius*)   Plate 9

Adult. (NJ) M 134–223 g, F 134–281 g; 24–30 cm (9.4–12 in.). Nest: modify nests built by other birds. **Nestling.** Semi-altricial. Bill: bluish, white egg tooth. Cere, L/F: yellow. *1st down*: short, above buffy or brownish & white, white below; *2nd down* d4–8: longer, grayish-brown above, white below. G&D: d0: 13 g (avg mass gain d0–17: M 12.0 +/- 0.5 g, F 14.8 +/- 0.4 g); d9–11: contour PinUn; d12–14: prim PinUn; d15–17: rect PinUn, stands upright. **Fledgling.** ~29 d post-hatching. d22–28: short flights. Bill: bluish, dark tip. Cere: yellow. LF: yellow. Iris: dark brown. Plumage like Ad F (Boreal ssp darker). Above: brown, nape has pale mottling, remiges brownish-black with pale tawny-rufous spots in barred pattern, dorsal bands of F tawnier (M more gray). Below: white throat, breast & belly tawny to whitish with heavy streaking. Independent: 1–4 wks post-fledging.

Gyrfalcon (*Falco rusticolus*)

Largest of the falcons; breeding range is circumpolar arctic and subarctic regions. **Adult.** M 960–1304 g, F 1396–2000 g; M 48–61 cm (19–24 in.), F 51–64 cm (20–25 in.). Nest: scrape on cliffs or stick nests (other birds). **Nestling.** MC: pinkish, then greenish or greenish bluish-gray. Bill (paler in H overall): pale horn (white morph), bluish-horn (gray morph), blackish (dark morph). Cere & LF: pale yellowish-pink, then greenish blue-gray. *1st down*: varies relative to shade of juv plumage (or morph): light morph with thick, pure white down; darker morph has white down with dark tinge on head &/or back; fully covered (sparser on ventral). *2nd down* ~d8: lighter, does not replace 1st down. Iris: black-brown. G&D (captive-raised gain 59 g/d between d6 & d27): d0: 52.1 g (avg), eyes almost open; d5: weight doubles; d7: egg tooth lost; d8: 2nd down emerges; d11: primary no. 7 emerges; d35: FF, flight feathers not fully grown. **Fledgling.** 1st flight: 45–50 d. Plumage with 3 morphs: white, gray/intermediate & dark. Color abbreviations: Vandyke brown (VDB), olive-brown (OB), horn (HN), sepia (S). *White morph*: background color white to pale HN; forehead, crown, nape fine streaking of VDB; mantle mod to heavy markings (arrowhead tips) & barring (incomplete) OB to S; breast & belly with VDB teardrop streaks; tail barring varies from none to heavy OB to S; primaries with subterminal band VDB to S, barred on inner web. *Gray/intermediate morph*: forehead, crown, nape & cheeks pale HN with heavy streaks of OB to VDB; nape: some with pale HN eyelike patches; facial stripe VDB; mantle pale HN (some salmon), barred OB to VDB; breast & belly pale HN with OB to VDB streaking; tail pale HN to light gray, barring (heavy to mod) with OB to VDB; prim pale HN with barring (irregular & incomplete) of VDB to S. *Dark morph*: forehead, crown, nape, mantle & cheeks VDB; breast & belly pale HN with heavy streaks of OB to VDB; tail light gray, heavy barring of OB to VDB; prim pale HN with VDB to S bars on inner webs, spots/speckles on outer webs; sec barred OB to VDB.

Peregrine Falcon (*Falco peregrinus*)   Plate 9

Adult. M 410–1060 g, F 595–1600 g; M 36–49 cm (14–19.3 in.), F 45–58 cm (17.7–23 in.). Nest: on ledges. **Nestling.** Semi-altricial, eyes may be closed or open as slits. Bill: pale blue-gray, white egg tooth gone d2–3; cere & lore pale bluish. Skin: bright pink. *1st down*: off-white. *2nd down* d6–8: dense, long, very white. LF: grayish-white. G&D: d0: 35–40 g (mass doubles by d6); d6–8: 2nd down appears; d14: rect sheath ~2mm, usually 9th prim has emerged; d17: contour PinUn; d20: PinUn wing margins, tail & around eyes; d30: appears half in juv feathers, half in down; d35: appears FF with downy patches on legs, underwings & crown; d40: nearly FF, traces of down. **Fledgling.** d40: weak flight. Cere & lore: yellowish (deep to pale) to greenish or bluish. LF: variable by ssp, some yellowish, some may be bluish-gray to bluish-green. Iris: dark brown. Plumage varies geographically. *F. p. anatum*: above usually sepia or fuscous some with buff edges, buff ocelli markings may be distinct or partly covered with darker nape feathers; pale streak through eye; moustache stripe variable but dark & usually narrower than Ad. Background color below: tawny to pale buff, streaked (teardrop or arrowhead shaped) with shades of fuscous (sepia). Rectrices: upper surface sepia with 5–7 pale rufous or bluish patches that may have bars, large spots, or no markings.

Prairie Falcon (*Falco mexicanus*)

Adult. (ID) M 484–661 g, F 779–1133 g; 37–47 cm (14.6–18.5 in.). Nest: mostly cliffs, also trees, buildings & other structures. **Nestling.** Bill, cere, eye ring & LF are bluish. *1st down*: pure white. *2nd down*: white tinged buff. G&D: d0: 24.5–34.3 g, eyes slightly open; d1–2: eyes fully open; d10–12: juv PinEm; d21 (1 captive bird): 679 to 885 g (d36). **Fledgling.**

686 g (avg.), M 558 g, F 807 g. Fledges ~38 d post-hatching. Care: young fed ~30–35 d post-fledging. Plumage similar to Ad but darker dorsally, below more streaked, less spotted. Iris dark brown.

PARROTS AND RELATIVES

(Order: Psittaciformes)

Many species in this order have become naturalized in southernmost areas of FL, TX, and CA. **Adult.** Size varies greatly: lovebirds and budgerigars 16.5 cm (6.5 in.); Blue-and-yellow Macaw (*Ara ararauna*) 86 cm (34 in.). Nest: tree cavities or boxes. Eggs: subelliptical to oval, white. Bill: short, stout, culmen strongly decurved and sharply hooked, cere fleshy and bare, with open nostrils. Mouth: fleshy tongue. Plumage: variety of colors, mostly greens; accents of other brilliant colors; some with crests. Wings: most rounded and broad. Tail: variable, some short, most long and pointed. Legs/feet: zygodactyl, fourth toe reversible; tarsi reticulate, shorter than longest toe; grasping claws. **Young.** Altricial. Down: some naked, some with down. Fledge: variable. Biparental care, dependent several weeks post-fledging.

PASSERINE BIRDS

Passeriform (passerine) birds are the most diverse and largest known order, with over 5,000 species in the world. Over 900 occur in North America north of the Mexican border. There are two suborders of passerines: suboscines (Tyrant Flycatchers) and oscines (songbirds). General characteristics of all passerines include anisodactyl feet with a special leg tendon that enables perching or clinging; a muscular syrinx (except suboscines) for producing complex sounds; and wings and tails that are variable in shape with 9 or 10 primaries (with the 10th full length or rudimentary), 9 secondaries, and 12 rectrices. Nests are cup shaped, either "open" in vegetation, on structures such as building ledges, or enclosed in a cavity within a tree, rock crevice, or man-made structure. Nest height may be anywhere from on the ground to in a tree canopy. Eggs are highly variable in size, shape, and color and with or without markings. All passerines have altricial young. Most chicks have varying amounts of down; some hatch with no down. Nestling period and dependency after fledging is fairly predictable by species, but if the nest is disturbed, nestlings may fledge prematurely. The smallest of North American passerines are the Verdin and Golden-crowned Kinglet and the largest is the Common Raven.

TYRANT FLYCATCHERS

(Family: Tyrannidae)

Tyrant Flycatchers have several differences that separate them from true songbirds, including a simple syrinx and songs that are innate (not learned). Distinguishing flycatchers to the genus level will help to narrow down choices (see comparisons in Table D-2, Appendix D). The most challenging group to identify, even when in hand, are members of the genus *Empidonax*, as they are of similar size and plumage. Wing and tail formulas (Pyle 1997a, 1997b) may be the only methods to differentiate some of these species. Nest type and location can be a determining factor. **Adults.** Small to medium passerines. Diet: insects during breeding season, catch flying insects in midair or perch-to-ground flights; some eat fruit in other seasons and larger species take small vertebrates. Nest: open cup most common, sometimes on ground or ledge under overhang; also in tree cavities or spherical arboreal nests with inner chamber. Eggs: subelliptical, short to long, white or off-white, with spots or streaks. Bill: size varies by species, related to size of prey captured; straight, wide, and flat; rictal bristles at base; triangular in outline (dorsal view); culmen somewhat decurved toward tip forming slight hook in many species; commissural point just below midpoint of eye; wide gape. Nostrils: circular.

Plumage: sexes colored similarly; above colors olive-greens or greenish, black, browns, grays; below are lighter colors of whites, grays, pale to bright yellow, a few rust, and males of one species with red on head and undersides. Wings: 10 primaries (10th full length), pointed, outermost primary usually longer than secondaries. Tail: usually square, sometimes forked. Legs/feet: small, weak; tarsi short (seldom longer than middle toe), irregularly scutellate, rounded behind. **Young.** Mouth: colors range yellow to orange. Gape flanges: white to yellow or pinkish-yellow. Bill: see species. Skin: pinkish to bright orange. Down: hatchling Tyrannids have the greatest number of neossoptiles and downy regions among passerines (Wetherbee 1958); a few species hatch naked and develop down by day 2; colors of light gray, white, buffy-white or dusky, concentrated on crown and back. Fledge: 14–17 days post-hatching.

Ash-throated Flycatcher (*Myiarchus cinerascens*)   Plate 15

Adult. 21–37.8 g, 21.6 cm (8.5 in.). Diet: primarily arthropods, small fruits (winter). Nest: cavity or cup inside plant matter; snakeskins incorporated. **Nestling.** MC: bright yellow, orange throat. GF: white, thick (Fig. G3 j, Illustrated Glossary). Bill: d3 pale gray-brown; d6 dark brown, broad, flat, pointed (Fig. G3 i, Illustrated Glossary). Skin: pinkish-gray. Down: thick, dark grayish-brown. LF: grayish. G&D (CA): d0: 3.7 g (wing chord 8 mm); d1–2: 5.3 g; d3–4: 10; d4–5: 12.7; d6: eye slits; d6–9: 18–23 g; d10–15: 25–27. **Fledgling.** 27.4 g (wing chord 60–77 mm), fledges d16–17. Bill: slate. Plumage above: head large with bushy crest, crown darker; dark olive-brown mantle; wing covs brown, edged buff; indistinct pale wingbars; rem & rect dark, edged rufous. Below: pale yellow & duller than Ad, pale gray throat & breast. Iris: dark. Dependent 21 d post-fledging. *Sim sp: Brown-crested Flycatcher (in s.w. US) and Great Crested Flycatcher (in TX); when distinguishing, identification should be based on multiple characters.*

Great Crested Flycatcher (*Myiarchus crinitus*)   Plate 15

Adult. 26–40 g, 22.2 cm (8.75 in.). Diet: insects mostly (+ small fruits). Nest: the only cavity-nesting flycatcher of eastern N.A., snakeskins incorporated. **Nestling.** MC: yellow or orange-yellow. GF: white or cream. Bill: dark. Skin: dark pinkish, turns darker. Down: hatches naked, soon develops down; crown dark brown, back brown to mouse-gray, sides & flanks yellowish. G&D: d0: 3 g (avg); d2–3: 5–8; d4: 11; d4–8: eyes open; d6–8: 18–24 g, prim well-developed; d9–10: 29–30 g. Voc: constant *peeping* (like tree frogs) from d1, then *twee-et* or *twee-eet*. **Fledgling.** 30 g, fledges d14–15, sustained flight ~23 meters. Plumage above: dark brown & olive-brown, rufous-buff, or buff-yellow wingbars; rect dark, edged rufous. Below: chin to breast pale gray, primrose-yellow belly. LF: sepia-brown. *Sim sp: see "Ash-throated Flycatcher."*

Cassin's Kingbird (*Tyrannus vociferans*)   Plate 14

Adult. (AZ) M 45 g, F 41.4 g, (TX) 37.4–50 g; 23 cm (9 in.). Diet: insects (& some berries). Nest: substantial cup. **Nestling.** MC: orange. GF: yellow, then cream. Bill: gray upper, yellow-pink lower. Skin: pinkish-cinnamon. Down: sparse, buffy-white, grayish-yellow, or pinkish-buff. G&D: unknown (unhatched egg 4.0 g). **Fledgling.** Fledges d14–17. Plumage above: grayish-brown crown (no red) & hindneck; brownish wash, covs edged cinnamon-buff. Below: paler & duller than Ad, chin white, gray breast, yellow crissum. Tail: grayish-white outer rect. *Sim sp: Western Kingbird.*

Western Kingbird (*Tyrannus verticalis*)   Plate 14

Adult. 35–44 g, 20.3–24 cm (8–9.5 in.). Diet: insects (bees). Nest: open cup (untidy with plant materials & mammal hair). **Nestling.** MC: (bright) yellow to orange-yellow. GF: bright yellow, thick. Bill: H mandible copper to smoke-gray; wide, flat, tapers to point, rictal bristles. Skin: reddish & pink. Down: white. Iris: brown. LF: H copper to smoke-gray. G&D: d0: 2–3 g; d1: 3.4; d4: eyes open, PinEm; d7: 16–17 g (62–70 mm); d8–9: PinUn most tracts; d15–16: FF. **Fledgling.** 36 g, fledges d13–19. Plumage above: dark grayish-olive with brownish wash, no red crown, dark "stripe" from corners of bill through eyes. Below: pale gray throat & breast, pale yellow belly. Tail: short, outer web to outer rect whitish. Dependent for 2–3 wks post-fledging. *Sim sp: Cassin's Kingbird.*

Eastern Kingbird (*Tyrannus tyrannus*)   Plate 14

Adult. (PA) 35.8–41 g; (NY) 33–47 g; 21.6 cm (8.5 in.). Diet: insects; some fruit & some small fish. Nest: open cup in tree. **Nestling.** MC: yellow to deep orange-yellow. GF: cream to yellow, thin. Skin: orange or orange-yellow, dark gray by d2. Down: gray to white (capital tract buff or brown). G&D (Murphy 1981): d0: 3.6 g; d1–2: 5.5–7.8, wing PinEm; d3: 10.4 g, dorsal PinVis; d5: 16.6 g, most PinEm; d8–10: 25.2–27.6 g; d10–12: mostly FF; d13–14: 29.8 g. **Fledgling.** 33–38 g, fledges d13–17, limited flight. Bill: dark. Plumage above: crown dark, lacks central patch; dark gray back, prim edged pale buff, covs & sec edged cinnamon. Below: chin & belly white, breast grayish. Dependent several wks post-fledging,

Scissor-tailed Flycatcher (*Tyrannus forficatus*)   Plate 14

Adult. 35–56 g, 22–37 cm (8.7–14.6 in.). Diet: insects (grasshoppers, beetles, crickets). Nest: loose cup in tree or shrub. **Nestling.** MC: (bright) orange-yellow. GF: yellow. Skin: reddish-brown. Down: sparse, short, white, or yellowish-white. Bill: paler copper than Ad. LF: nestling copper-red, turns dark. G&D: (mass comparable to Western Kingbird), d0: 2.9 g (2 chicks, KS), d9: most PinUn; d13: FF. **Fledgling.** 30 g, fledges d14–17. Plumage above: no red in crown, may have dark "stripe" through eye; brownish wash, scaly appearance on upperwing covs from pale whitish edges. Below: pale cream-buff, no color on side of breast, no salmon-pink yet (as in Ad). Tail: short.

Olive-sided Flycatcher (*Contopus cooperi*)

Adult. (PA) 26.7–42.2 g; 17.8–20.3 cm (7–8 in.). Diet: flying insects. Nest: open cup, high in tree. **Nestling.** MC: yellow. GF: bright yellow. Bill: broad, upper mandible darker than lower, dark tip. Down: sparse drab-gray. LF: brownish-black. G&D: unknown. **Fledgling.** Fledges d15–21. Bill: stout, dark upper, paler lower. Plumage above: large head, small crest; brownish wash, white cotton-like tuft on side of rump (not always visible), pale buff indistinct wingbars. Below: throat, center of breast & belly white; gray sides & flanks (may be streaked) giving appearance of unbuttoned vest. Dependent 7 d post-fledging. *Sim sp: Greater Pewee (Contopus pertinax), sympatric in AZ.*

Western Wood-Pewee (*Contopus sordidulus*)   Plate16

Adult. (w. US) 11–15 g; 14–16 cm (5.5–6.3 in.). Diet: flying insects. Nest: open cup. **Nestling.** MC: yellow. GF: bright yellow. Bill: broad, pointed, flattened (ducklike), dark upper, yellow-orange lower. Skin: reddish-pink. Down: well covered above, whitish then pale gray by d3; d6 mottled appearance, dark gray PinUn underneath down. G&D: unknown. Voc: H high-pitched weak *peep*. **Fledgling.** 8.3–8.8 g (up to10 g), fledges d14–18. Two fledglings* (fully feathered with gape flanges present): 12.5 & 12.6 g, length 78 mm, tail 13 mm. Plumage above: head with high crest, sloped forehead; overall dark gray with brownish or cinnamon wash, little down left, dull buff-gray or cinnamon wingbars, (juv) wing tips reach to tips of undertail covs. Below: grayish breast appears like a vest, buttoned at top only, light belly. No tail flicking. *Nearly identical to Eastern Wood-phoeb (sympatric in a few small areas). *Data from two MVZ fluid specimens (1959, Rabbit Creek, AK).*

Eastern Wood-Pewee (*Contopus virens*)   Plate 16

Adult. (PA) 12–15.4 g, 15.2 cm (6 in.). Diet: flying insects, sometimes from ground. Nest: shallow cup, high in canopy, well concealed, some feathers in lining. **Nestling.** MC: yellow. GF: pale yellow or yellow. Bill: broad, pointed, upper dark, lower pale orangish (variable). Down: sparse, whitish-gray or white with olive-brown tips. G&D: unknown. **Fledgling.** 12–14 g, fledges d15–18. Plumage above: head with high crest, sloped forehead; back with brownish wash, down mostly gone, wingbars edged cinnamon to buff, upper less distinct. Below: similar to Western Wood-Pewee but vest not buttoned. No tail flicking.

Yellow-bellied Flycatcher (*Empidonax flaviventris*)

The only Empidonax with a yellow throat. **Adult.** 11.5 g, (PA) 9.2–15.5 g, 12.7–15.2 cm (5–6 in.). Diet: insects & other arthropods (occasionally fruit). Nest: on or near ground. **Nestling.** MC, GF & bill: yellow. Skin: orange-pink. Down: brownish olive-green or dark gray. LF: H yellow, turn gray or brown. G&D (4 chicks, MI): d0: 1.25 g; d2: 2.4; d4: 5.5, prim PinEm, tail PinEm; d6: 8.4 g, tail length 2 mm; d8: 12.7 g; d9: stretching, wing flapping. May weigh slightly more in Alaska. **Fledgling.** Fledges d13–15, flies weakly at first. Bill: small, short, wide, dark upper, pinkish-yellow lower. Plumage above: head big, rounded, slightly crested, large eyes, complete white eye ring; overall dull olive, more brownish, distinct buffy wingbars (more yellowish in juv). Below: throat & breast yellow. *Sim sp: Cordilleran Flycatcher (sympatric ranges).*

Acadian Flycatcher (*Empidonax virescens*)

Bill is longest & broadest of all Empidonax. **Adult.** (PA) 11.3–13.8 g, 14.6 cm (5.75 in.). Diet: insects & other arthropods. Nest: cup in fork, often over water. **Nestling.** MC, GF, bill: yellow. Skin: dark pink or pinkish-yellow. Down: white or buffy in occipital, spinal & humeral. LF: pinkish to yellowish-pink. G&D: d0: 1.9 g; d3: eye slits; d4: PinEm; d6: wingbars & ventral PinUn, tail length 2 mm; d8–9: flaps wings; d12: 12.3 g. Voc: weak *cheeping*, by d11 *seet* or *pseet* calls. **Fledgling.** Fledges d12–18. Plumage above: flat forehead, peaked, eye ring pale yellow & narrow; overall brownish-green edged buff, wide rich buffy wingbars. Below: dull white throat & flanks, lemon-yellow wash on belly (some with gray-green band) & undertail covs. LF: gray. Dependent 14 d post-fledging.

Alder Flycatcher (*Empidonax alnorum*)

Adult. 13.5 g, (ON, Canada) 10–14.5 g, 14.6 cm (5.75 in.). Diet: insects, some fruit in winter. Nest: open cup in fork. **Nestling.** MC: yellow. GF: deep yellow. Down: pale gray, head brownish-gray. G&D: egg 1.7 g; d5–8: PinUn; d6 eyes open. **Fledgling.** Fledges d12–15. Bill: upper dark gray, lower yellowish or pinkish. Plumage above: may have dark markings on crown; brownish wash, pale wingbars. Below: belly may have yellow wash. *Sim sp: Willow Flycatcher.*

Willow Flycatcher (*Empidonax traillii*)   Plate 16

Endangered status. **Adult.** (OR) 12.5–12.7 g; (CT) M 12–15.7 g, F 11.3–16.4 g; 14.6 cm (5.75 in.). Diet: primarily flying insects, occasionally fruit. Nest: compact cup in fork of slender branch. **Nestling.** MC: bright orangish or bright orange-yellow. GF: H pale yellow, turns brighter yellow, not prominent. Bill: buffy-orange, upper turns pink. Skin: pale pink & yellowish-apricot. Down: long, gray, crown olive-brown or mouse-gray. G&D: d0: 1.4–2.2 g; d1: 2–3; d2: PinVis; d3–4: 3.3–7.2 g; d4–5: egg tooth gone; d6: eye slits; d7: 11.5 g; d9: FF; d13: 13.3 g (WA & s.w. US). Voc: H faint *weep-weep* without opening bill, then *peep*. **Fledgling.** 12.5 g, fledges d13–16. Bill: brownish-black upper, lower lighter. Plumage above: crown "spots" small; brownish wash, well-defined buffy-brown wingbars. Below: olive-gray breast, brownish sides.

Least Flycatcher (*Empidonax minimus*)   Plate 16

Adult. (PA) 7.8–12.2 g, 13.3 cm (5.25 in.). Diet: insects, fruits occasionally. Nest: compact cup in fork of tree. **Nestling.** MC: yellow or orange. GF: pale yellow. Skin: orange-yellow. Down: light gray, including ventral tracts. G&D: d1: 1 g; d2: PinVis; d4–6: 4–5.5 g, all PinEm; d6–8: eyes open; d8: most prim & sec PinUn; d10: 9 g, rect PinUn; d12: 10 g. **Fledgling.** 10.8 g, fledges d13–14. Plumage above: white eye ring (complete); (bright) grayish-green back washed brownish, darker olive-brown wings & tail, broad buffy wingbars; rect tipped buffy. Below: whitish throat, dusky vest, pale yellow breast & belly. Dependent 2–3 wks post-fledging. *Sim sp: Yellow-bellied Flycatcher.*

Hammond's Flycatcher (*Empidonax hammondii*)

Adult. M 9.2–12 g, F 8–13 g; 14 cm (5.5 in.). Diet: insects (caterpillars, butterflies, moths). Nest: loose cup high in conifer. **Nestling.** MC: presumed yellow. GF: unknown. Down: on sides of head, dorsal & scapular tracts. G&D: mass unknown; d7: FF dorsum & alar; d11: FF. **Fledgling.** Fledges d17–18, remains in nest area ~20 d. Bill: tiny, short, may be mostly dark, orange near base. Plumage above: white eye ring, teardrop shaped (thickest behind eye); brownish wash, broad buffy wingbars. Below: abdomen yellowish or white; (juv) may have "vested" look due to contrasting dark flanks. *Wing formula may distinguish from other Empidonax (p10 usually ≥ p5; p9–p5 ≥ 5.0 mm)* (Pyle 1997b). *Compared to sympatric Dusky & Gray Flycatchers, Hammond's bill slightly shorter & narrower, prim extension longer & seems more pointed. Tail: mod forked (in Dusky slightly longer & double-rounded, nearly square-tipped in Gray).*

Gray Flycatcher (*Empidonax wrightii*)

Adult. (NV) 11.3–13.6 g, 15.2 cm (6 in.). Diet: insects; young fed grasshoppers, wasps, moths. Nest: bulky cup in fork of tree. **Nestling.** MC: presumed yellow. GF: unknown. Down: sparse. G&D: mass unknown; d7: eye slits, PinUn back & wings; d11: FF. **Fledgling.** Fledges d16. Bill: small, dark upper, pale pinkish to yellowish lower. Plumage above: pale band on forehead; olive washed, buffy wingbars. Below: brownish breast, lemon wash on belly. Wags tail. *Sim sp: see "Hammond's Flycatcher."*

Dusky Flycatcher (*Empidonax oberholseri*)   Plate 16

Adult. (AZ) 9.3–11.4 g, 14.6 cm (5.75 in.). Diet: mainly flying insects. Nest: compact cup low in mountain scrub. **Nestling.** MC & GF: bright yellow. Bill: orange-yellow, 75% gray (d9), darker (d12). Skin: bright reddish, thin & transparent. Down: white, long; on crown, occipital, ocular, dorsal & ventral. G&D: d0: 1–1.3 g; d1: 3.8, alar PinVis; d4: 4–5.8 g; d4–5: alar PinEm; d5–6: eye slits; d6: 7.7–9 g; d8: all PinUn; d9: 9.3–11.5 g, eyes fully open, prim PinUn; d10–11: appears FF, buffy wingbars; d11: perch on nest rim flap wings; d13: 10.2–12.2 g. **Fledgling.** 10 g, fledges d15–17. Plumage above: indistinct eye ring; dull olive fringed brownish; broad buff wingbars. Below: grayish-brown wash. Dependent for 3 wks post-fledging. *Sim sp: see "Hammond's Flycatcher."*

Pacific-slope Flycatcher (*Empidonax difficilis*)   Plate 16

Adult. (CA) 7.7–14.3 g, (AZ) 7.6–13 g; 14 cm (5.5 in.). Diet: mainly flying insects, young fed large, winged insects. Nest: cup in tree fork or in
cavity. **Nestling.** MC: bright yellow-orange to orangish. GF: yellow or
white. Bill: H yellow, very flat, pointy tip, wide "arrowhead" look. Skin:
bright pink or yellowish. Down: long, sparse, all tracts, densest on capital; color may be similar to Cordilleran. G&D (CA): d0: 1.5 g; d2–4:
2.8–5; d6–8: 7.3–9.4; d10–12: 10.2–11. Voc: insistent crowlike *squawk*
becoming froglike. **Fledgling.** 10.5 g, fledges d14–18, flies poorly until
5–7 d post-fledging. Bill: upper dark, lower lighter or yellow. Head: see
"Cordilleran Flycatcher." Plumage above: brownish wash, buff to cinnamon wingbars. Below: paler yellow or buff. LF: long, thin, delicate, dark
blue-gray (unique in Empidonax); white nails. Tail flicks upward when
perched. Dependent 14–22 d post-fledging. *Sim sp: Cordilleran Flycatcher.*

Cordilleran Flycatcher (*Empidonax occidentalis*)

Adult. 8–15 g; 14 cm (5.5 in.). Diet: mainly flying insects; young fed
leafhoppers, spiders, flies, wasps, bees. Nest: open cup or in cavity or
crevice. **Nestling.** MC: presumably yellow. GF: unknown. Down: dull
wood-brown or yellowish-gray, most tracts. G&D: mass unknown (egg
1.3–1.4 g), may be similar to Pacific-slope Flycatcher. **Fledgling.** Fledges
d14–18. Bill: lower all yellow or pinkish-yellow. Plumage above: head
peaked, bold teardrop eye ring widest behind eye; dull brownish wash
(also on sides), dark buffy or cinnamon-buff wingbars. Below: pale yellow. *Sim sp: in field, indistinguishable from Pacific-slope Flycatcher except in
Ad by M song.*

Black Phoebe (*Sayornis nigricans*)   Plate 15

Adult. (CA) M 16.9–22 g, F 14.6–20.8 g; 17.8 cm (7 in.). Diet: mainly
flying insects & other arthropods. Young (1–5 d) fed regurgitated bees,
wasps, flies, butterflies, moths. Nest: cup, mud pellets cemented to vertical wall or on covered ledge in outbuilding (barn, etc.). **Nestling.** MC:
bright yellow-orange. GF: creamy yellow, thin at first. Bill: N mandibles light yellow, turn darker by d8, wide, flat, pointed. Skin: orange &
yellowish-pink. Down: sparse, medium gray, dense on capital. LF: long,
thin orange & yellowish-pink; nails light yellow, turn black. G&D: d0:
1.5 g (CA), d2–4: 2–5 g, d5–12: eyes open; d5 to fledging: 7–15 g. Voc:
weak *peeping*. **Fledgling.** 18.7 g (approx. Ad mass), fledges d18–21, strong

flyers. Plumage above: browner than Ad; dark wing covs, prim & sec edged cinnamon; cinnamon wingbars. Below: white belly. Wags tail. Dependent 7–11 d post-fledging. *Sim sp: Say's Phoebe.*

Eastern Phoebe (*Sayornis phoebe*) — Plate 15

Adult. (PA) 16.5–23.2 g; 17.8 cm (7 in.). Diet: insects + 11% small fruits when available. Nest: cup of mud pellets, outside shapes to niche where built. **Nestling.** MC: deep yellow to orange or orange-red. GF: cream to yellow. Bill: flat, wide, upper mandible pale yellow, bill darker & longer by d6. Skin: deep pink or yellow-pink, fading after d1. Down: very long, sparse hair-brown or mouse-gray. G&D (Murphy 1981): d0: 1.7–2 g; d1–2: 2.8–4; PinVis spinal, humeral & ½ wings; d4: 7 g, capital & ventral PinVis, prim no. 9 emerges; d5–6: 9–11 g, cinnamon tufts on wing covs; d7–8: 12.5–14.2 g, more prim PinEm (d8–11); d9–10: 15.4–16.3 g; d11–13: 17–17.5; d13: flaps wings on nest rim. Voc: *tee-tee* (begging). Tail wags when fed. **Fledgling.** 17.5 g, fledges d16–18, flies well. Bill: pinkish-gray, tip dark. Plumage above: head & mantle drab-olive or dull brown; gray-brown with brownish-cinnamon wash; wings with some olive, buff-yellow edges, indistinct cinnamon-buff wingbars. Below: breast pale brownish, belly yellowish wash. LF: vinaceous. Dependent 2 wks post-fledging.

Say's Phoebe (*Sayornis saya*) — Plate 15

Adult. (NM) 15–24 g, 17 cm (6.7 in.). Diet: flying insects (beetles, grasshoppers, crickets, bees, wasps). Nest: bulky cup on sheltered ledge or nest of other species. **Nestling.** MC: reddish-orange. GF: yellow, thin. Bill: dark, wide, flat, tapering to point. Skin: deep yellow. Down: very sparse, light gray. G&D: d0: 2 g (est.). **Fledgling.** Fledges d13–21. Plumage above: browner than Ad, cinnamon or cinnamon-buff wingbars. Below: pale cinnamon, buffy chin, upper breast with brown. *Sim sp: Black Phoebe.*

Vermilion Flycatcher (*Pyrocephalus rubinus*)

Adult. M 11–12.4 g, F 12.6–14.8 g; 12.2–13.8 cm (4.8–5.4 in.). Diet: insects, other arthropods. Young fed butterflies, moths, spiders, grasshoppers, crickets, flies. Nest: flattish cup, in fork or crook of tree. **Nestling.** MC: bright yellow-orange. GF: yellow to bright yellow. Bill: somewhat broad, dark grayish-brown. Skin: blackish on back. Down: dense, short, light gray or creamy white. G&D: d0: 1.5 g (est.), d4: eye slits. **Fledgling.**

Fledges d14–16. Plumage above: grayish-brown, feather edges whitish or pale buff have scalelike appearance. Below: dusky spotting. Juv M may be tinged pink or have a few pink or red feathers in breast.

SHRIKES

(Family: Laniidae)

Two species of shrike raise young in N.A.; only one is covered here. Diurnal predators; known to impale prey on spikes. **Adults.** Diet: insects and small vertebrates (reptiles, birds, small mammals). Nest: bulky cup, fork of tree with dense cover. Eggs: subelliptical, smooth, white with grayish or buffish tinge, heavily marked. Bill: strong, short, culmen slightly decurved, hooked. Plumage: gray above, lighter below; broad black mask. Wings: 10 primaries (10th reduced), broad, rounded, black and white. Tail: long. Legs/feet: medium-long, black; sharp claws. **Young**: see "Loggerhead Shrike."

Loggerhead Shrike (*Lanius ludovicianus*)   Plate 17

Endangered status. **Adult.** (CA) 41–58.5 g, (FL) 35–58.5 g; 20.3–23 cm (8–9 in.). Large head. **Nestling.** MC: scarlet, orange tongue. GF: buffy-yellow, lined yellow-orange. Bill: H orange-yellow, turns dark & hooked, white egg tooth with pair of black spots each side, dark ring around bill near tip. Skin: gray but with pinkish-orange appearance. Down: all tracts, sparse, short, tan & white. LF: pink-orange; nails pale yellow, acute. G&D: d0: 2.8–3.5 g; d1–2: 4–6; d4: 12.5; d8: 21; d12–19: 46.5–48. Voc: d1 *tsp tsp*; d11 begging *tcheek*. **Fledgling.** ~48 g, fledges d17–21. Plumage above: facial mask dull, relatively indistinct; overall light grayish-olive with smoky-gray tips. Below: chin & throat white; breast & sides whitish or smoke-gray & finely barred; light buff wingbars. Biparental care: fed 3–4 wks post-fledging. Behavior: young shrikes must learn how to impale their prey (see Chapter 1, "Other Types of Imprinting"). *Song learning not studied.*

VIREOS

(Family: Vireonidae)

Thirteen species raise young in N.A., all in the genus Vireo. The most common species with the largest ranges are Red-eyed, Warbling, White-eyed & Yellow-throated Vireos. Five species with plain wings and contrasting pale eyebrow stripes are Warbling, Philadelphia, Red-eyed,

Black-whiskered & Yellow-green. Those with very small breeding ranges are Gray, Yellow-green (Mexico only), Black-whiskered (Florida only), Hutton's, and Black-capped (s-central OK, c. TX, and Mexico). **Adults.** Most are small to medium passerines. Diet: foliage gleaners of invertebrates (mostly insects), fruits (fall and winter), and some seeds. Nest: small, pendulous, open cup, suspended in fork or small branch, adorned with mosses, spider webs, lichen, few leaves. Eggs: subelliptical to oval, usually white with sparse dark speckles. Bill: usually short, rather straight, somewhat compressed at base, hooked, upper mandible with two tooth-like notches; rictal bristles present but not evident; nostrils somewhat concealed by bristle-like feathers. Plumage: sexes alike (except in Black-capped), with plain olive, olive-green, or gray upperparts; lighter (whitish, light yellow, or olive) below; some with wingbars, eye rings, or eyelines. Wings: 10 primaries (10th is reduced, minute [shorter than prim covs], or vestigial), long, variable in shape, usually somewhat rounded or notched. Tail: usually much shorter than wings; slightly rounded or notched; rectrices narrow. Legs/feet: strong, rather short, tarsi longer than middle toe with nail and scutellate; anterior toes basally adherent. **Young.** Mouth: most yellow, few orange, some yellow-orange. Gape flanges: white to yellow. Bill: many have long bills even at hatch with raised nostrils about midway from tip. Down: most hatch fully covered; some hatch with no down and do not develop any, while others grow down within a few days. Legs/feet: colors may be blue or bluish-gray in some species. Eyes: large and rounded. Fledge10–12 days; very short tails at first. Can be aged by plumage, tail shape, eye color & condition of primary coverts. Compared to similar songbirds, fledgling vireos have larger feet, longer legs, thicker bills (with hooked tip); gape flanges in recently fledged may still be prominent and slightly down-curved at corners; some have spectacles or eyelines; some with one or two wingbars; most have olive-green edgings on primaries and secondaries. Similar sympatric vireo sp: Warbling & Philadelphia, Warbling & Bell's, White-eyed & Yellow-throated. Song learning: juvenile vireos most likely learn some form of adult song within 30 days post-fledging while still in their home range; some are capable of singing adult song at 10–11 weeks old, many mimic songs of other species, which suggests they learn new songs for some time.

Black-capped Vireo (*Vireo atricapilla*)   Plate 17

Endangered status. **Adult.** (OK & TX) M 9.2 g, F 8.8 g; 11.4 cm (4.5 in.). Diet: insects (+ seeds in winter). Nest: pendulous cup in bushes or woody plants. Bill: black, short, tip slightly hooked & notched. Sexually dichromatic (female duller). Incomplete, prominent white spectacles on black face. Eye: red to brownish-red. **Nestling.** MC: yellow. GF: yellow. Skin: pink tones. Down: none. Bill & LF: H pink. G&D: d0: 1 g (gains ~1 g/d until 7–8 g); d3: capital & ventral PinEm; d5: eyes open; d7–8: PinUn. Voc: H faint *peeping.* **Fledgling.** 6.5–9.5 g, fledges 10–12 d (as early as d8). Plumage above: pale gray-green with pale gray cap, distinct white wingbars; spectacles less distinct than Ad; p10 reduced. Below: dull white or pale gray-white. Bill: brown-horn. LF: dull blue-gray to blackish. Iris: becomes red in December.

White-eyed Vireo (*Vireo griseus*) Plate 17

Adult. (FL) 10–14.3 g; 10.7–12.7 cm (4.2–5 in.). Diet: insects, some fruit (winter). Nest: pensile cup of plant matter & spider web. Bright yellow spectacles. **Nestling.** MC, GF: light yellow. Bill: upper mandible tip is gray. Skin: yellow-pink. Down: none. G&D: d0: 1.3–1.4 g; d1: 1.5–4; d2: 2.8–5.5, PinVis; d3–4: 3.8–8 g; d5: PinEm; d5–6: 6–9.5 g, eyes open; d7: 8–10.5 g, PinUn; d8: fear response, preen; d9–10: almost FF. Voc: H faint *peeps.* **Fledgling.** 10.4 g, fledges d9–10, weak flight, stays with parents ~23 d. Plumage like Ad but slightly darker; M Fl brighter overall. Above: overall dull brownish olive-green, rem & rect darker with bright olive-green edges, broad distinct buffy wingbars; beige auriculars, spectacles pale yellow with brownish loral streak through eye; p10 reduced. Below: dull grayish-white, buffy-yellow sides & crissum. LF: bluish-gray. Iris: brownish or grayish, turns lighter in November. *Sim sp: see family description, also mistaken for Empidonax flycatchers. Song learning: learns own song and mimics brief call notes of other species.*

Bell's Vireo (*Vireo bellii*) Plate 17

Endangered status. **Adult.** (AZ) 7.4–9.8 g, 11.4–12.7 cm (4.5–5 in.). Diet: insects (some fruit after July). Nest: pensile cup. Plumage: western sp drabber with little to no yellow (Least Bell's). Tail: relatively long. LF: eastern sp more bluish. Bill: blunt-tipped. **Nestling.** MC, GF: yellow. Bill & LF: N buffy-yellow. Skin: pinkish-yellow. Down: none. G&D: d0: 1 g; d2: 3.3; d5: 6.4 (eyes open); d7: 8.2 g, PinUn; d8: fear response; d9: 8.7 g. Voc: small nasal *peek.* **Fledgling.** 8.1g, fledges d11–12, parents feed 20–30 d post-fledging. Plumage above: gray with brownish wash, wingbars distinct; p10 reduced. Below: white. Least Bell's (CA): pale white cheeks & forehead, wings & tail edged green. *Song learning: juv sings subsongs of Ad ~40 d post-fledging.*

Gray Vireo (*Vireo vicinior*)

Adult. (Sonora, Mexico) 11–15 g, 12.7–14.7 cm (5–5.8 in.). Diet: large arthropods & other insects, some are frugivorous where fruit of elephant tree is available. Nest: pensile basket (cup shaped). LF: plumbeous or grayish-blue. **Nestling.** MC: yellow. GF: yellow. Down: H naked, then dull white above. Bill: pale brown. G&D: egg 1.92 g; 1 chick (TX): d5: 6.5 g; d6: eye slits, rem & rect PinEm; d8: 8.3 g. **Fledgling.** 10.85 g, fledges 13–15 d, flies weakly. Plumage above: crown dark, brownish wash, fairly distinct wingbars; p10 reduced. Below: dull, pale buff. Bill: blackish-blue. *Song learning: may be from father & neighboring M on nesting grounds.*

Hutton's Vireo (*Vireo huttoni*)   Plate 18

Adult. 9–15 g, 12.7–14.7 cm (5–5.8 in.). Diet: arthropods (including spiders); young fed spiders, small worms & insects. Nest: pensile cup, globular, well hidden. **Nestling.** MC: orange. GF: bright yellow. Bill: pinkish. Skin: pinkish or apricot. Down: little to none, becomes thicker (first appears on back), pale yellowish, grayish, or brown. LF: sky-blue to pale gray, turn gray, whitish toepads. G&D: egg 1.5–2.16 g; d0: 1–1.65 (est.); d2: mass doubles; d5: PinEm; d6: PinUn buff tips, head & neck PinEm, wingbars begin to appear; d8–9: ~7 g, eyes open, all PinUn, rect PinEm; d14: FF. **Fledgling.** 9–15 g (est.), fledges 14–15 d, fed 21 d post-fledging. Bill: dark. Plumage like Ad, except duller. Above: head rounded, high forehead with raised crown, indistinct broken eye ring; overall washed olive-gray or brownish, wingbars yellow-buff; p10 reduced. Below: yellowish wash over gray. Voc: begging calls, wheezy quality of Ad.

Yellow-throated Vireo (*Vireo flavifrons*)   Plate 18

Adult. (PA) 15.6–21.4 g, 12.7–14.7 cm (5–5.8 in.). Diet: insects (fruits & seeds fall & winter). Nest: cup (suspended). **Nestling.** MC: yellow. GF: yellow. Down: grayish above & below. LF: pinkish-buff. G&D: unknown. Voc: d4–5 persistent, high-pitched *ceeeeeee*. **Fledgling.** 13 g, fledges d13–14. Bill: light brownish to dark. Plumage like Ad but softer brownish-gray. Above: grayish-olive with soft brownish wash; buffy indistinct supraloral stripe, whitish or pale yellow eye arcs, dark gray lores. Wings: buffy wingbars, dark prim & sec edged buff-white; p10 minute. Below: pale buffy-yellow throat & upper breast, pale gray sides. LF: pale blue-gray. Independent 14 d post-fledging.

Cassin's Vireo (*Vireo cassinii*)

Adult. (w. US) 10–20.6 g, 11–13.8 cm (4.3–5.4 in.). Diet: insects (+ some berries). Nest: suspended cup. **Nestling.** MC: yellow-orange. GF: cream to light yellow. Skin: pinkish. Down: possibly hatch naked, then very sparse whitish. G&D: unknown. **Fledgling.** Fledges d13–14 (CA). Plumage above: white spectacles; brownish-gray overall; p10 minute. Below: dull white, yellowish tinge on flanks & crissum. LF: grayish-blue.

Blue-headed Vireo (*Vireo solitarius*)

Adult. 13–19 g, 12.7–14.7 cm (5–5.8 in.). Diet: insects (+ some fruits in winter). Nest: cup suspended by its rim in trees or shrubs. **Nestling.** MC: yellow-orange. GF: cream to light yellow, brightest at corners. Skin: pinkish. Down: sparse, white or grayish (depending on ssp). Bill: H brownish. LF: H pinkish-tan, then dusky (d8). G&D: d3: 4.6 g (length 51 mm), alar PinEm; d5: 8.4 (60 mm); d5–6: eyes open; d6–8 PinUn capital, dorsal, humeral, femoral & ventral; d7–8: appears FF, wing PinUn; d8:17 g (71 mm); d12: 14 g (94 mm). **Fledgling.** Fledges d12–13, flies weakly. Plumage above med gray with brownish wash, rect edged yellow-olive, outer rect edged whitish, rem dusky to black (edged olive-green), 2 buff-white wingbars, p10 minute; supraloral area & eye ring white (& narrow). Below: white patch extends up sides of neck; buffy-white, lemon wash over gray, lemon-yellow undertail covs. LF: become bluish-gray (juv). *Song learning: may learn basic song elements in 1st wks of life.*

Plumbeous Vireo (*Vireo plumbeus*)   Plate 18

Adult. (CO) 12–20.3 g, 12.5–14 cm (4.9–5.5 in.). Diet: insects & other arthropods (almost exclusively). Nest: cup (suspended). **Nestling.** MC: orange. GF: yellow. Skin: pinkish. Down: white, dorsal areas. G&D: d0: 1 g; d1: 2; d4: prim & tail PinEm; d6: most PinEm, eye slits; d9: appears FF; d10: 14–15 g. Bill: yellow, yellow-brown (d3), dark brown (d9). **Fledgling.** Fledges 13–15 d. GF: still present & yellow. Bill: black. Plumage like Ad but more brownish with whiter sides & flanks, buff-yellow wingbars; p10 minute. LF: grayish-blue.

Philadelphia Vireo (*Vireo philadelphicus*)

Adult. (PA) 9.8–12.7 g, 10.7–12.7 cm (4.2–5 in.). Diet: insects (some fruit fall & winter). Nest: cup suspended by rim, high in tree. **Nestling.** MC,

GF: yellow. Bill: white. Skin: light yellow-orange. Down: short, pale gray, above & below. Nails white. G&D: unknown. **Fledgling.** Fledges d13–14. Plumage above: wood-brown, wings & tail dark brown with olive-green edgings; p10 vestigial or minute. Below: pale yellow; buff-yellow auriculars, supercilium & eye ring; dusky lores & postocular streak. LF: slaty. Independent 24 d post-fledging.

Warbling Vireo (*Vireo gilvus*)   Plate 18

Adult. (western ssp) 9.9–15.8 g, (eastern ssp) 11–18.4 g; 12–12.7 cm (4.7–5 in.). Diet: insects & other arthropods (some spiders) & few berries. Nest: deep cup hanging below level of branch (in fork). **Nestling.** MC: orange. GF: H thin, yellow, become more prominent. Skin: dark yellow or yellow-orange. Down: sparse, light brown or white tufts above & below. G&D: little info. d0: 1.13–1.32 g, d6–8: eyes are open. **Fledgling.** Fledges d10.5–19 (avg 14–16 d). GF: still somewhat prominent. Bill: smaller & darker than eastern ssp. Plumage like Ad (grayish tinged olive-green back, crown darker) but browner, paler & duller. Above: white supercilium mod distinct, grayish eyeline, loral area dark, pale eye ring; faint buffy wingbars, dark brown prim & sec edged with olive, p10 reduced. Below: white with varying degrees of pale yellowish wash on sides, belly & crissum. LF: blue-gray or plumbeous. Voc: (eastern birds): little info; nestlings make calls resembling Ad calls. Dependent 2 wks post-fledging.

Red-eyed Vireo (*Vireo olivaceus*)   Plate 18

Adult. (PA) 14–21 g, 12–12.7 cm (4.7–5 in.). Diet: insects (mostly), caterpillars in spring and summer; fruit in winter. Nest: open cup, fork of branch. **Nestling.** MC: pale yellow to bright orange. GF: white to yellowish. Bill: hook on tip not evident until 10–12 d. Skin: pale pinkish-apricot to yellow-orange (darker dorsally). Down: sparse, long, gray or brownish-gray above, white on ventral (d1), middorsal region is flared. LF: H pinkish-buff. G&D: d0: 1.5–1.8 g; d1–2: PinVis; d4–6: eyes open; d5–6: 10.9–12.5 g; d7: prim PinF 8 mm, PinEm 1 mm all other tracts; d9: flight & contour PinUn; d11: appears FF except ventrally. **Fledgling.** 13.8 g, fledges d10–12. GF: yellow & still mod prominent. Bill: pale grayish & pinkish. Plumage above: grayish-olive with brownish tinge; buffy supercilium, grayish lores, dusky eyeline; dark dusky-brown wings edged olive-green, indistinct buff wingbars, p10 vestigial or minute. Below: white with pale yellow sides, flanks & crissum. LF: pinkish-gray, then plumbeous. Iris: brown or gray-brown until end of 1st winter (then red).

Song learning: some degree of learning occurs; can mimic other species. Juvs not observed singing on breeding territories. Ad M sings 30–40 different songs (Kroodsma 2005).

CROWS AND RELATIVES

(Family: Corvidae)

Adults. Medium to very large songbirds ranging from 26 cm (10.25 in.) to 63.5 cm (25 in.). Weights (given for most species as ranges) are variable by sex (M>F), geography, and individuals. Diet: all are omnivorous; some scavengers. Nest: base is basket of sticks with cup-shaped woven lining inside, some use mud; magpie nests are covered. Eggs: subelliptical (short to long), bluish-green with brown markings. Bill: all-purpose, powerful, medium to long, stout, culmen decurved toward tip, somewhat acute; nostrils covered with rictal bristles pointing forward. Plumage: most species predominantly black or blue, some boldly patterned, some have long tails and some have crests. Wings: 10 primaries (10th reduced), rounded in most sp. Legs/feet: large and strong, tarsi longer than middle toe with claw, scutellate. **Young.** Mouth: red or pink. Gape flanges: pink or yellowish, not prominent (thin), except magpies. Bill: long relative to head; nares flat. Down: most hatch with none, then grow very short, sparse, downy-like coat beginning a few days post-hatching. In most species, feathers of head, body, and legs grow directly out of the skin (no quills emerge). Wing and tail feathers emerge in quills. Head shape: high, rounded crown. Fledge 16–30 days (smaller species), 28–44 (larger species). Young are dependent on adults long after fledging. Behavior: some among the most intelligent of all animals with remarkable spatial memories; some make and use tools; most have highly complex social systems. Voc: a complex variety of sounds, including mimicry; do not "sing" as other oscines; calls are noisy, loud & harsh. *Aging corvids: feather growth and behavioral events are essential in addition to weight ranges. Song learning is important in some species, such as Pinyon Jay, because of high degree of sociality.*

Canada Jay (*Perisoreus canadensis*)

Adult. (ON, Canada) M 68.5–84 g, F 57.8–75.4 g; 27.4–30.5 cm (10.8–12 in.). Diet: arthropods, small vertebrates, carrion, berries, fungi, insects. Young fed arthropods. Nest: open cup, fork of small branch. Early nesting: mid-Feb. to end of Mar. **Nestling.** MC: pink to red. GF: white & thin. Skin: pinkish-brown, then pale orange. Down: sparse, long, above only. Bill: H pinkish-brown. LF: H pale pink, then gray. G&D (grows

5 g/d between d4 and d10): d0–1: 4.5–7 g; d4: purplish-blue feather tracts; d7: prim PinF ~8 mm, PinEm 1 mm all other tracts; d8–11: eyes open; d9: flight & contour PinUn; d10: 50 g; d11: appears FF except ventrally; d13: 60 g; d14: stretches & flaps wings; d17: 65 g; d20: brancher. **Fledgling**. Fledges 22–24 d, flies well. Plumage above: slate-gray. Below: lighter. Tail 1/3 full length (full length 44 d). Bill: black, small compared to other jays. LF: black. Family stays together > 3 wks post-fledging.

Pinyon Jay (*Gymnorhinus cyanocephalus*)

Extremely social. **Adult.** (AZ) M 111 g, F 99; 26.7 cm (10.5 in.). Diet: especially seeds of piñon-pine. Nest: bulky cup often high in piñon pines & junipers, in colonies. **Nestling**. MC: bright salmon-red; turns bright orange-red (d8). GF: yellow. Bill: yellowish-pink (white egg tooth), d15 grayish to purple. Skin: pink, then dark purplish-blue. Down: none. LF: pinkish-tan. G&D: d0 (AZ): 6.26 g; d7–8: eyes open; d8: dorsal & alar PinEm; d10: 54% of Ad body mass; d15: appears FF. **Fledgling.** 81.3 g (or 75% of Ad mass), fledges 22–23 d, flies poorly, remains in dense vegetation in trees & shrubs. Bill: d22 culmen yellowish-horn. Plumage mostly mouse-gray, except whitish feathers of bib; blue-gray tail. Voc: begging *chirr* (d15).

Steller's Jay (*Cyanocitta stelleri*)   Plate 19, Photos 22-25

Adult. (AZ) 98–117 g, (W. Canada) 111–142 g; 33 cm avg. (13 in.). Diet: arthropods, nuts, seeds, acorns, berries, small vertebrates, eggs, human food, garbage. Nest: open cup (bulky) on branch close to trunk; mud used. **Nestling.** MC: red to bright red, fades to pinkish-gray in juv. GF: white or pale yellow, thin. Bill: N pinkish-gray, turns dark (Fig. G2, Illustrated Glossary); long & pointed (d1 culmen 9–10 mm). Large gape. Skin: H pink, then darker. Down: none; 1st feathers grayish-brown, "downy-like." Crest on head forms early. LF: long, pinkish-gray, grappling toes. G&D: d0: 6–8 g; d1*: 13.3, 17.2 & 18.9 g, length 63–69 mm; d5: 30 g (Fig. G6 a & G7, Illustrated Glossary); d6: 30–40, PinEm wings; d7: eye slits, PinEm upper covs; d8–9: 55–65 g (length 120–125 mm); d9–10 PinUn all over, crest may appear; d10–11: 75–80 g, FF with 1-in.-long tail still mostly in sheaths, not longer than wings; d12: ~80 g; d14–15: ~90; d16: ~100 (tail 2/3 length of Ad); d24–28: 108 g; d40: 116. Voc: squeaky 2-noted "whine," louder d10. **Fledgling.** 90–100 g (CA), fledges d16–21 (not well known), weak flights 15–22 d, sustained flight ~30 d. Bill & LF: dark. Plumage above: crest very short; body, head & mantle sooty-gray;

gray overall; indigo blue or cobalt blue prim & sec; venetian or phthalocyanine blue greater covs; indistinct barring on outer web; wings & tail very short. Below: light gray (no white). Family groups may remain together until fall. *Sim sp: California Scrub-Jay nestling, no crest, weighs less, white underparts. *d1 data from three MVZ fluid specimens (1919, Yosemite Valley, CA).*

Blue Jay (*Cyanocitta cristata*)   Plate 19

Adult. (PA) 73–101 g; 25–30 cm (9.8–11.8 in.). Diet: arthropods, acorns, nuts, carrion, small vertebrates, soft fruits, human food. Nest: open cup, fork of tree or branch; uses mud. **Nestling.** MC: deep pink or light salmon; tongue lavender-gray. GF: white & thin, downcurved. Bill: H pinkish-white, N grayish upper. Big gape. Skin: H gray tinged pink on dorsal, (d2); d2 skin turns brick-red brown with olive on spinal tract and bluish-green on wings & around eyes. Down: none (see d6). LF: lavender-gray, long, smooth (d9). G&D: d0: 5–6 g (51 mm avg); d3: 13.5 g (62 mm); d5: 20–28 g (76 mm), eye slits, most PinVis, wing PinEm; d6: 30 g, appears FF with short, fuzzy greenish-yellow feathers; d7: 35 g, eyes fully open (d7–11), wing cov PinUn; d8: 40 g, blue evident on head, wings, some underparts; d9: 45 g; d10: 50; d11: 55; d12: 58, dull white around eyes, head pattern begins to show; d13: 61 g; d14: 56, PinEm sides of head & throat; d16–19: 58–60 g; d21: 62. Voc: harsh, squeaky, almost like yelping puppies; 2-noted (same time), makes sounds while food is placed in mouth. **Fledgling.** 58–62 g, fledges 17–21 d. Plumage pattern resembles Ad. Above: brownish-gray or dull blue-gray mantle, fluffy medium gray downy coat on back & rump, crest dull & short. Blue in wings & tail: depending on lighting, may reflect colors of Prussian blue, cobalt blue, or phthalocyanine blue. Below: dull white.

California Scrub-Jay (*Aphelocoma californica*)   Plate 19, Photo 26

Adult. 70–100 g, (CA) M 96.6 g, F 82.7 g; 29.2 cm (11.5 in.). Diet: 27% animal, 73% vegetable matter. Young fed insects & other arthropods. Nest: open cup (bulky), low in tree or bush. **Nestling.** MC: deep rose to reddish. GF: white, very thin. Bill: N pinkish-gray, long & pointed, tip hooked. Big gape. Skin: pinkish-brown (ruddy). Down: none, then grayish-brown downy-like plumage, "furry" head. LF: pinkish, then pink-gray, long legs & grappling toes; white nails. G&D*: d0: 5–6 g; d1: 6–9; d3–4: skin dark, wing & tail PinEm, white ventral PinEm; d5: 18–38 g, body PinEm; d7: 22–50 g, PinUn, blue evident on wings; d10: 32–55 g, head fuzzy, short, fluffy gray feathers all over, ventral grayish-white; d14: 45–76 g, FF, wings & tail mostly in shafts; d18: 67–76 g; d20: 80; just prior to fledging: 74 g (est. ~80% of Ad mass). Voc: hatchling makes short repeated *peeps*, later single *squawk*. **Fledgling.** 74 g (variable: 50–80 g), fledges d16–26, unable to fly well for 3–4 wks. Plumage above: super-cilium narrow, indistinct, grayish-blue overall, wings & tail dull pale blue & gray. Below: grayish-white. Bill: pink at base, gray toward tip. LF: brownish-black. Dependent > 30 d post-fledging. *Sim sp: see "Woodhouse's Scrub-Jay" (has thinner, pointed bill); nestling jays may be hard to distinguish until feathers appear. *Nestling weights from Ritter 1984, Butte County, CA.*

Woodhouse's Scrub-Jay (*Aphelocoma woodhouseii*)

Little information on young, 2016 taxonomic split from California Scrub-Jay. **Adult.** (w. US) M 69–98 g, F 65–83 g; 28 cm (11 in.). Diet: arthro-pods & fruit (spring/summer), seeds of masting trees (fall/winter). Nest: open cup, placement variable in trees or shrubs, often heavily concealed. **Nestling.** MC: pink. GF: presumed white. Down: none. LF: pinkish-tan. G&D: unknown, probably similar to California Scrub-Jay. **Fledgling.** Plumage above: head grayish-brown or brown; dull brown-back, blue tinge to wings. Below: dull gray, throat patch less pronounced. LF: brownish-black. *Sim sp: see "California Scrub-Jay."*

Clark's Nutcracker (*Nucifraga columbiana*)

Adult. (MT) M 137 g, F 123 g; 30.5 cm (12 in.). Diet: primarily conifer seeds + insects, spiders, small animals & carrion. Nest: open cup on outer branches of tree. **Nestling.** MC: salmon-red. Down: none, then sparse white all tracts. LF: gray; turn black by 6 wks. Iris: gray. G&D: d0: 7.1–8.2 g; d4: dark PinVis all tracts; d5–11: eyes open; d11: PinUn; d16: 95 g. **Fledgling.** 88–105 g, fledges d20–22. Plumage above: mouse-gray tipped ochreous-buff, less white around bill, eyes & face. Iris: brown.

Black-billed Magpie (*Pica hudsonia*)   Plate 20

Adult. (CO) M 159–209 g, F 135–197 g; 48.3 cm (19 in.). Diet: arthro-pods, seeds, carrion. Nest: domed, mud cup & anchor, sometimes in ag-gregated colony. **Nestling.** MC: deep pink to red. GF: pale pink to white. Bill: H pink, turns bluish-pink, egg tooth remains until fledging. Skin: pink, turns yellowish & grayish, then black by 5–10 d. Down: none. G&D (egg weight: 9.4 g, N gains 10 g/d 1st 15 d): d0: ~7 g (est.); d1:

14 g (~65 mm long); d2: 25–28 g (~75 mm long), upper mandible shorter than lower, egg tooth evident; d3–8: 35–76.4 g (eyes open d7); d7: prim & sec PinEm, PinUn on head, throat, back & femoral tracts; d9–15: 63–128.6 g; d16–24: 155–183.6. Voc: begging trill & soft whistle. Highly competitive for food with siblings. **Fledgling.** 180.5 g, fledges d24–30. Plumage like Ad but less iridescent, more green than blue on wings & tail, wing tips rounded, bare skin around eyes. LF: soles of feet light, legs black. Iris: brown. Care: remains near nest 3–4 wks; independent 6–8 wks post-fledging.

Yellow-billed Magpie (*Pica nuttalli*)   Photo 27

Endemic to California, found in open oak woodlands. **Adult.** (CA) M 158–189 g, F 126–158 g; 43.2–53.3 cm (17–21 in.). Diet: arthropods, grain, acorns, carrion. Nest: large, globe shaped & domed, high in large trees, loose colony. MC & bill: yellow. **Nestling.** MC: H pale pink, then deep pink. GF: creamy yellow outside, yellow inside & prominent but not thick; wide gape. Bill: H pink; d16 tip turns yellow, egg tooth remains until fledging. Skin: pink, then turns yellowish & grayish. Down: none, PinVis alar & spinal. LF: very long pink legs & grappling toes, light brown nails, tarsus 11 mm. G&D: d0: 5.7–8.1 g; d0–1*: no mass taken, 65 mm long, beak culmen 10 mm, tarsus 8 mm, no down, no wing pins; d5–7: 40–55 g (captive); d7: all PinEm except capital; d8–9: 106 g, d11: nape blacker than crown, PinEm on rem some with tufts; white tufts ventral & femoral; d14–16: 115–135 g (captive birds). Voc: squawk, whistle. **Fledgling.** 123–141 g, fledges ~30 d; clambers around on branches in nest tree 4–5 d. Plumage (complete 50–60 d post-hatching): like Ad but duller, less iridescence on tail, face more naked, bright yellow around eyes. Dependent 50–55 d post-fledging. *d0-1 data from one MVZ fluid specimen (1971, Monterey County, CA).*

American Crow (*Corvus brachyrhynchos*)   Plate 20, Photo 28

Adult. Mass varies by ssp & sex: (CA) M 390–457 g, F 347–405 g; (OH) M 436–637 g, F 416–610 g; (IL) 474–538 g; 47 cm avg (18.5 in.). Diet: 72% plant matter, 28% animal. Nest: open, bulky, isolated trees in fields. **Nestling.** MC: medium-pink to red; d25–30 deep pink or bright red. GF: white or pinkish, small. Skin: pink or pinkish-tan, turns darker (black d5–10). Bill: H gray; N medium length, heavy, wedge-shaped; d25–30 dark gray to black with pink overtones. Down: none, then sparse grayish-brown on capital, alar & ventral. LF: H pinkish, then grayish-black; long toes, heavy legs; d25–30: toe pads yellow-horn to dark gray. G&D (see Table C-4, Appendix C): d0: 10–13 g. Iris: gray-blue. Voc: H weak, d4 variety of *churrs*, *cheeaps*, *wa-eeks*, *yumyumyum* sounds when fed. **Fledgling.** 300–370 g, fledges d28–38. MC: pinkish palate until fall when it may change to black (variable). GF: still visible (until ~d90). Bill: gray-black (~d25). Plumage like Ad but duller. Iris: gray-blue (~d60), then gray, eventually brown. Care: fed by parents & helpers 2–4 mos. Post-fledging, remains in natal territory ~9 mos., may stay in family group > 5 yrs. *Note: If fledgling lands on ground during first 10 d post-fledging, it often cannot get back into the tree & is vulnerable. Sim sp: comparison to Common Raven: crow weighs at least half at same age, bill smaller, bristles shorter, tail square or rounded. Large adult M American Crow may weigh close to small F Common Raven in some areas.*

Fish Crow (*Corvus ossifragus*)

Adult. (FL) M 260–332 g, F 195–304 g; 30.5–40.6 cm (12–16 in.). Diet: carrion, marine invertebrates, eggs & nestlings of many bird species, insects & fruits. **Nestling.** MC: deep rose pink. GF: vinaceous. Egg tooth remains several wks. Down: hatch naked or with light drab-gray or grayish-brown. LF: neutral gray, turn black by 3 wks. Iris: grayish-blue, then neutral gray. G&D: d10: prim PinEm; d18: head FF. **Fledgling.** Fledges 21 d or more (32–40 d in FL), flies short distances at first, leaves nest area 70 d. Plumage: brownish-black, wings & tail lustrous black with greenish iridescence.

Chihuahuan Raven (*Corvus cryptoleucus*)
(previously White-necked Raven)

Adult. M 442–667 g, F 378–607 g; 49.5 cm (19.5 in.). Diet: omnivore: grasshoppers, beetles, mammals, carrion, eggs, sorghums, fruits, human food. Nest: cup in tree or man-made structure. **Nestling.** MC: ruby. GF: pale yellowish-pink. Bill: H reddish-orange, turns grayer d2, then dark upper with lower mostly pinkish. Skin: reddish-orange. Down: short, grayish-white, buffy-white to yellowish or pinkish-white. LF: H pinkish to reddish, then grayish-black. Iris: bluish. G&D: d0: 13–16 g; d6: wing PinEm, eye slits; d8: rect & dorsal PinEm; d10: ventral PinEm; d11: capital PinEm; d12–13: eyes open; d15–19: begins to stand & to move wings. Voc: d21 adult-like *croak*. **Fledgling.** Brancher at 30 d, fledges 37–40 d. Bill: black. Plumage: dull black (except crown) with metallic bluish or greenish tinge. *Sim sp: American Crow & Common Raven. White at base* *of neck feathers (gray-based in Common Raven). Longer wings & heavier bill than crows. Chihuahuan Raven occupies limited range in desert & grassland habitats.*

Common Raven (*Corvus corax*)   Plate 20

More solitary (or in pairs) than other Corvus except in nonbreeding season may be found in large foraging flocks. N susceptible to hyperthermia and dehydration; adults bring water to feed & cool N. **Adult.** Mass varies by ssp & sex: 689–1625 g, (CA) 585–985 g, (AK) 1097 g; 56 to 69 cm (22–27 in.). Diet: omnivore & scavenger, of various vertebrates & invertebrates, carrion, vegetable matter, human food. Nest: open cup (bulky), well hidden in crotch usually near trunk of tree. Massive bill: culmen length 59–86 mm, culmen height 21–30 mm, long bristles over nares. Elongated throat feathers. Tail: wedge-shaped. **Nestling.** MC: deep pink to red. GF: yellowish (possibly white). Skin: H deep rose-pink, turns pinkish-salmon (d1), then darker. Down: H none to little, (d14) sparse, long & gray & black on dorsum; (d21) thicker & fuzzy, short gray or dull brown all over. Bill: H pinkish-tan, turns gray (d10). LF: pink, then gray, then black with gray soles d9–10 (Fig. G5 a, Illustrated Glossary). Iris: N light blue, turns gray ~60 d. G&D (mass varies by ssp & sex): d0: 19–34 g (est.); d4: ~7 cm, eyelids closed; d8–12: eyes cannot focus on objects < 0.5 m away (Whitemore and Marzluff 1998); d9–10: 180–355 g; d9–10* (age est.): 219 g, 165 mm (6.5 in.), culmen 25 mm, tarsus 38 mm; prominent flanges; short feathers growing all over, including ventral, none in sheaths except all flight feathers with tips exposed; feathers on head very short, longest on spinal tract, especially upper back; d10–11: 347–405 g, gray bill & LF; d12–14: eye slits; d14: 20 cm long, rectrices ~1 cm, cannot stand; d21: wing & tail PinF ~2 cm; d24: 997–1190 g; d28: covered with contour feathers, defensive response; d30: 1031–1327 g, begins moving in & out of nest; d35: FF, active, exercises wings. **Fledgling.** Fledges 35–49 d, many factors affect departure, weak flights until ~4 d post-fledging, remain near nest up to 28 d post-fledging. MC: transition from pink to black variable in 1st year; dominant individuals tend to change to black sooner. Plumage: overall dull brownish-black, little to no metallic greenish/purplish on wings & tail, throat feathers shorter than Ad. Post-fledging care: may stay with parents through 1st winter (some observed begging & being fed); in some areas young may disperse sooner. *See "American Crow" for comparisons. *d9–10 data from one (CAS) fluid specimen (2002, Bakersfield, CA).*

LARKS

(Family: Alaudidae)
Of two species occurring in N.A., only one is covered here.

Horned Lark (*Eremophila alpestris*)   Plate 21

Adult. Small to medium-size songbird, (TX) 32.2 g, (WA) 30.2 g; 16–20.3 cm (6.3–8 in.). Diet: mainly seeds & insects. Nest: shallow cup on ground, variety of open habitats. Bill: short, conical, slender & acute; feather tufts conceal nares. Plumage (highly variable geographically): head crested or "horned," conspicuous black patches on head & neck; dull-colored with mostly brown, white & black. Wings: 9 (visible) primaries (10th vestigial), long, pointed. Tail: nearly square & shorter than wing, outer rectrices white or edged white. LF: moderate size, tarsi longer than middle toe, rounded behind & scutellate. Nails: hallux nail is very elongated, equals hallux in length (similar to pipits). **Nestling.** MC: bright orange-yellow. Tongue spots: 5 black spots (1 at tip of upper & lower mandible, 1 at tip of tongue & 2 on back of tongue). GF: yellow, thin; N: whitish outside, yellow inside, turned down at corners. Skin: black above, orange below. Down: very long, well covered dorsally; olive-buff, creamy buff-yellow, or cream-buff on capital & all dorsal, a few on abdominal. Bill: N buff-yellow, darkens to neutral gray. Egg tooth remains until fledging. LF: pinkish-tan, then light neutral gray. Hallux nail nearly Ad length d8 (8.75 mm). G&D: d0: 2.4 g; d1: 4.3; d2–4: eyes open; d3–4: 9.1 g, PinVis head, back, wings, breast, abdomen; d4: 12.7 g, prim PinEm; d5: 16.3 g; d6: body PinUn, rect PinUn (continue to d12); d7–8: 18.2–20.7 g; d10: FF; d12: down gone. **Fledgling.** 22.3 g, fledges 9–12 d, flies poorly until 3–5 d post-fledging. Plumage: overall color varies (by ssp) light gray, nearly white to nearly black; mottled appearance from light gray triangular tip on each feather; dusky-brown occipital feather tufts. Tail: outer rect white (lateral webs). Below: duller white & some birds may have an incomplete ashy-brown or faint gray pectoral (chest) band. Independent 21–30 d post-fledging.

SWALLOWS

(Family: Hirundinidae)
Distinguishing features: wide gape is twice the length of the culmen; broad, rounded head shape; short neck, short tarsi, long pointed wings, and vocalizations. **Adults.** Most are small passerines except the largest

swallow, the Purple Martin. Diet: aerial hunters, insects almost exclusively taken in flight. Nest: varies by species; in tree cavities, nest boxes, or earthen tunnel; may be entirely mud, composed of mud and feathers, or a mud structure fixed to rocks or walls (with sheltered overhang). Eggs: subelliptical or oval, usually white or finely spotted or speckled in some species. Bill: short, wide at base (triangular in outline when viewed from above), depressed, slightly hooked, culmen somewhat decurved toward tip. Plumage: compact, boldly patterned, generally darker above, some are iridescent with blues or greens with contrasting patches of white or rust; others are in shades of browns and grays. Wings: 9 (visible) primaries (10th vestigial), primaries long and pointed, secondaries generally very short. Tail: shorter than wing (except Barn Swallow) either notched, squared, or forked; lateral rectrices sometimes filamentous. Legs/feet: tarsi short (usually shorter than middle toe, including nail) and scutellate, toes tiny and weak, hallux nail shorter than hallux. **Young.** Mouth: yellow range. Gape flanges: creamy-white to yellow. Down: mostly grays & browns, some white; feather tracts split at upper back (upside-down Y). Fledge: see species. Biparental care.

Bank Swallow (*Riparia riparia*)   Plate 21

Smallest swallow. Thin wings. Slender, notched tail. **Adult.** (CA) M 10.9–16.4 g, F 11–18.8 g; 13.3 cm (5.25 in.). Diet: insects. Nest: excavates burrow or uses kingfisher burrow. **Nestling.** MC: pale, lemon-yellow. GF: pale yellow, relatively thin. Bill: small, yellowish-gray, dark tip; wide gape. Skin: pink to bright reddish-pink. Down: sparse, pale gray or gray-white, turns gray-brown d10. LF: pinkish-gray. Eyes: large. G&D: d0: 0.86–1.16 g; d9: able to move out of nest by shuffle-walking; d10: PinF appear like spines; d15–17: moves to burrow entrance to be fed. **Fledgling.** (WI) 10–16.2 g, (CA) 9.3–19.4 g, fledges 18–24 d, flies well, returns to burrow for ~5 d post-fledging, juv gather in groups. Plumage above: white curls up around dark auriculars. Below: drab, throat dusky, wide brown breastband.

Tree Swallow (*Tachycineta bicolor*)   Plate 21, Photo 29

Broad wings, tips reach tail tip. Short, notched tail. **Adult.** 17–25.5 g, 13.5 cm avg (5.3 in.). Diet: insects, berries when insects scarce. Nest: cavity in tree or box, lined with feathers (from other species) that curve up over eggs. Eggs: pure white. **Nestling.** MC: pale yellow, then bright yellow-orange throat (d10). GF: creamy-white to yellow, thick, prominent, straight. Bill: small, short, wide gape. Skin: pale pink or apricot. Down: very sparse, white, mouse-gray to gray-brown. LF: H tarsus short, pinkish, white toenails turn dark. G&D (see Table C-5, Appendix C): d0: 1.3–1.8 g. Voc: begging call variable, 1st week 1–3 short, pure tone sounds, *eep* or *tseep*. **Fledgling**. 20–21 g, fledges 8–25 d, flies well. Plumage above: dusky-brown (no trace of greenish-blue until completion of preformative molt by early winter), white on throat extends to nape contrasting sharply with dusky crown. Below: dull white often with gray-brown wash across chest (not distinct as in Bank Swallow), small white crescent on sides of rump. Tail: slightly forked. *Sim sp: Northern Rough-winged Swallow and Bank Swallow.*

Violet-green Swallow (*Tachycineta thalassina*)   Plate 21

Tips of wings project beyond tail tip. Short, notched tail. **Adult.** (CA) M 13–16.3 g, F 12.5–15.2 g; 12 cm avg (4.7 in.). Diet: insects. Nest: lined with feathers, in cavity, may use old nest of another species. **Nestling.** MC: orange-yellow. GF: creamy-white to pale yellow, very thick, prominent. Bill: small, short, somewhat pointy, wide gape. Skin: pale pink, darkens by d5. Down: silvery white & light gray, turns dirty brown. LF: short & pink. G&D: d0: 1.5–1.7 g; d1: 2+ (40–45 mm); d8–9: rem PinUn, eyes open; d10: 16 (= Ad), rem 1/8 beyond sheaths, white feathers appear on sides; d10–13: PinUn; d17: 20–21 g. Voc: faint single-note *peep* at 7 d, faint double-note *peep* 9–10 d. **Fledgling**. Fledges 23–24 d. Mass decreases to 15–16 g (11.8 g CA). Plumage above: gray-brown overall, white from cheek partly encircles eye, large white rump patches nearly meet at base of tail. Below: dull white. *Comparison: Violet-green Swallow has more down covering than Tree Swallow.*

Northern Rough-winged Swallow (*Stelgidopteryx serripennis*)   Plate 22

Broad wings. Barbs on primary feather edges. Square tail. **Adult.** (PA) 10.3–18.3 g; 14 cm avg (5.5 in.). Diet: insects. Nest (cup): in burrow excavated by other species or crevices/holes in man-made structures. **Nestling.** MC: yellow. GF: cream or pale yellow, thick, somewhat prominent. Skin: bright pink. Bill: H pink; N broad with hook at tip, nares roundish. Down: sparse, fluffy, pale gray & cream, then brownish. LF: short, yellowish-pink, purplish 4 d, dark 7 d. G&D: d0: 1.1–1.9 g (length 17 mm); d5–10: eyes open; d7–10: PinUn; d9–10: 16 g; d12: 17.6; d14: able to perch; d15–20: FF. **Fledgling.** 14–14.5 g, fledges 19–21 d, can fly fairly well, clumsy landing. Plumage above: sepia with washed cinnamon; dark grayish-brown wings & tail, buff or cinnamon wingbars, margins on sec feathers dull buff, edge of first prim not rough as in Ad. Below: dull white; throat & breast dusky with vinaceous or pale cinnamon.

Purple Martin (*Progne subis*)   Plate 22

Largest of swallows with longer head & larger bill. Wings angular, pointed. Tail forked. **Adult.** (TX) M 53.5 g, F 54 g; (ME) 49 g; 19–20.3 cm (7.5–8 in.). Diet: insects. Nest: cavity in birdhouse, dead tree, cactus, crevice in cliff or building. **Nestling.** MC: lemon-yellow (pink in juv). GF: white or cream outside (light yellow inside), fairly prominent. Bill: short with wide gape. Skin: reddish-pink or pink, as tracts darken skin appears more bluish-purplish in those areas. Down: none to very sparse, by d9 may be short, fuzzy & grayish. G&D (weights from Hill 1994): d0: 2.8–3.5 g; d3: 11; d5–6: 20–25, PinVis cast bluish-purplish tint; d6–10: eyes open; d7: wing PinEm; d8: 34 g; d11–12: body PinEm, white feathers begin to appear at midback; d13: 51 g; d14–15: fear response; d15: 54 g; d17–20: 60 g; d23: large white flank patches. **Fledgling.** 46–56 g (d28), fledges 27–36 d (typically 28–29), but may perch for 1–4 d before flying. Plumage above: drab brown & hair-brown, little to no metallic bluish-green. LF: short. *Starlings nest in martin nests, which may cause a mistaken identification.*

Barn Swallow (*Hirundo rustica*)   Plate 22

Tail long, forked ("swallow-tailed"). **Adult.** 17–20 g avg, 17 cm (6.75 in.), outer tail length M 8–10.6 cm, F 6.8–8.4 cm. Diet: insects. Nest: cup, mud & grass, lined with feathers. **Nestling.** MC: orange-yellow or lemon-yellow. GF: cream, or pale to lemon-yellow; fairly prominent, not thick like other sp. Bill: H cream, upper turns dull brown & lower dull pink; wide gape, flat at base, pointed. Skin: pink. Down: light to med gray & grayish-brown. Head shape: flattened forehead. LF: short. G&D: N gains 1.9 g/d; d0: (KS) 2.2 g, (NY) 2.4; d5–11: eyes open; d9: outer prim & tail PinUn; d10: 19.5 g; d14: forehead PinUn. **Fledgling.** 17.5 g, fledges 18–23 d, may roost on nest several days & continue to be fed by parents midair. Plumage similar to Ad DefB except much duller. Above: chestnut or fawn just above bill, blue-black on crown & dorsal areas (faint greenish iridescence), wing covs edged reddish-brown. Below: breastband blackish mixed with cinnamon, vinacaous-cinnamon chin & throat. Tail not forked, short, white band. LF: reddish-brown. Voc: faint *chirp* or *chup*, 6–7 times while begging. *Gape much wider than in Tree Swallow.*

Cliff Swallow (*Petrochelidon pyrrhonota*)   Plate 22

Highly social. Tail short, square-edged appearance. Rusty or buffy rump. **Adult.** 19–34 g avg, (CA) 17.5–26.7 g; 12.7 cm (5 in.). Diet: insects. Nest: colony of jug-shaped mud clusters under building or bridge overhangs. **Nestling.** MC: pale yellow-orange. GF: creamy-white to yellow, somewhat prominent at first. Bill: H is pale, may have dark tip, flat, pointy, wide gape; N clove-brown upper. Skin: light pink or bright reddish-pink. Down: dull, creamy to light gray above, 50% of hatchlings show scant down on abdominal tracts, may have some on scapular & femoral tracts. G&D (NY): egg 1.4–2.4 g, N gains 2.36 g/day; d0: 1.6–2.2 g; d5–6: 11–13, begins vocalizing; d7–8: 18.5, mostly FF, cinnamon evident on dorsal areas; d9: 22.5, outer prim PinUn; d10: 24.3 g; d12–17: 26.5; d19–26: weight decreases. LF: short, chubby toes, vinaceous-pink, turn dark; nails white. Voc: barking type *chirp*. **Fledgling.** 21–22 g (NE), 28.3 g (NY); fledges d 20–21 (NY), d 23 (CA), 21–24 d (varies geographically), may return to roost for several days. Plumage highly variable. White speckling on (dull russet to gray) forehead, chin/throat varies individually. Above: dark crown & auriculars, hindneck pale cinnamon to russet; back & upperwing covs fuscous with buffy to pale cinnamon edges (scalloped effect); buffy to pale cinnamon on lower back & rump patch; uppertail covs edged white, or buff to cinnamon. Below: chin & throat mixed with white, dark gray, and/or pale chestnut; whitish undersides washed buff to light brown on sides/flanks. Tail unmarked, square-edged; spotting on undertail covs. Care: 3–5 d post-fledging. *Sim sp & hybrids: Cave Swallow; juveniles may be difficult to distinguish, Cave has pale throat & forehead.*

Cave Swallow (*Petrochelidon fulva pallida*)   Plate 22

Adult. (NM) 24.2 g, 14 cm (5.5 in.). Nest: of mud in caves, culverts, bridges. **Nestling.** MC: white to pale yellow. GF: pale. Down: buffy-white. LF: light pinkish. G&D: unknown. **Fledgling.** Able to fly 20–22 d, some may remain in nest to 26 d. Plumage similar to Ad DefB but duller, little metallic sheen, forehead duller, lighter & narrower; feathers edged with dull buff, cinnamon, or white. Below: may have variable degree of small fuscous spots on chin, throat & breast. *Sim sp & hybrids: see "Cliff Swallow."*

CHICKADEES AND TITMICE

(Family: Paridae)

Small, energetic, sociable. In N.A., chickadees are slightly smaller than titmice and are not crested. **Adults.** Chickadees: 12–14 cm (4.75–5.5 in.). Titmice: 14.6–17 cm (5.75–6.75 in.). Diet: mainly insects, especially eggs & larvae of insects and spiders; + seeds in winter. Nest: cup shaped within cavity (tree, artificial box, wall, or bank). Eggs: elliptical, short, white, some with reddish-brown spots. Bill: short, straight, stout, conical, compressed, pointed, for picking up and pounding on seeds and nuts. Nostrils concealed by dense tufts of stiff feathers; rictal bristles sometimes evident. Plumage: most are boldly patterned with black or brown caps and throat patches (chickadees) or small crests (titmice). Wings: 10 primaries (10th reduced), rounded. Tail: either as long or shorter than wings, slightly rounded. Legs/feet: short, strong for hanging upside down or holding a seed while pounding it open; tarsi longer than middle toe with nail, scutellate. **Young.** Mouth: orange to yellow. Gape flanges: most are yellowish, few white, thick, wide. Bill: flat, pointed; nares tiny. Down: sparse, grayish, head & back mostly. Fledge 14–28 days. Biparental care.

Carolina Chickadee (*Poecile carolinensis*) Plate 23

Adult. (PA) M 10.1 g, F 9.9; 12 cm (4.7 in.). Diet: insects & spiders, plus seeds & fruits in winter. Nest: in cavity, both Ad excavate in tree. Gape is pink. **Nestling.** MC: light yellow. GF: white or ivory, prominent. Skin: pinkish-white to salmon. Down: gray or dark gray on capital (thickest), alar & rump. G&D: d0: 0.75–0.86 g; d1: wing & spinal PinVis; d2–3: prim PinEm, dorsal tracts darker; d4: ~25 mm long, white ventral PinEm; d6–7: PinUn crown, dorsum & wings, bill darker; d8: mantle PinUn (brownish-gray); d9: eyes open, crown PinUn; d10: wing PinUn; d11: ~7.5 cm long; appears FF. **Fledgling.** Fledges 16–19 d. Plumage like Ad with duller black cap; cheek patch white blending to pale gray toward rear. Dependent 14–21 d post-fledging. *Song: learned. Sim sp: sympatric (in small area) & may hybridize with extremely similar Black-capped Chickadee; distinguish fledgling Black-capped by its more contrasting markings & entirely white cheek patch.*

Black-capped Chickadee (*Poecile atricapillus*) Plate 23

Adult. (PA) 8.6–12.9 g; 13.3 cm (5.25 in.). Diet: insects mainly (+ seeds & fruits). Nest: in cavity of moss (base), fur & soft plant fibers. **Nestling.**

MC: pale, turns bright yellow with pink lining. GF: creamy (yellow inside). Bill: H pinkish upper, turns blackish. Skin: rose-pink or pinkish-orange. Down: sparse, pale mouse-gray or brown-gray above. LF: pinkish-gray. G&D: d0: 0.71–0.93 g; d2–3: PinVis, wing & tail PinEm; d4–5: dorsal PinEm; d5–6: PinUn all over; d7–12: eyes open, mostly FF; d12: tail lengthened; d15: 11 g. Voc: H makes faint calls. **Fledgling.** Fledges 16 d. Plumage like Ad with black cap & bib, white cheeks, short tail. Dependent 21–28 d post-fledging. *Song: learned. Sim sp: see "Carolina Chickadee."*

Mountain Chickadee (*Poecile gambeli*)   Plate 23

Adult. (CA) M 10.0–14.3 g, F 9.4–14.5 g; (UT) 10.8 g; M 12.8 cm, F 12.3 cm (4.8–5 in.). Diet: insects (& seeds of conifers). Nest: in cavity, coniferous mountain forest. **Nestling.** MC: yellow. GF: bright yellow, very prominent. Bill: N top dark greenish, flat, pointed. Skin: pink & apricot. Down: grayish, capital & spinal. G&D: d0: ~1 g (~36 mm); d2: PinVis all tracts; d6: eyes open; d10–11: FF; d11: white superciliary stripe evident. Voc: H makes faint *peep*. **Fledgling.** 13.4 g (avg), fledges 18–21 d. Plumage like Ad DefB except black cap duller. Dependent 21–28 d post-fledging.

Chestnut-backed Chickadee (*Poecile rufescens*)   Plate 23

Adult. M 8.5–11.4 g, F 7.5–12.1; 12 cm (4.75 in.). Diet: insects mainly (& seeds). Nest: in cavity, coniferous forest. **Nestling.** MC: orange-yellow. GF: yellow, prominent, becomes creamy & thin. Bill: flat, wide. Down: gray, H head only, back & wings d3. LF: long, N pale bluish-purple. G&D: unknown, d0: ~1 g, d10: chestnut on back appears. Voc: squeaky *cheep*. **Fledgling.** 9 g* (avg.), fledges 18–21 d. MC: has turned gray. Plumage like Ad DefB but duller overall. Above: black head, buff-white narrow cheek patch. Below: buff abdomen. Dependent 21–28 d post-fledging. *Weight from 16 specimens (2014–19, Santa Rosa, CA, Mario Balibit).*

Boreal Chickadee (*Poecile hudsonicus*)   Plate 23

Adult. (ON, Canada) 7–12.4 g; 12.5–14 cm (4.9–5.5 in.). Diet: seeds & arthropods. Nest: in cavity, in conifer or mixed forest, moist habitat (bog or muskeg). LF: bluish-gray. **Nestling.** MC: presumed yellow. GF: very prominent. Down: sparse, long, mostly on crown, chestnut or dark sepia*. G&D: little known; hatch weight may be closest to Black-capped Chickadee; length* 31–44 mm (est. 5 d). **Fledgling.** Fledges: 18 d, flies weakly. Plumage: similar to Ad but duller & paler, above brownish mouse-gray, crown & nape pinkish drab-gray, wings with white edges. Below: dingy white, sides & crissum washed pale cinnamon, chin & throat dull black. *Data from 6 MVZ fluid specimens (1959, Bear Creek, AK).*

Oak Titmouse (*Baeolophus inornatus*)   Plate 24

Open oak or oak-pine woodlands (s.w. OR through CA to n. [& extreme] s. Baja California). Sedentary; defend territories year-round; permanent pair-bonds. **Adult.** Mass varies from CA to OR: M 13–20 g, F 11–21.5 g; 14.6 cm (5.75 in.). Diet: insects mainly (+ fruit & seeds). Nest: in cavity. **Nestling.** MC: light orange-yellow. GF: creamy-white (or pale yellow), very prominent, wide gape. Bill: N yellowish, turns tan d4 with dark tip, pointed. Skin: salmon-pink. Down: small tufts, gray or "tilleul-buff" on capital, occipital & rump. LF: gray nails. G&D: d0: +/- 2 g (est. based on egg weight); d3*: 3–5 g; d4: 5; d6: 7; d8–9: 10; d11: 11–12; d14: 13; d17: 15 (max). **Fledgling.** 10–13+ g, fledges 16–21 d, flies short distances. MC: yellow to pinkish-gray. Plumage like Ad, upperwing covs paler & greater covs with buffy tips. Below: light brownish-gray. LF: may be bluish-gray. *Sim sp: Juniper Titmouse (ranges barely overlap in n. CA near Lava Beds National Monument at Siskiyou-Modoc County border).* Dependent 5 wks post-fledging. *d3–17: 4 captive-raised birds (2019, Nevada County, CA, Janice Barbary).*

Juniper Titmouse (*Baeolophus ridgwayi*)

Poorly studied. Similar to Oak Titmouse. Recently Plain Titmouse was split into Oak Titmouse & Juniper Titmouse. The two sp occupy different habitats; Juniper Titmouse inhabits juniper & piñon-juniper woodlands of the intermountain region; see overlapping range above. **Adult.** (CA) 13.4–19.5 g; (AZ & NM): M 13.5–16.8, F 14.5–18; 14.6 cm (5.75 in.).

Tufted Titmouse (*Baeolophus bicolor*)   Plate 24

Adult. (PA) 17.5–26 g; 16.5 cm (6.5 in.). Diet: arthropods (66%), + seeds & fruits (34%). Nest: cavity or box, moss (base), fur & soft plant fibers. **Nestling.** MC: yellow. GF: yellowish, very prominent. Bill: N brownish-black, wide gape. Skin: pale pinkish-salmon. Down: sparse pale blue-gray on capital, occipital, humeral & spinal. G&D: d4–8: eyes open, wing PinEm, gray dorsal feathers, rusty-brown flanks; d10: well covered; d14: FF. **Fledgling.** Fledges 15–16 d. Plumage like Ad except black patch on forehead smaller, paler & less distinct. Above: brownish mouse-gray, darker on wings with slight greenish tinge on secondaries. Below: dull grayish-white, flanks tinged rust-brown. LF: bluish-slate. Dependent 6 wks post-fledging.

Black-crested Titmouse (*Baeolophus atricristatus*)

Adult. (TX) 10.5–26 g, 15.2 cm (6 in.). Nest: cavity in tree or man-made structure. **Nestling.** MC: orangish-yellow. GF: cream. Skin: pink. Down: dark bluish-gray. G&D: H unknown; d5–7: eyes open; d10: FF. **Fledgling.** Fledges d15–18. Plumage similar to Ad F except crest shorter & black on crest reduced or lacking; above with brownish wash, buffy eye ring, wings & tail darker with olive-gray wash, sec with greenish tinge. Below: grayish-white, pinkish-buff tinge on flanks. LF: dull bluish-gray.

PENDULINE TITS (VERDIN)

(Family: Remizidae)

Of 11 species of penduline tits worldwide, only one occurs in N.A. nesting in desert regions of the s.w. US (& n. Mexico).

Verdin (*Auriparus flaviceps*)   Plate 24, Photo 30

Adult. (s.w. US) 5.5–8.5 g; 11.4 cm (4.5 in.). Diet: insects & spiders mostly, some berries & nectar. Nest: spherical with side opening, conspicuous. Eggs: subelliptical, greenish-blue, reddish-brown markings. Bill: short, conical, acute. Nostrils concealed by feathers. Plumage: grayish with yellow head, chestnut at bend of wing (concealed). Wings: 10 prim (10th minute), rounded. Tail: rounded, shorter than wing. LF: bluish-gray, tarsi longer than middle toe with nail, scutellate. **Nestling.** MC: bright yellow. GF: thick, cadmium-yellow. Bill: H arrow-like shape to mouth (above view), N dark brown upper. Skin: pale flesh-pink. Down: none. LF: pale gray. G&D: d0: 0.7–1.0 g (~27 mm); d2: PinVis, long reddish-brown feathers appear on crown & occipital (> 50 strands, 11 mm longest strand)*; d3: 3.8 g, PinEm; d5: PinEm all tracts & tail; d7: eyes open, length ~44 mm, wings 5–10 mm, rect 2.5 mm; d9: 4.7–6.2 g, wing & tail PinUn, plumage dark reddish; d12: 7.3 g; d14–15: 6.4–7.8, FF. Voc: (H) faint *peep.* **Fledgling.** 6.5–7.6 g, fledges 17–21 d, able to fly short distances. Plumage above: gray (no yellow), lesser wing covs olive to olive red-brown (gray in juv F). Below: pale brownish-gray. GF: cadmium-yellow. Bill: yellow at base. LF: pale gray. Voc: *tseet* response to parent. Dependent 18 d post-fledging. *d2 *description of crown feathers from four MVZ fluid specimens (1910, Imperial County, CA).*

LONG-TAILED TITS (BUSHTIT)

(Family: Aegithalidae)

Of 11 species of long-tailed tits worldwide, only one occurs and raises young in N.A. Distinctive characteristics: tiny, highly energetic, short bill, tail longer than wing, F has pale blue eyes; highly social, occur in flocks (nonbreeding).

Bushtit (*Psaltriparus minimus*)   Plate 24

Adult. (CA) 4.5–6 g; 11.4 cm (4.5 in.). Diet: gleans insects & spiders from plant surfaces; occasionally berries & seeds. Nest: unique, long, pendulous, sock-like, entrance near the top. Bill: very small, short, compressed (stubby), decurved culmen. Nostrils concealed by feathers. Plumage: sexes alike, mostly dull black, brown & white, dark ear covs variable. Wings: 10 prim (10th reduced), rounded. Tail: longer than wing; rounded or graduated. LF: tarsi longer than middle toe with nail, scutellate. **Nestling.** MC: orange-red to orange-yellow. GF: cream or pale yellow, somewhat prominent. Bill: light, may have dark spot near tip. Skin: bright pink. Down: H bare (Fig. G6 b, Illustrated Glossary), then (d3) scanty grayish-white. LF: long, delicate, light pink toes. G&D: d0: < 1 g; d0*: 0.75 g, length 21 mm, beak culmen 2.5 mm, tarsus 3.5 mm); d1*: 1 g, PinVis; d2: 2–3 g, ~38 mm; d3+: 3.5–4; d7–8: eyes open, many PinUn, dark above, buffy-white below with dark streaks on throat, first crown feathers stand upright. Irises: dark; in F lighten after d13. Behavior: stands up & leans back while gaping. Voc: 3 syllable "locator" call. **Fledgling.** 5.5–6.4 g (62–65 mm long), fledges 18 d, does not return to nest. GF: reduced but still evident. Bill: dark, very narrow (like tweezers), culmen ~5 mm. Plumage like Ad, very dark gray basally (especially breast & belly), dark feathers around eyes, amount of black in the auricular varies by ssp. Below: light whitish, grayish or buffy. Wings: lighter than body with dark buffy edges. LF: tarsus ~14–15 mm, very dark; nails long & dark, undersides flat. Iris: M brown, F pale gray, white, or yellow ~d46 post-hatching. *d0 data from one MVZ specimen with umbilicus attached (1919, Berkeley, CA; d1 data (2020, Nancy Barbachano).*

NUTHATCHES

(Family: Sittidae)

Distinctive feature of nestlings: "arrowhead" look (dorsal view) of bill and gape flanges, and features of hallux and nails. Family unit: nest helpers. **Adults.** Small songbirds. Diet: mainly invertebrates, nuts, and seeds. Nest: cup in cavity in dead wood. Eggs: subelliptical, white with rufous spots or streaks. Bill: medium-long (as long as head), strong, straight, compressed, acute (strongly pointed), underside upturned toward tip (chisel shaped), nostrils covered more or less with stiff rictal bristles that project forward. Plumage: compact, coloration of grays and blacks above, most are gray and white or gray and reddish, with black or brown crown and whitish eye stripe. Wings: 10 primaries (10th reduced), long, pointed. Tail: much shorter than wings; nearly square, usually with white markings, rectrices broad with rounded tips. Legs/feet: short, sturdy, strong, tarsi as long as middle toe with nail, scutellate in front, long hallux. Nails: acute, laterally compressed; hallux nail very long. **Young.** Mouth: yellow. Gape flanges: creamy-white, flared at corners. Bill: N long (not longer than head), pointed. Head: flattened appearance from side view. Down: whitish to gray, sparse on head and back. Fledge: 14–26 days.

Red-breasted Nuthatch (*Sitta canadensis*)   Plate 25

Adult. (NJ) 8.0–12.7 g; 11.4 cm (4.5 in.). Diet: insects & seeds. Nest: in cavity. **Nestling.** MC: yellow. GF: bright white to creamy-yellow, prominent, thick. Bill: pointed, yellowish-orange. Skin: bright pink. Down: dark gray, sparsely covered. G&D: unknown. At 4.5 g wing & tail PinF just emerging. **Fledgling.** Fledges 18–21 d. Plumage: sexes distinguishable, Fl M resembles Ad M but duller, white supercilium, chin & cheeks may be faintly speckled dark. Fl F has grayish cap, dull head markings, no black. Dependent 2 wks post-fledging.

White-breasted Nuthatch (*Sitta carolinensis*)   Plate 25

Adult. (PA) 18.3–23.2 g; 14.6 cm (5.75 in.). Diet: insects, spiders, nuts & seeds. Nest: natural cavity, occasionally in box. **Nestling.** MC: bright yellow. GF: creamy-yellow, prominent (Fig. 2.6), thick at corners (more yellow inside). Bill: H acute, short, yellowish; d5 much longer, pointed (Fig. 2.6) & yellow-orange. Skin: light pink. Down: H white, then light gray (back), crown darker. LF: long legs, big toes compared to body, pinkish-gray (d7). G&D: H: 1.52–1.8 g; d6: cap visible; d3–4: 10.5–12.5 g,

PinUn crown, wing & tail covs, throat & breast (whitish), eyes fully open; d9–10: 18.2–20.8 g. **Fledgling.** (CA) 15.75 g (avg), fledges 19–26 d, flies poorly at first. Bill: "dusky" pinkish-buff & very long. Plumage: similar to Ad DefB but paler & duller. LF: pinkish-buff. Dependent several wks post-fledging.

Pygmy Nuthatch (*Sitta pygmaea*)   Plate 25

Adult. (w. US) 10.6 g; 10.8 cm (4.25 in.). Diet: insects & seeds. Nest: in cavity. **Nestling.** MC: yellow. GF: creamy-white, very prominent, wide gape. Bill: N light or horn-yellow, pointed. Skin: brownish-pink. Down: very sparse, pale smoky-gray. LF: H light pink. G&D: d0: 1.2 g; d6–9: all PinEm; d7–10: eyes open, PinUn; d8–10: prim PinUn; d9: 9.3 g. Voc: *pseep* at d1. **Fledgling.** 10.5 g, fledges 14–22 d, flies well 21 d. Bill: dark above, yellow base. Plumage like Ad DefB but back, rump & tail less bluish than Ad; crown & nape gray; yellowish-brown tips on wing covs. Below: white to buff, buff-brown sides & flanks. Tail: middle pair rect with white bases. Dependent 23–28 d post-fledging.

Brown-headed Nuthatch (*Sitta pusilla*)   Plate 25

Small nuthatch. Endemic to pine forests of e. US; prefers decayed snags for nesting (vulnerable to habitat alteration) but will use artificial cavities near pines; cooperative breeders (nest helpers may be closely related); tool use documented. **Adult.** (GA) 10.2 g; 10.8–11.4 cm (4.25–4.5 in.). Diet: insects & seeds. Nest: excavates cavity. **Nestling.** MC: bright yellow. GF: creamy-white (yellowish at corners), very prominent & thick. Bill: yellowish, turns gray or black d10–12. Skin: tannish-pink, then rose-pink. Down: light brownish-gray or mouse-gray. LF: H light pink. G&D (GA): d0: 1.2 g; d6–9: all PinEm; d8–9: 6.9–7.3 g, PinUn; d9–10: PinUn crown, spinal, femoral (cinnamon), ventral (whitish tips); eyes open; d11–13: 7.2–8.9 g; wing PinF ½–1/4 unsheathed, rect ¾ unsheathed but short, ventral creamy (center uncovered). Voc: H *pseep*. **Fledgling.** 10.1 g, fledges 18–19 d. Plumage much like Ad but duller & less bluish above, greater wing covs edged pale brownish-buff. Below: brownish-buff sides, flanks & undertail covs. Dependent 3–4 wks post-fledging.

TREECREEPERS

(Family: Certhiidae)

   Of 11 species of treecreepers worldwide, only one raises young in N.A.

Brown Creeper (*Certhia americana*)   Plate 25, Photo 31

Adult. Small songbird. (PA) 7–10 g, ~12.7 cm avg (~5 in.). Diet: insects by gleaning leaves & probing into bark; nuts & seeds in winter. Nest: "hammock" of twigs, bark fibers, mosses & spider cocoons usually behind bark of tree. Eggs: subelliptical, white gently spotted with brown. Bill: length varies by individual, may be shorter or longer than head, for probing in tiny bark crevices; decurved, slender, laterally compressed; nostrils entirely exposed. Plumage of M & F alike: cryptic, mottled brown upperparts speckled with white, buff & black; white below. Wings: 10 prim (10th reduced), somewhat long, rounded. Tail: rounded, as long as or slightly longer than wings; rectrices stiff & acuminate (pointed tips). LF: strong, tarsi shorter than middle toe with nail, scutellate; nails long & acute, especially hallux. **Nestling.** MC: yellow. GF: yellowish-white, prominent. Bill: H pinkish-buff & stubby, becomes pointed. Skin: yellowish-pink. Down: thick, long, grayish-black on superciliary and occipital regions of 3 rows, 6 neossoptiles each. Tail short with stiff feathers. Below: 1st feathers on anterior underparts flecked pale brownish-gray. LF: short legs, long toes & nails, pinkish-buff. G&D: unknown. d8: eyes open. Voc: older nestlings *ts-tssi*. **Fledgling.** Fledges 13–16 d, climbs well, flies weakly. MC: reported to be red in recently fledged birds (Davis 1978). Plumage like Ad but supercilium less distinct, paler overall with spots & streaks; narrower barring on central rect. Below: as nestling.

WRENS

(Family: Troglodytidae)

Adults. Small to medium songbirds. Diet: mostly insects; rarely plant material. Nest: cup in cavity, in rock crevices, or a woven globular structure. Eggs: subelliptical to oval, white, cream or pink, mottled with browns. Bill (for gleaning insects from ground and vegetation): varies in length by species, from half as long to about as long as head, usually decurved, thin or slender, and compressed. Head shape: "flattened" appearance. Plumage: brown predominates, paler below, wings and tail barred. Many have long, pale eyelines. Wings: 10 primaries (10th reduced), short, concave, rounded. Tail: varies from slightly longer than wings to two-thirds as long, rounded, rectrices soft with rounded tips; many carry tail upright (especially when excited or agitated). Legs/feet: strong, tarsi longer than middle toe with nail, scutellate in front, sometimes behind; anterior toes partly adherent; long claws. **Young.** Mouth: bright yellow or orange-red. Gape flanges: yellowish. Head: somewhat flattened forehead.

Down: varies by species, scant, dark to whitish-gray, on head and back, thickest on crown. Fledgling has mottled underparts and short, stiff tail feathers sometimes held upright. Biparental care. *Wrens are nervous birds that require specialized care so must be quickly identified. Do not have a true crop. May not be tolerant of boldly patterned clothing. Should not be housed in same room with species that prey on them. Wrens sometimes confused with other cavity-nesting sp, such as titmice.*

Rock Wren (*Salpinctes obsoletus*)

Adult. (CA) 16.5 g; 15.2 cm (6 in.). Diet: insects & other arthropods. Nest: cavity or crevice of rocky cliff or hillside. **Nestling.** MC: yellow or pinkish-yellow. GF: yellow. Skin: pink-yellow. Down: dull tilleul-buff, vinaceous buff, or drab (Oberholser 1974). G&D: d0: 1.5 g; d3: 4.5; d6: 8; d9–11: 12–15. **Fledgling.** 14–17.5 g, fledges 14–16 d. Plumage like Ad DefB but paler & more buffy above, few markings on crown, buffy rump, few to no markings on underparts. Dependent: 7–18 d post-fledging.

Canyon Wren (*Catherpes mexicanus*)   Plate 26

Adult. 9.9–14.8 g, 14.6 cm (5.75 in.). Diet: insects & spiders. Nest: cup in rock crevice on cliff or bank. **Nestling:** MC, GF: yellow. Bill: pinkish. Skin: pink. Down: mouse-gray, mostly head & upper back. G&D: unknown (egg size: 18mm x 14mm). **Fledgling.** Fledges 12–17 d. Bill: long, thin, downcurved. Plumage: fewer white specks than Ad, dull chestnut on belly with narrow dusky bars. Dependent 5–10 d post-fledging.

House Wren (*Troglodytes aedon*)   Plate 2

Adult. (PA) M 9.8–12 g, F 9.8–13 g; 12 cm (4.75 in.). Diet: insects (& other invertebrates). Young fed insects; d3 fed bits of grit & shell. Nest: in natural cavity or box; coarse platform of twigs, lined with downy feathers & fine fibers. **Nestling.** MC: H pale yellow, then orange-yellow (throat dark). GF: cream to pale yellow, mod prominent by d3, somewhat wide. Bill: pointed, buffy, turns dark with growth. Skin: H pinkish, darkens d3. Down: long tufts, "mouse-gray" or drab brown thickest on crown, occipital & upper back (Fig. G6 c, Illustrated Glossary). Wing chord: H 3 mm, Fl (d15–18) 44–48 mm. LF: H pale pinkish-tan, N dark gray. G&D: d0: 1.04–1.17 g (length 2.5–3 cm); d1: 1.7 g; d2: 2.5; d3: 4, PinVis; d4: 5 g; d5: PinEm wings, tail, upper back; d6: 8 g; d6–7: rufous PinUn (tips) dorsum & flanks; d8: 9.5 g; d9–10: 10. Voc: H makes *peep*, d5 distinctive begging chatter. **Fledgling.** 10.2 g, fledges 15–17 d, can fly. Plumage similar to Ad DefB but barring & superciliary line less distinct, pale orbital ring, mottling on breast darker, rufous on flanks, very little down still attached to tips of feathers.

Pacific Wren (*Troglodytes pacificus*)

Adult. (BC, Canada) M 8–11.6 g, F 7.6–11 g; 7.6–12 cm (3–4.7 in.). Diet: insects, spiders & other invertebrates. Nest: variable, uses existing cavity, creates hole (in bank or dead snag), or domed-shaped hanging. **Nestling.** MC: yellow. GF: yellow, somewhat prominent, wide. Down: sparse drab brown. G&D: d4: prim PinEm; d7: head PinEm; d10: eyes open; d17: FF, down on head still present. **Fledgling.** Fledges 15–18 d, stays with parents for some time. Plumage like Ad but no whitish bars on upperparts, indistinct supercilium, darker below.

Winter Wren (*Troglodytes hiemalis*)

Adult. (PA) 7.5–10.5 g, 7.6–12 cm (3–4.7 in.). Diet: insects, spiders & other invertebrates. Nest: variable (as Pacific Wren). **Nestling.** MC: bright yellow. GF: pale lemon-yellow. Skin: pinkish-tan with yellowish tinge. Down: short, dark gray or drab brown. G&D: little info, same or similar to Pacific Wren as once were considered same species. d0: 1 g, d10: eyes fully open. **Fledgling.** Fledges 16 d. Plumage similar to Ad DefB but without barring on back, scapulars, rump, or uppertail covs, indistinct head stripes. Below: darker than Ad with less distinct barring.

Sedge Wren (*Cistothorus stellaris*)

Adult. M 7.8–9 g, F 7–9 g; 11.4 cm (4.5 in.). Diet: insects & spiders. Nest: globular ball in sedges or fine grasses. **Nestling.** MC, GF: presumed yellow. Bill: tiny, yellowish lower mandible. Down: no info. LF: pinkish. G&D: d0: 1.34 g; d3: 1.46; d4: eyes open; d6: 3.6 g; d9: 6.5; d11: 7.4. **Fledgling.** 7.4 g, fledges 12–14 d. Plumage above: buffy supercilium & stripe below eye, dull black back, streaked pale cinnamon or white; rump & wings barred; tail mottled with black. Below: pale buff. Pale cinnamon crissum & flanks. *Song learning: Sedge Wrens of N.A., because of their nomadic lifestyle, do not learn by imitating older males; rather, they improvise a large repertoire of songs (Kroodsma 2005).*

Marsh Wren (*Cistothorus palustris*)   Plate 26

Adult. (NY) M 10.5–13.5 g, F 9–13.5 g, (w. US) M 10.9–12.7 g; 12.7 cm (5 in.). Diet: insects (some snails). Nest: spherical, of reeds/grass over water. **Nestling.** GF: yellow. MC: deep yellow. Skin: pinkish. Down: whitish, head & back only. LF: pinkish-tan. G&D: d0: 0.87 g; d3: 2–3; d9: 10; d12: 11.08. **Fledgling.** 11.5 g, fledges 13–16 d. Plumage like Ad but back has few black markings & no distinct white streaks, indistinct or no supercilium, wings barred faintly. Dependent 2 wks post-fledging, remain in family groups through summer.

Carolina Wren (*Thryothorus ludovicianus*)   Plate 26

Adult. (AL) M 18.5–27 g, F 16.2–21.5 g; 12–14 cm (4.7–5.5 in.). Diet: insects & spiders. Nest: natural cavity or box; bulky, of variable debris, usually domed with tunnel-like entrance. **Nestling.** MC: bright orange-yellow to reddish-orange. Tongue: 2 dark spots on tip & spurs of tongue. GF: pale lemon-yellow to yellow, wide, somewhat prominent. Bill: H pale with dark tip, upper mandible turns dark, lower paler with shades of brownish-tan; d9 longer & blacker, distinctly pointed. Skin: salmon-pink or reddish-pink, duller d1. Down: pale gray, dark gray or grayish-brown on capital, spinal, humeral & femoral. G&D: d0: 1.8–2.6 g, 35.5 mm long, wing chord 6.29 mm; d1: 3.13 g; d2: 4.7, PinVis; d3: 6.2 g; d3–4 skin more pinkish-tan, alar PinEm; d4: 8.2 g; d4–8: eyes open; d5: 9.9 g; d6: 11.9, ventral & femoral pins cinnamon; d7: 13.53 g, black eyes & face, wing & tail PinUn, may cower in nest; d8: 14.85 g, 71.5 mm long, wing chord 32.93 mm, all PinUn, very alert; d8–9: dorsal feathers buffy-rust colored, ventral PinUn (2 broad cinnamon strips). Voc: H soft *peep*, then *seee*, d8 *chip* locater call. **Fledgling.** 79% of Ad mass, fledges 10–16 d, short flights. Bill: 50% of Ad length. Plumage: paler than Ad, duller supercilium & postocular stripe, wing covs tipped buff, no barring on undertail covs, chin & sides of head with dull black flecking or barring. *Source of G&D: Jonsomjit et al. (2007).*

Bewick's Wren (*Thryomanes bewickii*)   Plate 27

Adult. (AZ) 7.8–11.8 g, (CA) 9.6 g; 13.3 cm (5.25 in.). Diet: insects. Nest: in cavity; twig/leaf base lined with soft materials including snakeskin or cellophane. **Nestling.** MC: orange or bright yellow. GF: yellowish with orange tinge, wide, somewhat prominent. Bill: yellow (dark?), narrow, distinctly pointed. Skin: pinkish. Down: 9 tufts, sparse, long, hair-brown or gray, mostly on crown, occipital & upper spinal. LF: very long tarsi & toes; pinkish legs darken to gray d8. G&D: d0: 1.4 g; d2: dorsal PinVis; d3: weak food cry; d4: prim & sec PinEm ~1 mm; d5: eye slits; d7: wing & tail PinUn (tips); d10: nearly FF; d11 superciliary line evident; d13: FF. Voc: *cheeping* or *chirping* (d8). **Fledgling.** ~10 g, fledges 14–17 d, flies well d16. Plumage: sepia & raw umber; dusky-edged underparts, narrow eye stripe. Voc: begging call *sc-i-i-t sc-i-i-t*. Dependent 14 d post-fledging. *Song learning: learns father's song 1st 4–5 wks post-hatching, then 1st spring as an adult, learns dialects of neighboring M near his natal grounds (Kroodsma 2005). Ad M sings 14–20 different songs.*

Cactus Wren (*Campylorhynchus brunneicapillus*)   Plate 27

Adult. (AZ) 33.4–46.9 g, 21.6 cm (8.5 in.). Diet: insects mainly, some fruits. Nest: large globular nest chamber of grasses & fibers usually in cactus. **Nestling.** MC: bright yellow to orange-red. GF: yellow. Bill: N light brown d4. Skin: pinkish, then darker. Down: white or ash-gray, long & fluffy, all tracts. LF: light gray. Iris (d16–17): pale grayish or yellowing. G&D: d0: 2.6–4.1 g; d6: alar & dorsal PinEm, fuzzy tufts all over; d6–8: eye slits; d8–10: prim PinUn, sepia with buff tips; d10: superciliary line whitens; d14: FF; d15: 28–34 g. Voc: sharp *peeps* d2. **Fledgling.** 31.3 g, fledges 17–23 d. Plumage like Ad but buffier above, mottled appearance on wings & back, tail with white markings & full subterminal white bar across webs. Below: few to no spots. Dependent 17–25 d post-fledging.

GNATCATCHERS

(Family: Polioptilidae)

Adults. Very small songbirds. Diet: insects gleaned from shrubs. Nest: variable, most are tiny woven, compact cups. Eggs: subelliptical; colors vary whitish, bluish, pink, buff, or greenish; may be heavily mottled or variably speckled with brownish dots that may form a "wreath" around larger end. Bill: short, straight, slender, pointed, somewhat depressed at base, culmen decurved toward tip, upper mandible notched near tip, surrounded by rictal bristles. Plumage: bluish-gray upperparts, white underparts. Wings: 10 primaries (10th reduced), long, rounded. Tail: longer than wing, usually square or rounded, outer rectrices white. Legs/feet: moderately strong, tarsi longer than middle toe with nail, and scutellate. **Young.** Mouth: yellow. Gape flanges: yellow. Down: none to some, not well known. Leave nest: 10–16 d, remain with adults several weeks.

Blue-gray Gnatcatcher (*Polioptila caerulea*)   Plate 27

Adult. (CA) 5.4–6, (PA) M 5.2–7 g, F 4.8–8.9 g; 10–11.4 cm (4–4.5 in.). Diet: insects & spiders; young fed soft-bodied insects (grasshoppers, moths). Nest: cup with high walls. **Nestling.** MC varies pale cream to sulphur-yellow; 2 black spots on tongue. GF: pale yellow. Bill: N grayish, pointed. Skin: light to dark bluish-gray. Down: none. G&D: d0: 0.63–0.87 g; d2: wing PinEm; d3: body PinEm; d5–8: eyes open; d6–7: PinUn; d9: 6.2 g. **Fledgling.** 5.9 g (CA), 5.5 g (FL), fledges 10–15 d. Plumage like Ad F, head mouse-gray, white eye ring, grayish-brown back & rump. Tail: black with white outer rect. Below: lighter. LF: light gray. Dependent 19 d post-fledging.

California Gnatcatcher (*Polioptila californica*)

Endangered status. **Adult.** (CA) M 5.2–7 g, F 5.3–6.4 g; 11.4 cm (4.5 in.). Diet: wide variety of arthropods. Young fed large crane flies & praying mantis. Nest: deep cup in dense shrub usually on slope. **Nestling.** MC: yellow, 2 black spots on tongue. GF: presumably yellow. Bill: H short. Skin: pink. Down: none. G&D: egg wt 0.82–1.1 g, d0: < 1 g; d2: skin darker, wing PinVis; d3: spinal PinVis, wing PinEm; d4: head & body PinVis in rows; d6: eyes open, back PinEm; d7: wing PinUn; d8: wing PinF 2 mm, tail PinF 6 mm; d10: all PinUn (50–75%), appears "downy"; d11: FF, short tail. **Fledgling.** Fledges 10–15 d. Bill: fine, narrow with slightly hooked tip. Plumage like Ad but wing covs edged with gray or brown, dorsum with pale brown wash. Dependent for 21–35 d post-fledging. *Additional source: Grishaver et al. 1998.*

Black-tailed Gnatcatcher (*Polioptila melanura*)

Adult. (s.w. US) 5.4 g; (AZ & CA) M 5.1 g, F 5.3 g; M 11 cm (4.3 in.), F 9.7 cm (3.8 in.). Diet: wide variety of arthropods (+ small berries). Nest: cup in fork of dense shrub or tree. **Nestling.** MC: yellow. GF: yellow. Down: none. G&D: little info, may be similar to California Gnatcatcher. d3–4: prim PinEm. **Fledgling.** Fledges 9–15 d. Bill: dark gray (paler at base). Plumage sim to Ad DefB, brownish-gray crown & back, may have black in supercilium (by mid-June). Dependent 3 wks post-fledging.

DIPPERS

(Family: Cinclidae)

Of five species worldwide of these exclusively aquatic songbirds, only one raises young in N.A.

American Dipper (*Cinclus mexicanus*)   Plate 27

Adult. Medium-size songbird: (OR) M 55–58.5 g, F 44–61 g; (UT) M 57–66 g, F 43–65 g; 14–20.3 cm (5.5–8 in.). Diet: exclusively aquatic invertebrates & small fish; occasionally fruits & small vertebrates; plunges under fast-moving streams or rivers, picks or probes on surfaces or vegetation along shore. Nest: spherical "hut" of interwoven moss usually behind waterfall or on midstream rock. Bill: short, straight, slender, laterally compressed, culmen decurved toward tip, gonys (ventral ridge of lower mandible) recurved toward tip, upper mandible notched near tip. Plumage: firm with soft undercoat of down, compact, coloration with predominating brown or gray. Wings: 10 prim (10th reduced), very short, concave, square or slightly rounded. Tail: held upright; short, more than half as long as wings, square to slightly rounded, rectrices broad with rounded tips. LF: strong, tarsi longer than middle toe with nail & booted. Nails very curved. **Nestling.** MC: orange-yellow, darker interior. GF: white. Bill: pointed, narrow, very long, partly yellow. Skin: pink on dorsum, orange on flanks & belly. Down: H sparse, slaty-gray; d6 long, thick pale gray capital & dorsal areas. G&D: d3: 10–25 g; d16: 40–55; d20: FF. **Fledgling**. 53 g, fledges 24–26 d. Bill: salmon base. Plumage: slate-gray, feathers edged paler gray or whitish to cinnamon. LF: legs pinkish-tan. Biparental care.

KINGLETS

(Family: Regulidae)

Of six species of kinglets worldwide, two raise young in N.A. The Golden-crowned Kinglet is the smallest passerine in the US with a diagnostic white eyebrow stripe and orange crown patch on the adult male. The Ruby-crowned Kinglet is a bit larger with pale eye ring and red crown patch on the adult male. **Adults.** Tiny songbirds. Diet: arthropods and their eggs, occasionally fruit and sap. Nest: cup shaped, deep, in tall conifers in high elevations. Eggs: short elliptical to subelliptical, dull white to cream with faint spots. Bill: short, small, slender, straight, somewhat depressed at base, culmen decurved toward tip, upper mandible notched near tip, for gleaning insects and collecting spider webs for nest building. Plumage: mostly greenish-gray with some black patterns. Wings: 10 primaries (10th reduced), long, rounded. Tail: shorter than wing, variable in shape, notched, rectrices broad and acuminate at tips. Legs/feet: moderately strong, tarsi longer than middle toe with nail, and booted. **Young.** Mouth: orange, orange-red, or bright red. Down: none or few tufts depending on species. Fledge 14–19 days. Biparental care.

Golden-crowned Kinglet (*Regulus satrapa*)   Plate 27

Adult. (PA) M 4.9–7.7 g, F 4.5–7.8 g; 10 cm (4 in.). Diet: arthropods (soft-bodied) + limited seeds & fruit. Nest: pendulum with deep cup. Restricted to nesting in dense coniferous woodlands. **Nestling.** MC: orange or orange-red. GF: pale yellow. Bill: N yellow. Skin: pinkish-tan. Down: few tufts, fine, gray occipital & crown. LF: large & strong relative to body size. G&D: d0: size of a bumblebee, 0.73–1.03 g; d1: 1.5–2; d2: alar & dorsal PinVis; d5–11: eyes open; d6: wing PinEm; d8: crown darkens; d9: wing & tail PinUn; d11: appears FF. Voc: *tsip.* **Fledgling.** Fledges 14–19 d. GF: yellow to orange. Plumage above: grayish-brown crown (lacks yellow), broad white supercilium, dark "mask," brownish-olive overall. Below: gray. Dependent 15–17 d post-fledging.

Ruby-crowned Kinglet (*Corthylio calendula*)   Plate 27

Adult. (PA) M 5.8–7.1 g, F 5.4–6.8 g; 10.8 cm (4.25 in.). Diet: arthropods (small), some berries & sap. Nest: usually pensile. **Nestling.** MC: bright red. GF: yellowish-orange. Bill: yellow, tinged brown. Down: little to none; bare head. G&D: unknown. **Fledgling.** Fledges 12 d (or 16 d another source). GF: yellow to orange. Plumage above: no red on crown, dull white eye ring, crown & back with dusky mottling, buffy wingbars. Below: dull white tinged gray-brown. LF: toe pads orangish.

SYLVIID WARBLERS (WRENTIT)

(Family: Sylviidae)

Of 68 species in Sylviidae worldwide, only one sylviid warbler raises young in N.A.

Sedentary species, nonmigratory, young remain in territories within 1300 feet from where they hatched. **Adult.** Small songbird. (CA) M 12.4–18 g, F 12.4–17 g; 14–15.2 cm (5.5–6 in.). Diet: insects & spiders gleaned from bark or from ground, + fruits & seeds. Nest: cup of cobwebs suspended in horizontal fork. Eggs: subelliptical, whitish, blue, or green; markings vary from unmarked to heavily spotted. Bill: short, straight, stout, compressed, culmen very decurved, nostrils entirely exposed. Wings: 10 prim (10th reduced). Tail: much longer than wings, rect narrow & graduated but broader toward rounded tips. LF: short & strong, tarsi much longer than middle toe with nail. **Nestling.** MC: sulphur-yellow or deep orange. GF: yellow, thin. Skin: light pink or purplish. Down: none, most tracts dark with PinVis, light spots on ventral tract. Bill: H ochre with orange-yellow; N pointy, straw-colored, darkens by d7. Eye area dark. LF: straw-colored, turn darker. G&D*: d0: 1.5–1.6 g (3.3 cm); d1: 2.4 g; d2: 3.5; d3: 4.4; d4: 5.6; d5: ~7, PinEm dorsal, ventral & wings; d6: 7.74 g; d6–8: eyes open; d7: 8.75 g (4.8 cm), PinF only; d8: ~10 g; d9: 10.7; d10: 11.4, wing PinUn, appears FF. Voc: *wheat.* **Fledgling.** Fledges 15–16 d, cannot fly. Plumage like Ad, first contour feathers gray-brown, cinnamon underparts. Irises turn pale by 15 d. Independent 18–41 d post-fledging. *Injured adults & young must be returned to exact territory where found. *G&D information: Jonsomjit et al. 2007.*

THRUSHES

(Family: Turdidae)

Adults. Small to medium songbirds. Diet: arthropods and worms (spring), fruit (fall/winter); pecking in soil or plucking from vegetation. Nest: open cup lined with sticks and grasses, sometimes reinforced with mud; typically in tree or shrub but sometimes man-made or natural cavity (bluebirds), occasionally on ground. Eggs: subelliptical to oval, pale blue to occasionally white, some with brown flecks. Bill: variable in length (short to medium), straight, relatively slender, compressed, not pointed, culmen decurved toward tip, upper mandible notched near tip, adapted for eating soft foods (insects, worms, berries). Eyes: large and dark, eye rings (complete or incomplete). Plumage: terrestrial species colored more in earth tones. More arboreal species are brightly colored. Sexing by plumage is reliable in some species, such as bluebirds. Wings: 10 primaries (10th reduced); in most species, wings are long, pointed. Tail: usually shorter than wings, square to slightly rounded. Legs/feet:

strong, tarsi usually longer than middle toe with nail, and booted. **Young.** Mouth: orange-yellow. Gape flanges: most are fairly prominent, pale yellow, and turned down at corners. Down: sparse, whitish, gray, or drab. Fledge 8–19 days depending on species. Fledglings are always spotted or speckled above and below. Song learning: some components are innate and some are learned.

Eastern Bluebird (*Sialia sialis*)   Plate 28

Adult. (e. US) 28–32 g; 15.2–20.3 cm (6–8 in.). Diet: insects, spiders, occasionally small vertebrates, small fruits (winter). Nest: cup of fine grass or needles in cavity; open nest rare. Eggs: powder blue (sometimes white). **Nestling.** MC: ochre-yellow. GF: cream to yellow, color & prominence varies. Bill: H pink, black (older N), rictal bristles. Skin: yellowish & bright coral-pink. Down: dark gray, dark drab, or black on capital, scapular, humeral & lower ½ of dorsal (2 rows along spine). LF: yellowish, turn black. G&D: d0: 1.7–3.1 g; d1–2: 4–5.5; d3–4: PinVis; d4: 14 g; d5: PinEm wings; d5–8: eyes open; d6–8: 20 g; d10–12: 25–25.5, PinUn all over; d12–13: almost FF, white eye ring; can be reliably sexed—wings & tail show blue (M), dull gray-blue (F). **Fledgling.** 27.2 g, fledges 17–19 d, flies strongly. Plumage browner than Ad, dusky spots & streaks above & below, amount of blue varies in individuals, M with dull Prussian blue in wings & tail; F secondary wing covs dull blue, outer tail covs white. Voc: *tu-a-wee*. Remains in family groups 4 wks.

Western Bluebird (*Sialia mexicana*)   Plate 28, Photos 32-35

Adult. (CA) M 26–32.5 g, F 22.5–30.5 g; (AZ) M 23.5–28.5 g, F 24.5–31.5 g; 17.8 cm (7 in.). Diet: insects (spring/summer), small fruits & some seeds (winter). Nest: in cavity of tree or box. **Nestling.** MC: orange-yellow. GF: cream to pale yellow, yellower at corners, somewhat prominent (Fig. 2.4). Bill: H pink or light horn, pointed. Skin: vinaceous-pink, ochreous-buff to orange-ochreous on wings, legs, joints & caudal region; body color fades by d6. Down: dark bluish-gray above, dense on crown. By d12–14, M may have blue in wings & tail (F, little to no blue). LF: pink, then yellow-orange, darken to gray; N pinkish-buff, orange toe pads. G&D (see Table C-6, Appendix C): d0: 2.1–3.0 g. Voc: d1 faint *peeping*, stronger by d5. **Fledgling.** ~28 g, fledges (OR) 16–23 d, (AZ) 18–25 d. Plumage above: white eye ring, grayish-brown head & neck. M: back & scapulars fuscous, heavily mottled or streaked white; rump dark grayish-brown; wings & tail smalt blue (amount varies by individuals), outermost rect with narrow white streak along distal edge. F: dull ultramarine blue in wings & tail. Below: heavily mottled/streaked white. Independent by 2 wks post-fledging but remains with parent(s) & begs for food. M may only be tending to young while F starts new nest.

Mountain Bluebird (*Sialia currucoides*)   Plate 28

Adult. (OR) 25.7–31.6 g; 16.5–19 cm (6.5–7.5 in.). Diet: insects (92%), vegetable matter (8%). Nest: in cavity—natural or box. **Nestling.** MC: yellow (orange-yellow throat). GF: yellow. Skin: salmon-pink to orangish, turns pale yellow. Down: sparse, dark gray. G&D: d0: 2.3–2.7 g. From study in MT (Power 1966): d0–2: 4.0 g; d3–4: 6.4; d7–8: 12.4, eyes open; d9–10: 21.2 g; d11–12: 23, appears FF; d13–14: 27.2 g; d15–16: 26.2; d17–18: 24.7; d19–20: 19.9. Voc: faint at first. **Fledgling.** 28 g (25.8 OR), fledges 17–24 d, flies short distances at first. Plumage above: brownish-drab to mouse-gray (F darker), amount of blue varies in individuals, M wings & tail pale cerulean blue to pale cobalt blue (F wings pale blue with more greenish or grayish), white eye ring; below pale white with pale buffy-brown & mouse-gray streaks; throat, breast & sides pale gray & spotted white. Independent 2–4 wks post-fledging.

Townsend's Solitaire (*Myadestes townsendi*)   Plate 28

Adult. (CA) 28.4–39 g, 21.6 cm (8.5 in.). Diet: insects (+ fruit in winter). Nest: cup usually on ground. **Nestling.** MC: yellow (bright orange throat). GF: yellow, not thick or prominent. Skin: pink. Down: sparse, blackish neutral gray. G&D: d0: 2.5 g; d2–3: wing PinEm; d4: body PinEm; d5: rect PinEm, eyes open; d6–7: PinUn; d8: 28.1 g (avg. in CA), FF. Voc: *trill* (d6). LF: salmon or pinkish-yellow. **Fledgling.** Fledges 9–15 d. Plumage above: overall black & heavily spotted with white or buff; grayish eye ring. Below: neutral gray or white with less distinct spots or mottling. As bird ages, plumage overall becomes more grayish & more mottled appearance below; wings have buff-yellow wingbar on proximal third; eye ring becomes more white and distinct. Dependent 2 wks post-fledging.

Veery (*Catharus fuscescens*)

Adult. (PA) 26.8–37 g, 17.8 cm (7 in.). Diet: insects & fruits (mostly in fall/winter). Nest: open cup, ground or low bush. **Nestling.** MC: orange-yellow. GF: white, not thick, not prominent. Down: sparse, dark gray to brownish. LF: pinkish. G&D: d3: PinVis; d4–5: eyes open; d5–6:

PinEm; d10: FF. **Fledgling.** 24.5 g, fledges 10–12 d. Bill: pink. Plumage sim to Ad DefB but appears spotted because of buffy tips on feathers of upperparts. Below: heavy spotting, dusky bars & mottling.

Swainson's Thrush (*Catharus ustulatus*)

Adult. (PA) 25–36 g, 17.8 cm (7 in.). Diet: insects (+ fruit/berries in winter). Nest: cup. **Nestling.** MC: yellow. GF: yellow. Bill: N burnt orange tinge. Down: short, dark burnt amber. LF: burnt-orange tinge. G&D: d2: wing PinEm; d3: body PinEm; d4: tail PinEm; d5–6: PinUn; d10: FF. **Fledgling.** Fledges 10–14 d. Bill: brownish-gray. Plumage above: eye ring distinct & buffy; olive-brownish overall, spotted above & below. LF: yellow-pink. *Sim sp: Hermit Thrush. Juv can be distinguished by primary 6 (p6): in Swainson's p6 is not notched, whereas it is notched in Hermit Thrush.*

Hermit Thrush (*Catharus guttatus*)   Plate 29

Adult. (PA) 26.6–37.4 g, 14–17.8 cm (5.5–7 in.). Diet: insects (65%), fruit & other vegetable matter (35%). Nest: cup. **Nestling.** MC: orange-yellow. GF: yellow or cream. Bill: gray. Skin variable: pinkish or dark yellow-pink, then brick-red. Down: sparse, dark grayish on capital, humeral & lower ½ dorsal. LF: dull pink. G&D: d0: 4.1 g; d1: 4.9; d2–3: 7.2–10; d3–5: eyes open; d4: 14.8 g, alar PinEm; d5: 17 g, all PinVis; d7: wing PinUn; d9: 25 g; d10: FF; d11: 25.6 g; d12: 24.8. **Fledgling.** Fledges 10–15 d, flies short distances. Bill & L/F: dull pink-buff. Plumage above: generally sepia or olive-brown spotted buff-white; rump, uppertail covs & tail rufous-cinnamon or rufous-brown. Below: white base, tinged buff on throat & breast, whiter on belly; black spotting or flecks on sides of neck, chest, flanks & vent. *Sim sp: see "Swainson's Thrush."*

Wood Thrush (*Hylocichla mustelina*)   Plate 29

Adult. (DE) M 37–62 g, F 40.5–76 g; 19.7 cm (7.75 in.). Diet: invertebrates mainly in soil, adds fruit late summer, fall & winter. Nest: open cup in tree or shrub. **Nestling.** MC: bright yellow. GF: whitish or cream, thin, not prominent. Bill: long. Skin: pinkish. Down: very sparse, mouse-gray or dark gray on capital, humeral & middorsal tracts. G&D: d0: 2.8–4.0 g; d2: PinVis dorsal, alar, femoral & humeral; d2–5: eyes open; d4: wing PinEm; d8: dorsal plumage brownish-olive; d10: 30–34 g; d12–14: 34.6. **Fledgling.** 12–15 d. Plumage above: olive-brown streaked pale tawny or ochreous; wings & tail olive-brown; narrow white eye ring.

Below: white, breast & flanks (lighter) with large, dark, indistinct blackish streaks/spots. Dependent 28–36 d post-fledging. *Song learning: 3 parts to the song, the first & last part may be innate or learned, the middle part is learned from conspecifics.*

American Robin (*Turdus migratorius*)　　　　　Plate 29

The largest of thrushes and one of the most common birds in rehabilitation centers. **Adult.** (w. US) 56–112 g; (OR) M 85 g, F 75.5 g; (PA) M 84.8 g, F 75 g; (NY) M 77.4 g, F 83.6 g; 25.4 cm (10 in.). Diet: omnivorous; arthropods, especially insects & earthworms, + berries. Hatchlings fed partly digested, regurgitated worms & insects, then whole foods (d5). Nest: cup shaped, bulky, of grass with mud. **Nestling.** MC: yellow to yellow-orange (pumpkin). GF: white exterior, pale yellow lining inside at corners, turned down at corners, somewhat prominent. Bill: H orange-pink or pink, large & adult-like Ad; egg tooth remains until fledging; large gape. Skin: light pink, then yellowish & orangish. Down: whitish, turns creamy and then gray. LF: long legs, delicate-looking toes. Voc: staccato *trill.* G&D (see Table C-7, Appendix C): d0: 4.1–6.7 g. Distinguishing features: d4–7: skin pink, orangish tint; beak longer, gray-black; whitish feathers begin to appear on flanks; begin to see rusty-tipped feathers outlining ventral tract; d9–10: white on throat with two dark vertical streaks; rufous (or rust) ventrally with dark sepia flecks or speckles on an underlayment of rust-yellow. Darker dorsally with numerous buffy streaks on upperwing covs & back. **Fledgling.** 13–16 d, 50–62 g, flies poorly, can run, spends 10–15 d under cover in vegetation. Bill: pinkish. Plumage above: often with buffy-whitish supercilium; rust-tipped feathers, pale speckling dorsal & ventral, tufts of down protrude from feathers mostly on head. Below: chin & throat white. Variation occurs in amount of dorsal speckling and in amount & intensity of orange or rufous on ventral areas (Fig. G8, Illustrated Glossary). Independent 4 wks post-fledging, remains in natal area up to 4 mos., then gathers in mixed flocks with Ad before & during fall migration. *Sim sp: often confused with mockingbird nestling: see comparison Table D-4 (Appendix D). Song learning: may be innate, may make up own repertoire.*

Varied Thrush (*Ixoreus naevius*)

Adult: M 70–100 g, F 70.5–99.5 g; 24 cm (9.5 in.). Diet: omnivorous; mainly arthropods; fruits & berries late summer & fall. Nest: cup. **Nestling.** MC: yellow. GF: presumed yellow or cream. Down: sparse vinaceous-buff or grayish. G&D: unknown. **Fledgling.** Fledges: 13–15 d. Plumage: orange supercilium indistinct, two conspicuous wingbars. LF: horn-colored nails.

MOCKINGBIRDS AND THRASHERS

(Family: Mimidae)

Adults. Small to medium songbirds. Diet: glean insects and small fruits from ground and low vegetation. Nest: simple, cup shaped, exterior of sticks and twigs, fairly low in dense vegetation, tree, cactus, or yucca. Eggs: buffy, blue, or green, most have speckles. Bill: varies from short and straight (mockingbirds) to long and decurved (thrashers), terete and decurved toward the tip, upper mandible notched toward tip (except in genus *Toxostoma*), rictal bristles evident. Plumage: various shades of brown and gray, some with streaking on pale underparts. Wings: 10 primaries (10th reduced), variable in length, usually short and rounded. Tail: variable in length, usually somewhat longer than wings, rounded, sometimes graduated. Legs/feet: rather long, strong; tarsi distinctly longer than middle toe without nail and scutellate in front, often booted behind; base of middle toe adherent to outer toe. Vocalizations: complex calls and songs, some species with a high degree of mimicry. **Young.** Mouth: yellow or orange-red. Gape flanges: whitish, not particularly prominent, long, down-curved. Down: sparsely covered. Fledges 12–15 days, sometimes longer. Dependent up to 3 weeks post-fledging. Song learning: little known, except primary song appears to be mostly learned, especially in species that incorporate a high degree of mimicry in their repertoire, e.g., Northern Mockingbird, Gray Catbird, and some thrashers.

Gray Catbird (*Dumetella carolinensis*)   Plate 30

Adult. (PA) 27.3–43.5 g, 21–25.4 cm (8.3–10 in.). Diet: insects & small fruits. Nest: open-cup, low in shrubs & small trees. **Nestling.** MC: bright yellow or orange-yellow. GF: creamy-white, blackish mouth edges & tongue tip. Bill tip: dark gray. Skin: blackish-gray with pinkish-tan on dorsum, lighter below. Down: clove-brown, dark gray, or black, above only. LF: yellowish-tan. G&D: d0: 2.52–3.41 g; d2: 6.9; d4: 12.8, some PinUn; d6: 20.6 g; d8: 25.3; d10: 26.5 (avg in MI); d12: 28.5 g. **Fledgling.** Fledges 8–12 d (up to 15 d). MC: mix of yellow, pink & gray. Bill: very pointed. Plumage like Ad but cap is dusky (not dark) & crissum buffy to pale rufous. Above: washed brownish, wing covs brownish-gray (may be edged brown); face may be bare. Below: gray with brown mottling, rust crissum. LF: dark pinkish-buff. Iris: cloudy grayish. Dependent 12–24 d post-fledging. *Sim sp: Brown Thrasher. Song learning: does not need exposure to normal catbird song (Kroodsma 2005).*

Curve-billed Thrasher (*Toxostoma curvirostre*)

Adult. (*T. c. curvirostre*) 85.23 g; (*T. c. palmeri*) M 80.9 g, F 75.8 g; 28 cm (11 in.). Diet: arthropods & gastropods, berries in fall. Nest: cup, in spiny shrub or cactus. **Nestling.** MC: ochre-yellow. May have black spots on tongue. GF: whitish (not prominent) or yellow. Skin: pinkish-red. Down: long gray to grayish-black above, pale gray below. G&D: d0: 3.6–6.8 g; d2: PinVis; d4–6: eyes open; d5: wing & tail PinEm; d8: all PinUn; d9: 30.2–55.4 g (AZ); d12: FF. **Fledgling.** 46.6 g, fledges 14–18 d. Plumage sim to Ad DefB & variable by ssp, rusty-brown upperparts contrast with brownish back; paler below with variable amount of spotting. Independent 22–26 d post-fledging. *Sim sp: Juv plumage of Curve-billed Thrasher indistinguishable from juv Bendire's Thrasher; nestling Bendire's without black spot on tongue. Song learning: 2–3 day-old chicks capable of making "wit-weet-wit" call of Ad.*

Brown Thrasher (*Toxostoma rufum*)   Plate 30

Sings repertoire of 2,000 to 3,000 different songs (Kroodsma 2005). **Adult.** (PA) 57.6–89 g, 23.5–30.5 cm (9.3–12 in.). Diet: insects & other arthropods, + berries & seeds. Nest: cup, in low dense shrubs (with thorns). Iris: orange-yellow. **Nestling.** MC: creamy-yellow, orange throat. GF: whitish (not prominent). Bill: N upper bluish-pink, lower rose-pink, for shape see Fig. G3 h, Illustrated Glossary. Skin: dark pinkish-tan, turns darker. Down: H plentiful grayish-white above only; N appears darker gray or umber. LF: H deep vinaceous. Iris: dusky-brown. G&D: d0: (KS) 4.8 g, (KY) 6.4 g (avg); d3: PinF appear, eye slits; d9: 45.5 g. (ON, Canada): d0: 4.2–6.3 g; d1: 10.2–11; d4: 19.7–22; d7–8: 34–38, PinUn; d9: 42; d11: 50; d14–16: 56–58; d21: 62. **Fledgling.** 41.5 g, fledges 9–13 d, runs well. Bill: length ~same as head, decurves toward tip; nares (at base) not covered. Plumage similar to Ad except buffy wingbars & subtle mottled appearance on mantle (some individuals). Above: mostly sepia with variable amount of umber to burnt umber on dorsum, rump & tail. Below: dull white with dull blackish streaks. LF: light brownish & pinkish. Iris: olive-gray. *Sim sp: Gray Catbird H has darker skin & weighs less at same age.*

California Thrasher (*Toxostoma redivivum*)

Adult. (CA) 78–93 g; 24–28 cm (9.4–11 in.). Diet: insects (+ berries in fall/winter), seeds year-round in drier habitats. Nest: cup, in dense shrubs. **Nestling.** MC: orange-yellow. GF: cream, N downcurved at corners. Skin: pinkish-tan. Down: long dark gray. LF: long legs. G&D: d0–3: 6–35 g, N (d4+): 40–60 g. **Fledgling.** Fledges 12–14 d. Plumage sim to Ad DefB but duller & features of face & throat less distinct. Above: browner than Ad, indistinct pale eye stripe, light cinnamon-brownish wing covs, tail edged rusty-brownish. Below: medium gray throat patch, dusky breast, crissum duller.

LeConte's Thrasher (*Toxostoma lecontei*)

Adult. (s.w. US) 54.5–75.5 g; 28 cm (11 in.). Diet: arthropods (almost exclusively), rarely drinks. Nest: cup, in dense thorny desert shrubs & cactus. **Nestling.** MC: orange-yellow. GF: yellow, thin. Egg tooth until d6. Skin: pink. Down: long, fluffy dull white. LF: H pink, then pinkish-horn. G&D: d0: 3.7–4.4 g, d4: PinVis (skin darkens) along tracts; d6–9: eyes open, PinEm; d14–15: wing PinUn, body FF. Voc: quiet until d5, then *peeping*. **Fledgling.** 38–47.7 g (70% of Ad mass), fledges 12–20 d, flies poorly, then fairly well 10–15 d post-fledging. Bill: ~half the length of Ad. Plumage above: paler than Ad, uppertail covs pale brown, buffy edges to wing covs. Below: pale buff crissum. *Song learning: mimics sounds of birds & mammals in its area 6–9 months post-fledging.*

Crissal Thrasher (*Toxostoma crissale*)

Adult. (AZ) 53–70 g, 29.2 cm (11.5 in.). Diet: arthropods. Nest: cup, in dense shrubs. Iris: brown. **Nestling.** MC: light yellow. GF: light yellow or creamy-white. Down: long, dense, blackish or charcoal-gray. LF: H light colored, turn dark. Iris: N gray, then ivory. G&D: unknown, egg mass 5 g (avg CA). **Fledgling.** 74% of Ad mass, fledges 11–16 d (12.5 d avg), flies poorly. Plumage paler & duller than Ad; rusty tone, including uppertail cov & tips of tail; shows black mustache stripe.

Sage Thrasher (*Oreoscoptes montanus*)

Adult. (WA) 39.6–50.2 g, 21.6 cm (8.5 in.). Diet: insects (+ berries). Nest: cup, in sagebrush. **Nestling.** MC: yellow to bright orange. GF: creamy-white. Skin: reddish-orange turning grayish-yellow. Down: blackish or dark bluish-brown. G&D: d0: 3.9 g; d1: 5.8; d4: 16.45; d6: all PinEm; d8: 34 g. **Fledgling.** 38 g (d12), fledges 8–13 d, can fly. Plumage like Ad with paler grayish-brown back & darker brown streaking. Dependent 7 d post-fledging.

Northern Mockingbird (*Mimus polyglottos*)   Plate 30, Photo 36

Adult. (CA) 39.7–57 g, (FL) 36.2–55.7 g; 21–25.4 cm (8.3–10 in.). Diet: insects (+ fruit). Nest: open cup, usually in shrub or tree. **Nestling.** MC: yellow, crescent markings on roof of mouth (Fig. 2.2). GF: off-white or creamy-yellow. Broad gape. Bill: N dark yellow & narrow, culmen decurved. Skin: pink with yellow tint or olive. Down: above & below, plentiful dark gray or sepia-brown, some white on lower abdomen. G&D (Horwich 1966): d0: 3.28–3.84 g; d1: 4; d2: 8; d3: 13.3, PinEm; d3–5: eyes open; d4: 17 g, PinUn spinal & femoral; d6: 26 g; d7–9: white breast feathers (with dark brown streaks) cover most of ventral apterium; d8: 33.5 g; d12: mass varies geographically (60–70% of Ad), range 20.4–39.6 g (CA), appears FF; d16: 40 g; d18: 41. Voc: H single faint *peep*; then sounds like throaty *bark*. **Fledgling.** 32 g (FL), fledges 10–15 d, can run, flies well 8 d post-fledging. Plumage sim to Ad DefB but breast is brownish-gray to whitish with brownish or dark grayish spots, speckling & streaks (mottled appearance), tail patterned & spotted, white patch on wing. Down remains until d20 on upper tracts. Iris: dark gray to gray-green. LF: long legs, H nails white, turn dark. Biparental care: 3 wks post-fledging (Ad F may begin 2nd nest). *Sim sp: often confused with American Robin nestling: see comparison Table D-4, Appendix D. Song learning: mostly mimicry; Ads sing 100–200 different songs (Kroodsma 2005).*

STARLINGS

(Family: Sturnidae)

  Two species of Sturnidae occur in N.A., but only one is covered here. European Starlings were introduced in the late 1800s, are not protected, and are detrimental to native cavity nesters.

European Starling (*Sturnus vulgaris*)   Plate 31

Adult. Medium-size songbird. M 73–96 g, F 69–93 g; (OH) M 84.7 g, F 79.9 g; (NY) M 79–100 g, F 76.3–92 g; 21.6 cm (8.5 in.). Diet: omnivorous, probes with bill on ground; mainly insects, other invertebrates, fruit, nectar, grain & bird eggs. Young fed insects & fruits. Nest: cavity in tree, building or artificial box, commonly in urban areas. Eggs: subelliptical; pale blue, tint varies, slightly glossy, usually without markings. MC: pink. Bill: as long as head, straight, slightly depressed toward tip (tapers to a point); commissure somewhat angulated; feathers of forehead partially divided by culmen; dark (juv & nonbreeding Ad), becomes bright yellow (breeding season). Plumage: feathers of head, neck & breast long & narrow; coloration metallic & somewhat iridescent. Wings: 10 prim (10th minute), long, triangular, pointed, outermost primaries rudimentary & acuminate. Tail: short, half the length of the wings, nearly square to slightly notched. LF: strong, tarsi longer than middle toe without nail & scutellate. **Nestling.** MC: bright yellow. GF: pale yellow to bright lemon-yellow; very prominent, described as "clown lips," lower mandible greatly enlarged, extends beyond upper (Fig. 2.1). Head: flat, sloping forehead; small eyes (not bulging as other songbirds). Skin: orangish-pink & rose-pink. Bill: H pinkish-horn, long, pointed, culmen decurved. Down: relatively large quantity of down compared with other cavity-nesting species (Wetherbee 1958); very long, thick, gray to grayish-white, darker (brownish) on head. LF: long tarsi & toes, pinkish-brown, white nails. G&D (see Table C-8, Appendix C): d0: 5–6.4 g. Voc: H single squeaky note. **Fledgling.** Fledges 21–23 d, 71 g. GF: prominent on newly fledged. Bill: black, turns yellow first winter. Plumage above: almost entirely brown or gray-brown, or dark brownish-olive; cinnamon tips on wings & tail. Below: paler, whitish chin, some with faint streaking on breast, some with broad streaks across abdomen. LF: reddish-brown, blackish claws. Biparental care: 4–5 d post-fledging. *Mistaken identification: starlings take over nests of other sp so can result in misidentification, such as woodpeckers or Purple Martin, both sp without down. Fledgling starlings have been mistaken as Ad F House Finch because of streaked underparts; distinguish by comparing size, weight, head (flatter forehead vs. rounded), & beak shape. Song learning: own song mixed with variety of mimicked sounds.*

WAXWINGS

(Family: Bombycillidae)

Two of three species of waxwings raise young in N.A. The Bohemian Waxwing (*Bombycilla garrulous*), not covered here, nests in northernmost territories of Canada and AK; Cedar Waxwing nests across Canada and northern states of the US. Distinguishing characteristics: waxlike tips on secondary feathers. **Adult.** Diet: frugivorous, insects fed to young. Nest: woven cup high in fork of tree. Eggs: subelliptical to oval, pale blue or blue-gray; markings of sparse dark and gray blotches. Bill: short, stout, straight, upper mandible slightly hooked, notch near tip, culmen de-curved (high ridge), gape deeply cleft & wide, nearly equal to length of exposed culmen. Plumage: soft, dense, predominantly brownish, prom-inent crest on head, black band from bill through eye, nostrils clearly concealed by feathers. Wings: 10 primaries (10th minute), long, pointed; outermost primary less than half as long as primary coverts. Tail: shorter than wing, square to slightly rounded; terminal band variously colored yellow, orange, or red possibly due to diet (Hudon and Brush 1989). Legs/feet: black, strong, short, tarsi shorter than middle toe without nail, and scutellate; middle and outer toes united basally; long claws.

Cedar Waxwing (*Bombycilla cedrorum*)   Plate 31, Photo 37

Adult. Medium-size songbird. (PA) M 25.5–40 g, F 26.4–39.6 g; 18.4 cm (7.25 in.). Diet: fruits, insects when feeding young. Nestlings fed insects for 2–3 d, then fruit gradually until 85% fruit when they fledge. **Nestling.** MC: deep pink or red with violet-blue markings inside mouth at cor-ners. GF: creamy-yellow, white & thin at corners & bill edges. Bill: upper mandible yellowish, distinct shape by d10; nares at middle of upper, not covered. Skin: pink, then purplish-pink on body & flanks, ochre-yellow on crown & neck. Down: none. G&D (ON, Canada): d0: 2.8–3.3 g; d1: 4.5–5.4; d3: 9–11.3, rect PinEm (just barely) with yellow waxy tips; d4: 11.6–13.5 g, able to grip, dorsal PinVis; d5: 14–16 g; d6: 16.3–18.5, dorsal & alar PinEm; d6–8: eyes open; d7: 18.4–19.5 g; d8: 19.6–20.8, all PinEm; d9: 20.5–21.8 g, can perch; d10: 20.7–22.3 g, waxy tips on sec evident inside shafts; d11: 22.5–23.2 g, waxy tips unsheathe as tiny points; d12: 24 g, throat & rect PinUn; d12–14: FF; d13: 25 g; d14: 26, black "mask" feathers last to unsheathe; d15: 27 g; d19: 30. Voc: H is quiet, d5 faint *buzzing* or *chirping*. Begging call: long, excited, high-pitched trill using several *chip* syllables, also described like a little bell. **Fledgling.** 25–30 g, fledges 14–18 d. Bill brown. Plumage like Ad but grayer overall, very small crest, black mask not as distinct, wings with fewer waxlike tips (see Ad tail). Below: pale with dusky-brown streaking. LF: legs brown. Iris: grayish-brown. Biparental care: 6–10 d post-fledging.

SILKY-FLYCATCHERS

(Family: Ptiliogonatidae)
    Of four N.A. species, one raises young in N.A. north of Mexico.

Phainopepla (*Phainopepla nitens*)   Plate 31

May nest in loose colonies; breed twice/year in two different areas. Strong association with mistletoe and mesquite. **Adult.** Medium-size songbird: (AZ) 18–28 g, 17.8–21 cm (7–8.3 in.). Diet: small berries (primarily) and flying insects, specialist on mistletoe berries (nonbreeding). Young fed insects 1st four days, then regurgitated fruit is gradually added. Nest: loose cup, central fork of tree or shrub, well concealed. Eggs: subelliptical, grayish or bluish, heavy mottling. Bill: short, stout, straight, upper mandible slightly hooked, with notch near tip; culmen decurved, gape deeply cleft & wide, rictal bristles evident. Plumage: soft & silky, distinct crest, M black, F gray (no markings). Wings: 10 prim (10th reduced), short, rounded. Tail: slender, variable in shape, usually equal to or longer than wings. LF: strong, tarsi usually shorter than middle toe with nail & scutellate, hallux very short. **Nestling**. MC: pink. GF: yellow. Bill: black. Skin: slaty or purplish-black. Down: white, long tufts on capital (center of crown bare), alar & caudal. LF: short, chunky, pinkish to gray. G&D: little info, d7: PinVis; d10–11: crest appears. N plumage very dark feathers with white edges; d17: 21.5 g. **Fledgling**. Fledges 14–19 d. Plumage like Ad F but browner; flight feathers dark gray. Begins pre-supplemental molt a few days after fledging: new feathers on median or greater wing covs grayer, some down remains on head.

SPARROWS, OLD WORLD

(Family: Passeridae)
   Two species raise young in N.A., only one is covered here.

House Sparrow (*Passer domesticus*)   Plate 31, Photos 38-40

Young of this family differ from New World sparrows. House Sparrows are more heavy bodied, have stouter bills, undersides are not streaked, have shorter wings & tails. House Sparrows aggressively compete with native species for nests & are not protected under any state or federal regulations. **Adult.** Medium-size songbird. (PA) 20–34.5 g, (CA June) F 25 g; 15.2–18.4 cm (6–7.25 in.). Diet: seeds, some insects & fruit; young fed 68% insects, 30% seeds & grains. Nest: cavity or domed often with tunnel-like entrance, in crevice or on ledge, usually in urban areas. Uses nest boxes & nests of other birds. May take over or add to existing nest with eggs or young & may kill the young & parents. Eggs: oval to long-oval, white to greenish or bluish-white, gray or brown markings may form a "wreath" around wider end. Bill: short, strong, conical, slightly decurved culmen, angulated commissure, relatively blunt tip, pinkish-buff except black in breeding M, rictal bristles present but not conspicuous, nostrils uncovered. Plumage overall: shades of browns & grays; M has black bib. Wings: 10 prim (10th reduced), long, pointed, outermost primary rudimentary. Tail: shorter than wings, somewhat square. LF: strong, pinkish-buff, tarsi shorter than middle toe without nail & scutellate. **Nestling.** MC: pinkish, pinkish-yellow, or red. GF: H white, then lemon-yellow (d4), prominent. Bill: H pinkish-tan, conical, short; d3: taking shape, base darker; d11–12 dark "ring" at base of culmen & lower mandible. Skin: pink or reddish. Down: none, does not develop. LF: short, chunky, pinkish to gray; nails whitish. G&D: d0: 1.5–3 g (see Table C-9, Appendix C). Voc: melodic single *chirp*. **Fledgling.** Fledges 11–16 d, 25–27 g, can fly but runs & hides well. MC: pinkish-yellow. Plumage: like Ad F; dark throat & breast in M begins (variable) 2–7 wks post-hatching, no streaking on underparts. Dependent 2 wks post-fledging.

PIPITS

(Family: Motacillidae)

Pipits and wagtails have long tails that bob (up and down) or wag (side to side). **Adults.** Small to medium songbirds. Diet: terrestrial and aquatic invertebrates gleaned from ground and edges of ponds and streams. Nest: open cup, or some domed; variety of locations from ground to buildings & cliffs. Eggs: subelliptical, white, pale green to dark olive, typically speckled gray or brown or may have fine dark streaks. Bill: short, straight, very thin, acute, culmen decurved toward tip, upper mandible notched near tip, rictal bristles present but not evident. Plumage: pipets have cryptic coloration mostly brown and streaked; wagtails boldly patterned with black, white & yellow. Wings: 10 primaries (10th reduced), long, pointed, tertiaries elongated, nearly equaling primaries in length. Tail: variable length, never shorter than wings; variable shape, rectrices narrow & acuminate (except possibly middle pair). LF: strong, tarsi usually longer than middle toe with nail, and scutellate, elongated hallux nail (as long as hallux). **Young.** Mouth: orange to reddish-orange. Down: thick on head & back. Fledge: 10–16 d. Biparental care: 14–18 d post-fledging.

American Pipit (*Anthus rubescens*)

Adult. (AK) M 19.2–25.5 g, F 18.6–23.2 g; 15.2–17.8 cm (6–7 in.). Diet: wide variety of invertebrates, few seeds. Nest: open-cup, ground.

Nestling. MC: reddish-orange. GF: very pale yellow. Skin: pinkish-tan. Down: long, fairly thick, brownish-gray above, shorter & lighter below. G&D: d0: 1.6–2.4; d4–5: eyes open; d8–9: appears FF. **Fledgling.** Fledges 13–16 d (avg 14). Plumage sim to Ad but more streaked. Above: drab (brownish washed), streaked black, fawn feather edges on back, buffy eye ring & eyeline, wings & tail fuscous edged fawn, buff-yellow edges on wing covs & sec. Below: cream, throat & breast streaked. Tail: r6 nearly all white, r5 extending 15–40 mm from tip.

Sprague's Pipit (*Anthus spragueii*)

Adult. (OK) 21.4–29.7 g, 16.5 cm (6.5 in.). Diet: arthropods & few seeds (winter); young fed arthropods exclusively. Nest: cup on ground, open grassland. **Nestling.** MC: orange. GF: yellow, downcurved. Bill: dull gray-pink (paler lower mandible). Skin: pink. Down: light gray, very long & thick above. LF: salmon, then pinkish, long relative to body. G&D: d0: 2.5 g; d1: 3.3, most PinVis; d3: 6.7 g; d3–5: eyes open; d4: 8 g; d4–5: prim & contour PinEm; d5–6: 11.7–13.4 g, all PinUn; d7–9: 15–19.3 g, prim PinUn; d10–11: 18.8–19 g. Voc: d3 soft clicking; d5–6 faint peeping. **Fledgling.** Fledges 9–12 d, flies poorly. Plumage above: head & back dark with dull black spotting. Below: brown chest, tan or white under. Tail: r6 mostly white, r4 & r5 extensively white.

FINCHES AND RELATIVES

(Family: Fringillidae)

Adults. Small to medium songbirds; variations in body size, bill size and shape, and plumage. Diet: seeds, other plant foods, some insects. Nest: open cup, in trees, shrubs or herbage, on ground, in crevice, holes in trees, or in covered nests of other birds. Eggs: may be white to bluish. Bill: short, stout, strong, conical (except crossbills, mandibles are crossed), culmen decurved, commissure abruptly angulated. Variation in bill structure related to primary type of seed preferred by the species: thin and pointed (Pine Siskin), more conical (goldfinches and rosy-finches), stubby and slightly hooked (Pine Grosbeak), crossed bills (crossbills). Plumage: nostrils concealed by tufts of bristle-like feathers; rictal bristles usually present; males brightly colored, females more drab, little to no color. Wings: 9 (visible) primaries, variable length and shape. Tail: variable length and shape, generally relatively short, square or slightly forked. Legs/feet: typically short to medium length, tarsi variable length but always scutellate (visible on d7 post-hatching). **Young.** Mouth: pink to red. Gape flanges:

yellowish or white, rose-pink at the rictus. Down: sparse on head & back, white or light grayish. Fledge: 9–25 d. *Goldfinches nest until late summer and early fall. Fostering: House Finches and probably other finch species will accept similar-age foster nestlings. Total nestlings should not exceed 4 at season's end or 6 at season's beginning, and there must be two parents tending to young. Table D-3, Appendix D, compares three nestling goldfinch species.*

Evening Grosbeak (*Coccothraustes vespertinus*)   Plate 32

Adult. (UT) M 55.7 g, F 54.6 g; (PA) M 52–71.7 g, F 48.2–71.7 g; 20.3 cm (8 in.). Diet: seeds & some fruits; insects only in spring/summer; young fed regurgitated larvae of insects. Nest: open cup. Bill greenish-yellow, off-white in nonbreeding; upper mandible very deep, thick, downcurved at tip. **Nestling.** MC: violet and red. GF: white. Bill: yellow. Skin: dark. Down: white, capital & dorsal. G&D: little info, d4–6: eyes open, d12 PinEm on dorsal tracts. Voc: *peep.* **Fledgling.** Fledges 13–14 d. Plumage like Ad F DefB but duller, browner. Above: buffy or dull gray, wings dark & edged white with small yellowish patch on greater wing covs, tail black with white tips. Below: paler than above (M more yellowish).

Pine Grosbeak (*Pinicola enucleator*)

Adult. 53–77.6 g; 20.3–25.4 cm (8–10 in.). Diet: seeds, buds, fruits (+ some insects & spiders in nesting season). Young fed mainly insects & spiders, mixed with some plant matter. Nest: cup. **Nestling.** MC: red or orange. GF: pale yellow. Down: very sparse, grayish-black above, d7 denser & turns paler. G&D: egg mass 4.3 g; d0: no info; d5: eyes open; d6–7: PinUn. **Fledgling.** Fledges 14–18 d. Plumage above: gray-brown, crown & rump tinged yellow; gray crescent under eye, pale buff wingbars. Below: light gray-brown front & sides. Dependent 21 d post-fledging.

Gray-crowned Rosy-Finch (*Leucosticte tephrocotis*)

Breed in remote alpine areas, few nests have been found. **Adult.** (CA) 24.6 g, 16 cm (6.25 in.). Diet: seeds mostly, + insects & emergent vegetation; young fed both plant & insect matter. Nest: cup on cliff or rocky area (high altitudes). **Nestling.** MC: not described. GF: yellow. Down: long, fluffy, gray or grayish-white above. G&D: little info; d6: 14 g, eyes open, PinEm; d9: 21 g; d14: 26. **Fledgling.** 40.5 g, fledges 15–22 d, can fly, follows parents. Plumage above: grayish-brown, wingbars pinkish-buff to tan.

House Finch (*Haemorhous mexicanus*)   Photos 41-46

Adult. (CA) 19–25.5 g, 15.2 cm (6 in.). Diet: granivore; young (d1–3) fed regurgitated plant food (mainly dandelion or other weed seeds). Nest: cup, just about anywhere; ring of poop around nest rim (from nestlings evacuating on rim). M head & bib colors variable because of genetic factors and/or diet. **Nestling.** MC: H pink, then (varies) dark pink to bright orange-red. GF: H pale yellow, then yellow with creamier "white" at corners. Bill: pinkish-tan on top with darker tip, short, broad, soft; d7 hardened, light brown, more conical. Skin: pinkish-apricot. Down: plentiful white or grayish-brown, all dorsal tracts. Whitish down on head (2 rows on crown & a small row above each eye). LF: short & stocky. G&D (see Table C-10, Appendix C): d0: 1.1–2.3 g. House Finch development is slower than other small passerines. Voc: none at hatch, d3 high-pitched *peeping*. **Fledgling.** 18 g (avg), fledges 15–17 d (variable), flies short distances. Plumage like Ad F but darker brown overall, no distinct facial markings, streaking heavier & narrower, buffy wingbars, down still on crown. Short wings & tails. Below: paler than Ad F, streaking not as defined. Dependent 2.5–3 wks post-fledging.

Purple Finch (*Haemorhous purpureus*)

Adult. (PA) 19.8–28.4 g, 15.2 cm (6 in.). Diet: seeds (+ some insects & fruit); young fed regurgitated seeds. Nest: open cup. **Nestling.** MC: red. Skin: pink. Down: dark gray. G&D: unknown. Egg weights: 1.86–2.09 g. **Fledgling.** Fledges d14. Bill: upper dark brown, lighter brown base & lower mandible. Plumage above: heavy brown mottling on crown, white superciliary & malar areas, auriculars dark brown, uniform brown on back & rump, little to no red. Below: heavy streaking (bright white on belly), throat streaked brown. LF: dark brown. Voc: *pee-wee. Western coast ssp M darker than e. ssp.*

Cassin's Finch (*Haemorhous cassinii*)

Adult. (AZ) 20.4–37.8 g, 16 cm (6.25 in.). Diet: buds, berries & other fruits, + seeds & some insects. Nest: open cup. **Nestling.** Down: plentiful, sooty-gray. G&D: unknown. **Fledgling.** First flight ~14 d. Plumage above: grayish or grayish-olive, blackish streaking on head & back; dusky wings & tail; grayish-olive lesser covs. Below: buff-white under with dusky streaks.

Red Crossbill (*Loxia curvirostra*)   Plate 32

Little known; nesting may occur mid-Dec. to early Sept. (or may skip a year), depending on conifer seed crops; may see juv all months except Jan./Feb. **Adult.** M 23.8–45.4 g, F 23.7–42.4 g; (AZ) 29.2–45 g; 14–20.3 cm (5.5–8 in.). Diet: seeds & buds (especially conifer) & insect matter in summer; young fed regurgitated seed "paste," dark in color so may contain some insects. Nest: open cup (twigs), well concealed, tree branch. **Nestling.** MC: red. GF: pale yellow, pinkish rictus. Bill: N large, uncrossed until fledge (complete by 30 d post-hatching); N yellow tip. Down: "mouse-gray" or dark gray, some on capital & dorsal. G&D: little info; d5: eyes open; d7: PinUn. **Fledgling.** 75% of Ad mass, fledges 15–25 d. Plumage: heavy gray-brown streaking all over, wings brown with olive-green or brown edges, rump yellowish, uppertail covs dark brown with olive edges, may have indistinct buffy wingbars. Below pale yellow or buffy-gray with heavy dark brown streaking.

Pine Siskin (*Spinus pinus*)   Plate 32

Adult. (PA) 11.7–14.7 g, 12.7 cm (5 in.). Diet: seeds, buds, nectar, sap. Young fed regurgitated thick "paste," greenish if it contains aphids. Nest: open cup (conifers). **Nestling.** MC: bright red. GF: yellow, bright red rictus. Bill: yellow. Skin: yellowish-orange. Down: dark gray. LF: pink. G&D: d0: 1.1 g; d3: PinVis; d3–4: eyes open; d4: prim PinF 3 mm; d6: dorsal PinUn; d7: body PinUn; d8: 10 g, perching; d10–11: FF; d12: 12.2 g. **Fledgling.** Fledges 13–17 d. GF: yellow, pink rictus remains. Bill: horn-colored culmen, pinkish at base, olive lower. Plumage above: olive-brown tinged with yellowish or buff edges, olive-brown tail & wings, 2 wingbars. Below: pale yellow with fuscous streaks. Dependent 3 wks post-fledging. *Additional source: Perry 1965.*

Lesser Goldfinch (*Spinus psaltria*)   Photos 47–49

Adult. (CA) 8.0–11.5 g, 11.4 cm (4.5 in.). Diet: seeds, mostly thistle, few insects. Young fed regurgitated seed material & insects (aphids mostly). Nest: open cup, fork in bush or tree or on branch; young evacuate on nest rim. **Nestling.** MC: H pink, then orange or orange-red. GF: pale yellow or cream outside lined yellow inside, pinkish rictus. Bill: N tan, darker on top; short, shape similar to House Finch. Skin: pink. Down: drab-gray to buff; one row of grayish down over each eye; another row crosses back of head forming a triangle. LF: short, pink, stubby. G&D: d0: < 1 g; d3–6: PinEm; d5–7 PinUn; d9: FF; d12: ~6 g, wings & tail short & stubby. Voc: N begging *cheep* (5 in a row). **Fledgling.** Fledges 11–15 d (not well known), fledge as group then join flocks. Plumage similar to Ad F DefB but above tinged buffy-brown, buffier wingbars, outer rect white patch indistinct in M (F with little or no white patch). Below buffy-yellow. Dependency post-fledge: may be wks. *Sim sp: see Table D-3 (Appendix D).*

Lawrence's Goldfinch (*Spinus lawrencei*)

Adult. (CA) M 8.8–12.5 g, F 9.8–14.3 g; 12 cm (4.75 in.). Diet: variety of plant seeds, few insects. Young fed regurgitated seed "paste." Nest: open cup. **Nestling.** MC: presumed red range. Skin: dark pink on head. Down: on capital & dorsal. G&D: unknown. **Fledgling.** Fledges 13–17 d. Plumage above: olive-brown crown, olive rump, blackish-brown tail; wings dark gray-brown with yellow edges. *Sim sp: see Table D-3 (Appendix D).*

American Goldfinch (*Spinus tristis*)   Plate 32

One of the last passerines to nest in temperate zone. **Adult.** (summer) 11–13 g, (winter) 13.5–20 g; 11.4–12.7 cm (4.5–5 in.). Diet: seeds (few insects), young fed regurgitated seed "paste." Nest: cup; young evacuate on nest rim. **Nestling.** MC: pink to pinkish-red. GF: creamy-yellow, pinkish rictus. Bill: H yellow-orange. Skin: yellow, appears bright pink. Down: pale grayish on dorsal (5–6 mm) & abdomen (2 mm); forms horseshoe shape from frontal tract to occipital, back to frontal. LF: H short, pinkish-tan. G&D: d0: 0.9–1.5 g, length ~29 mm; d1–2: 1.7–2.6 g; d3: 3.5, eye slits; d4: 5 g, alar PinEm; d5: 6 g, body PinEm; d6: 7 g, ventral PinUn; d7: 8.3 g, eyes open, most PinUn; d8: 9 g; d9: ~10; d9–12: feathering distinct from other goldfinches; d10–13: 10–12 g. Voc: faint, high-pitched *trill* begging calls, become louder by d8. **Fledgling.** Length ~6.3 cm, fledges 11–17 d (branchers), joins flocks. GF: pinkish rictus (newly fledged). Bill: short, somewhat pointed, tomia rolled inward at base. Bill & LF: dark grayish-brown. Plumage above: bare face; M darker than F, brown overall, olive-brown rump, blackish-brown wings & tail, peach to buffy wingbars & feather edges. Below: varying shades of yellow, throat of M becomes brighter with age. Dependent ~3 wks post-fledging. *Sim sp: see Table D-3 (Appendix D).*

LONGSPURS

(Family: Calcariidae)

Adults. Medium-size songbirds. Diet: invertebrates (insects and spiders) + seeds. Nest: cup in a hollow on or near ground. Bill: conical, angulated commissure, lower mandibular tomia usually rolled inward, rictal bristles usually present. Legs/feet: very long hallux nail, longer than middle toenail. Plumage: brown, gray, and white with chestnut, orange, or black accents, some with bold markings. Wings: 9 primaries, long. Tail: varies, short to fairly long. **Young.** Mouth: reddish to orange-red. Down: buffy or tan. Fledge: 7–15 days, can run well.

Chestnut-collared Longspur (*Calcarius ornatus*)

Adult. (SK, Canada) 19.3–22 g, 15.2 cm (6 in.). Diet: insects & seeds. Nest: depression on ground lined with grasses. **Nestling.** MC: orange-red. GF: white. Skin: appears yellow, orange, or dull red. Down: buffy-gray. G&D: d0: 1.7–2.8 g; d1: PinVis; d4: 6.43 g, alar PinEm; d6–7: eyes open; d7: 12 g; d8: appears FF. **Fledgling.** ~15 g, fledges 7–15 d. Plumage sim to Ad F DefB but crown appears streaked whitish, chin/throat with brown speckles, prominent streaks on breast & flank.

Thick-billed Longspur (*Rhynchophanes mccownii*)

Adult. (SK, Canada) M 26.7 g, F 24.7 g, 15.2 cm (6 in.). Diet: seeds & arthropods. Nest: depression on ground lined with grasses. **Nestling.** MC: bright pink. Skin: yellowish or reddish tinged. Down: buff. LF: pale yellow. Bill: grayish. G&D: d0: 1.6–2.9 g, ~3 cm long; d3–5: body PinEm, eyes open; d7: appears FF. **Fledgling.** Fledges d7–11, can run & flutter wings, d12 flies short distances.

SPARROWS, NEW WORLD

(Family: Passerellidae)

Identifying nestling and fledgling sparrows is challenging because of plumage similarities and indistinct facial markings, most have streaky breasts (whereas adults may have no streaks), multiple species occur in the same location, and individual variation. Newly fledged birds may be lacking facial feathers as they are the last to appear. Attention to subtle differences between species is key: feather fringing of flight feathers, tail markings, bill size, and developing crown & facial markings (Lai et al.

2017). **Adults.** Small to medium songbirds. Diet: in nesting season insects gleaned from ground or vegetation; seeds all other seasons. Many sparrows double-scratch in leaves for food. Nest: open cup on ground or in vegetation, a few domed. Eggs: subelliptical, white, tan, or bluish with markings that are plain or lightly mottled on wide end or densely covering egg. Bill: of various sizes depending on diet; conical, short, stout; commissure angulated; lower mandibular tomia rolled inward; nasal fossa somewhat triangular. Plumage: rictal bristles usually present; drab browns and mostly streaked, some boldly patterned, some with distinct head and facial patterns; sexes usually alike. Wings: 9 (visible) primaries, variable length & shape but generally rounded. Tail: extremely variable length and shape. Legs/feet: tarsi variable in length; hallux nail distinctly longer than middle toenail. **Young.** Mouth: pink to reddish. GF: cream to yellow. Bill: conical; nares small, nearest to base of forehead. Skin: pink to orange or reddish. Down: of various colors and amounts. Legs develop faster than wings in ground-nesting species. Head shape dorsal view: wide, rounded; side view somewhat flattened, bulging eyes. Iris: brown to black. Fledge: 7–15 days. Biparental care: independent 21–35 days post-fledging. Song learning known to occur in several species; young males learn songs from fathers and neighboring males.

Cassin's Sparrow (*Peucaea cassinii*)

Adult. (AZ) 14–23.5 g, 12.7–15.2 cm (5–6 in.). Diet: insects (nesting season), seeds other seasons. Nest: cup on ground or low shrub. **Nestling.** MC: dark red, 2 yellow lines on upper palate, 2 yellow spots on floor of mouth. GF: yellow, prominent. Bill: H brownish-gray. Skin: orange-pink, turns darker d4–5. Down: light gray or very dark. G&D: d0: 1–2 g; d1: 2–2.5; d3: 3–10, eyes open; d6: 11–12.5 g; d5–7: PinUn contour, head & flight feathers; d9: 11–12.5 g. Voc: high pitched *peeping*. **Fledgling.** Fledges 7–9 d, short flights at first. Plumage like Ad but buffier. Above: dull brown with dark brown central streaks & buffy edges, indistinct wingbars. Below: gray-buff to cream-buff with whitish belly; throat & flanks heavily streaked dusky-brown, darker on sides & flanks. Tail: nearly fully grown by d14.

Bachman's Sparrow (*Peucaea aestivalis*)

Adult. (AR) 20.8 g, (SC) 18.4 g; 12.5–15.2 cm (4.9–6 in.). Diet: seeds & insects. Nest: open cup on ground. **Nestling.** MC: red. GF: yellow. Bill: yellowish, darker on upper mandible. Skin: orange-pink. Down: light grayish or light drab. G&D: d0: 1.85–3.5 g; d1: 3.76, prim PinEm; d2–3: eyes open; d5: 11 g, PinVis; d4–6: most PinEm; d5: 11–12 g; d7–8: nearly FF; d9: 14.45 g. Voc: soft *peep*; *chitter* call. **Fledgling.** 70% of Ad mass, fledges 9–10 d, cannot fly. Plumage above: head & back gray-brown with tan or rust edgings (appears streaked). Below: whitish with heavy black streaking on throat & breast. Wings: gray-brown with white, buff, or rust edges. Outer prim & rect with narrow, tapered tips.

Grasshopper Sparrow (*Ammodramus savannarum*)   Plate 32

Populations have declined by nearly 70% since 1970. Florida ssp is endangered. **Adult.** (AZ) 13.4–28.4 g, (FL) 17.2–18.4 g, 12.7 cm (5 in.). Diet: insects, especially grasshoppers (summer), seeds (winter). Nest: cup, domed, ground. **Nestling.** Bill: Hook nose effect from flat-headedness. Down: grayish-brown above. LF: dusky to pinkish. G&D (MI): d0: 1.7–2.3 g, d2: 2.9; d4: prim PinUn; d6: 8.7–9.1 g; d6–7: contour PinEm, buffy crown stripe appears; d7–8: 9.7–10.5 g; d9–10: FF. **Fledgling.** Fledges 8–9 d, cannot fly, can run. Bill: upper black-slate, lower pink. Plumage above: back & rump brown, fringed buff; crown dark brown to fuscous; wing covs olive-brown to fuscous, tipped buff-white; wings light fuscous to dark brown, sides of head appear mottled with fuscous streaks/spots on buff-brown. Below: dull yellow or buff-white, throat & upper breast (broadly) streaked fuscous, lighter on sides & flanks. LF: pale pinkish-tan. Iris: hazel.

Black-throated Sparrow (*Amphispiza bilineata*)

Adult. (AZ) 10.2–16.4 g, 14 cm (5.5 in.). Diet: insects (spring/summer), seeds (other seasons). Nest: cup, low in bush or cactus. **Nestling.** MC: reddish or pinkish. GF: ivory. Down: sparse, white with some amount of gray, all tracts. G&D: d0: 1.7 g; d1: alar PinE; d2: down turns light gray; d3–5: eyes open, caudal & most contour PinEm; d6–7 FF. **Fledgling.** 11.3 g (80–90% of Ad mass), fledges 9.5–10 d, cannot fly, hops out of nest. Plumage: facial markings indistinct, more gray-brown, no black, white malar stripe, breast streaked, faint buffy wingbars. Dependent 14 d post-fledging.

Lark Sparrow (*Chondestes grammacus*)   Photos 50-51

Adult. (CA) 24.7–33 g, (UT) 23–31 g, (KS) 27–33 g; 16.5 cm (6.5 in.). Diet: seeds most seasons, arthropods (especially grasshoppers) spring/

summer. Nest: open cup usually on ground, sometimes in mockingbird or thrasher nests. **Nestling.** MC: dark pink to cherry-red. GF: whitish, (yellowish tinge), somewhat prominent. Bill: N dark, pointed. Skin: salmon. Down: gray-brown all over, including abdomen. G&D (based on one e. US bird): d0: 2.2 g; d1: 3.9; d2: 5.8, alar PinVis; d3: 8.7 g; d4: 11, rect PinUn, eyes open; d5: 12.9 g; d6: 14; d6: rem PinUn; d7: 12.5 g, FF (facial pattern dull, outer feathers with white); d8: 11.8 g; d9: 12.5. Voc: high pitched buzzing, soft continuous trills. **Fledgling.** 22.2 g, fledges 11–12 d (if disturbed d6, will not stay in nest), short flights 8–10 d. Plumage above: facial pattern distinct, but less striking than Ad (Fig. G9, Illustrated Glossary), crown dark with whitish streaks on center, rounded tail edged with white & white at corners on outer feathers (more visible on underside). Below: dark & light streaks on breast.

Chipping Sparrow (*Spizella passerina*)   Plate 32

One of the most common and widely distributed of N.A. sparrows. **Adult.** (PA) 10.3–14.5 g, 14 cm (5.5 in.). Diet: seeds mainly, insects (spring/summer). Nest: cup, fork in tree, rarely on ground, lines nest with hair. **Nestling.** MC: pinkish, then ruby-red. GF: cream to pale yellow, decurved at corners. Bill: gray tip, N light brown. Skin: orange or dark brick-red. Down: dark gray, plentiful, dense on crown. G&D: d0: 0.98–1.4 g, d1: 2, d7: eyes open, d10: 10.5 g. Voc: begging *zeee-zee-zee-zee.* **Fledgling.** 10.6 g, fledges 8–12 d, flies short distance, sustained flight d14. Bill: drab-brown upper, pale yellow lower. Plumage above: thin dark streaks on brown crown, eyeline faint buffy, brownish eyeline from lores to back of head, mottled breast & sides, fine cinnamon streaks on tertials & rump with rusty fringes, wingbars faint & buff. Below: buff-gray, streaking thin & dark from throat to flanks. LF: short pinkish legs, white nails. *Sim sp: Field Sparrow juv nearly identical anatomically with Chipping Sparrow.*

Clay-colored Sparrow (*Spizella pallida*)

Adult. (MN) 9–15 g, 14 cm (5.5 in.). Diet: invertebrates & seeds. Young fed variety of invertebrates. Nest: cup on or above ground. **Nestling.** MC: bright orange-red, darkens d3–4. GF: cream, turns yellow d3–4. Bill: H cream, then brown d1. Skin: pinkish-tan or dull orange. Down: sparse, gray (sparser & lighter than Chipping Sparrow, sparser than Brewer's Sparrow). LF: light tanish. G&D: d0: 1.2 g, d8: 10.3 (avg of 3 chicks). **Fledgling.** Near Ad mass, fledges 7–9 d, hops & climbs but cannot fly, flies d14–15. Plumage above: dull gray, light clay, or grayish cinnamon-buff; head may be more gray with narrow olive-brown or fuscous streaks on crown & nape; indistinct pale buff supercilium; back & scapulars streaked fuscous-black; yellowish wingbars. Below: washed light buff; throat, breast, sides & flanks streaked fuscous to dark reddish-black; dark buff crissum. Dependent 8 d post-fledging.

Black-chinned Sparrow (*Spizella atrogularis*)

Adult. 9–14.8 g, 14.6 cm (5.75 in.). Diet: insects (spring/summer), seeds (winter). Nest: cup. **Nestling.** Skin: pinkish-tan. Down: none or may have buffy-drab. **Fledgling.** Fledges 10 d. Plumage above: M like Ad F but brownish-gray crown & nape, back paler gray with broad black streaks, cinnamon-buff wing covs. Below: duller, dark gray streaks; chin & throat pale gray, white belly. **Juvenile.** F like M but crown & back darker & browner, wings more reddish, chin & throat tinged buff.

Field Sparrow (*Spizella pusilla*)

Adult. (PA) 11–15.6 g, 14.6 cm (5.75 in.). Diet: seeds (90% winter, 50% summer) & insects. Nest: open cup on/near ground, lined with grasses. **Nestling.** MC: bright red or deep pink. GF: pale yellow. Bill: pinkish, gray near tip. Skin: light orange or pale pinkish-tan. Down (variable): from light to dark gray, mouse-gray, or wood-brown, most tracts. LF: pinkish. G&D: d0: 1–1.4 g; d2: 3.5; d4: 6.44; d6: 9.7; d7: 10.2 (MI, weighs less in IL & PA). **Fledgling.** Fledges 7–8 d, flies short distances. Plumage like Ad but duller (no red-brown), narrow streaking on crown, white wingbars. Below: whitish tinged buff with narrow dusky streaking (may look like spots or flecks) on chest & sides. Tail: outer 2 pairs white. LF: light pink to salmon. *Sim sp: Chipping Sparrow juv nearly identical anatomically with Field Sparrow.*

Brewer's Sparrow (*Spizella breweri*)

Adult. (ID) 10.9 g (avg), 14 cm (5.5 in.). Diet: small insects & seeds. Nest: cup. **Nestling.** MC: red. GF: bright yellow. Skin: dark slaty-gray. Down: light drab to dark gray, most tracts. LF: pale pinkish. G&D: d0: 0.7–1.9 g. **Fledgling.** Fledges 6–9 d, cannot fly. Bill: thin. Plumage like Ad but less gray, crown streaked, rump spotted fuscous. Breast & flanks with fine brown streaks. Dependent several days post-fledging.

Fox Sparrow (*Passerella iliaca*)

Adult. (CA) 21.7–42 g, (PA) 30.2–39.2 g; 15.2–18.5 cm (6–7.3 in.). Diet: omnivorous, diet shifts seasonally, insects (mainly spring/summer); young fed insects only. Nest: cup on ground or in tree. **Nestling.** MC: presumed red. GF: yellow. Down: present but not described. G&D: unknown, d0: 2.4 g. **Fledgling.** 25–26 g, fledges 9–11 d, hops, cannot fly. Bill: upper & lower mandible color varies by age & ssp; in Sooty ssp, lower mandible orange or yellow at base; thick-billed ssp gray-green base. Color changes as birds age. Plumage like Ad DefB but browner & duller with paler edges. Below: buffy wash with dark heavy streaking. LF (varies by ssp): yellow, pink, or pinkish-brown (Fresno County, CA).

Dark-eyed Junco (*Junco hyemalis*) Photo 52

Considerable variation, 15 ssp divided into groups (Pyle 1997b, Billerman et al. 2020). **Adult.** (PA, *J. h. cismontanus*) 15.8–20.2 g; (PA, *J. h. carolinensis*) 18–23 g; (AZ, *J. h. mearnsi*) 15.5–23.5 g; (AZ, *J. h. dorsalis*) 18–26 g; (AZ, *J. h. caniceps*) 18–23 g. Overall length: 14.5–16.5 cm (5.7–6.5 in.). Diet (omnivorous): mostly vegetable matter (especially in winter), young fed winged insects & larvae/nymphs. Nest: cup. **Nestling.** MC: deep pink to red. GF: pale yellow to bright yellow, d30–50 flanges barely detectable at commissure. Bill: relatively thick, not prominent, darkens by d3; nares at base, not covered. Skin: pink to reddish with yellow-orange tint. Down: dark gray above, dense along spine & above each eye, long (~5–7 mm). Head shape: flat, wide. LF: N long (relative to body) pinkish-gray legs (tarsus front), cream (tarsus back), translucent nails. G&D: *J. h. mearnsi* d0: 1.68 to 1.94 g. *J. h. carolinensis* (weights d0–d11): d0: 2.3 g; d1: 2.7; d2–6: eyes open; d3: 7.4 g; d4–5: PinEm dark on crown, dorsal & breast; buff on ventral & femoral; d6: 14 g; d11: 17. Voc: several successive, very high-pitched rapid "trill," on the softer side, somewhat insect-like. **Fledgling.** 17–19 g, fledges 9–12 d (or sooner), flutter-flies, flies well d23–24. Plumage browner than Ad, some with pale wingbars; finely streaked breast. Tail: r6 all white, r4 & r5 less white than Ad. Care: self-feeding d25–26, joins flock after d30.

White-crowned Sparrow (*Zonotrichia leucophrys*)

Adult. Mass varies by ssp (CA, *Z. l. gambelii*): 21.0–28.5 g; (CA, *Z. l. nuttalli*): 27–35.5 g; (CA, *Z. l. pugetensis*): 21.4–29 g; (OR, *Z. l. oriantha*): 23.3–33.7 g; (PA, *Z. l. leucophrys*) 21.6–38.5 g; 17.8 cm (7 in.). Diet: insects (especially spring/summer) + seeds & berries. Nest: cup. **Nestling.** MC: pink then bright red. GF: yellow. Skin: yellowish. Down: brownish-gray. LF: pale pink; nails horn-gray. G&D (Banks 1959): d0: 2–2.8 g, PinVis some areas; d1: 2.8–4.4 g, stretch nearly erect; d2: 3.2–7.3 g, prim PinEm; d3: 4.4–10 g; d4–5: 6.5–15.6, crown PinEm, light-colored PinEm crural tract, PinUn upper ventral, eyes open; d6: 12.3–18.8 g: most PinUn, wing & tail cov PinEm, alarm call; d7: 13.2–21.6 g, prim & sec PinUn, ventral PinUn; d8: 14.8–23.5 g, light central stripe appears toward nape; d9: 14.5–22.7 g, appears FF, crown stripe evident, ventral feathers cover apterium. **Fledgling.** 19–21 g (geographic variations slight), fledges 8–10 d. Plumage above: crown stripe dull rufous-brown, pale buff midcrown stripe & buff supercilium. Below: breast & belly buff streaked black. *Song learning: must learn song from Ad, sensitive period between 8 and 50 days post-hatch (Marler 1970).*

Golden-crowned Sparrow (*Zonotrichia atricapilla*)

Adult. (BC, Canada) M 31.7 g, F 32.3 g; 15.2–17.8 cm (6–7 in.). Diet (omnivorous): seeds, fruits, buds, flowers, arthropods; young are fed insects. Nest: cup. **Nestling.** MC: red. GF: yellow. Down: gray. G&D: d0: 3.0 g; d3: prim PinUn; d3–5: eyes open; d5: tail PinEm; d7: all PinUn. Voc: faint *peeps* d4. **Fledgling.** 23 g, fledges 9–10 d, flies 2–4 d post-fledging. Plumage: head dark brown, light buff-yellow median crown stripe; dorsum streaked black & buffy-brown. Below: light brown throat tipped dark brown, breast & flanks with brownish streaks.

White-throated Sparrow (*Zonotrichia albicollis*)

Adult. (PA) 19.2–30.4 g, 17 cm (6.75 in.). Diet: insects (summer) + greens & fruit; mostly seeds & fruits (winter), some insects; young fed insects. Nest: cup on or above ground. **Nestling.** MC: deep pink to red. GF: yellow to bright yellow. Bill: pinkish base, gray upper, lighter below. Down: H pale clove-brown upper. LF: pale pink or pale brown. G&D: d0: 2.1–3.1 g; d3–4: eyes open, prim PinEm; d5: 12.4–16.8 g; d8: 16.4–20.8. Voc: soft insect-like buzzing. **Fledgling.** 20.3 g, fledges d7–12, flies poorly until 7 d post-fledging. Plumage above: forehead, crown, nape chestnut-brown (black streaked), buffy-white median crown stripe, lores gray or dusky, superciliary line buff-white to olive-gray, white eye ring, gray auriculars. Below: chin/throat white, dusky flecks; dark malar marks. Black streaking chest, sides, flanks; white belly. Dependent 2 wk post-fledging. *Sim sp: Song Sparrow slightly smaller than White-throated Sparrow; examine facial patterns carefully.*

Sagebrush Sparrow (*Artemisiospiza nevadensis*)

Adult. (OR) M 15.9–21.9 g, F 15.3–20.9 g; 15.2 cm (6 in.). Diet: omnivorous, diet shifts seasonally. Nest: cup, in shrub. **Nestling.** MC: bright red. GF: yellow. Bill: H yellow, turns grayish-brown d2–3. Skin: orange-yellow. Down: patches of light gray above. LF: orange-yellow. G&D (OR): d1: 2.3 g, d4–5: eyes open; d8: 14.7 g. Voc: begging calls 6–7 d post-hatching. **Fledgling.** Plumage like Ad but paler, grayer, head pattern indistinct.

Vesper Sparrow (*Pooecetes gramineus*)   Plate 33

Adult. (MI) M 26.5 g, F 24.9 g; 16 cm (6.25 in.). Diet: seeds, insects (breeding season). Nest: woven cup, ground, concealed in grasses. **Nestling.** MC: deep pink. GF: white or pale yellow. Bill: pinkish or lilac, upper turns dark. Skin: pinkish, then dark. Down: thick, long, whitish or buffy, well covered. G&D: d2: wing PinEm, faint calls; d2–5: eyes open; d6: PinUn; d7: crouches; d9: 18 g. **Fledgling.** 17.5 g, fledges 7–14 d (9 avg). Plumage like Ad but drabber brown, may be heavily streaked with more black & white, wing covs little or no rufous, wingbars wide & buffier; white eye ring. Below: heavily streaked (except midbelly). LF: short legs, pinkish. Independent 20–29 d post-fledging.

LeConte's Sparrow (*Ammospiza leconteii*)

Adult. (MN) 10–16 g, 12.7 cm (5 in.). Diet: seeds & arthropods. Nest: cup near ground in grasses. **Nestling.** Skin: pale pink. Down: sparse, wood or dull brown. G&D: d0: 1.7 g, d1: 2.5, d2: 3.4, d3: 4.5. **Fledgling.** Weight and fledge period unknown. Plumage above: pale buff with dark crown stripes & dark streaking on dorsum. Below: mostly white (throat pale buff), cinnamon-buff band across breast & down sides.

Seaside Sparrow (*Ammospiza maritima*)

Endangered status. **Adult.** (NY) M 19–28.7 g, F 19.4–28.5 g; 12.7–15.2 cm (5–6 in.). Diet: insects, spiders & amphipods (spring/summer); seeds, arthropods & mollusks (winter). Nest: cup in grasses just above high-tide line. **Nestling.** Down: grayish-brown above to whitish below. G&D: unknown, d2–6: eyes open. Voc: d5 "reedy vocalization," & *scree* call on d6. **Fledgling.** Fledges: d9–10, cannot fly, can run. Plumage above: olive-brown dorsum, streaked blackish; pale buff supraloral stripe, mustachial stripe dusky, no wingbars, crown with fine black streaks. Below: whitish throat & breast (with fine dusky streaks), longer streaks on sides.

Nelson's Sparrow (*Ammospiza nelsoni*)

Adult. (ND) M 13.7–16.2 g, F 12.8–15.2 g; (ON, Canada) 15.8–20 g; 12.7 cm (5 in.). Diet: insects mainly, seeds (mainly winter). Nest: cup, may be partially domed. **Nestling.** MC: reddish-orange. GF: yellow. Skin: yellowish-orange. Down: gray to brownish-black, blackish, or grayish-brown all dorsal tracts, whitish on ventral. LF: pinkish. G&D: d0: 1–2 g, PinVis; d2: PinEm; d7: PinUn. **Fledgling.** 86% of Ad mass, fledges 10–11, can run & climb, flies 3–5 d post-fledging. Plumage sim to Ad but crown dark with orange-buff median stripe & supercilium, dark postocular streak, orange-buff rump, prim edged pale gray to olive-gray. Below: orangish-buff, variably finely streaked black.

Saltmarsh Sparrow (*Ammospiza caudacuta*)

Endangered status. **Adult.** (NJ) M 18–23 g, F 15.3–19 g; 12–12.7 cm (4.7–5 in.). Diet: arthropods mainly, seeds (mainly winter). Nest: cup in grasses in tidal marshes. **Nestling.** MC: reddish-orange or pinkish-orange. GF: yellow. Bill: H pinkish, turns dark with age. Skin: yellowish-orange. Down: thin, grayish wood-brown or mouse-gray above; whitish ventral. LF: pinkish. G&D: d0: 1.37–1.89 g; d1: PinVis; d2–6: eyes open; d6: body PinUn; d7: streaks on breast appear, more distinct d8. Voc: single *peep*. **Fledgling.** Fledges d9–10, flies poorly. Plumage like Ad but buffier tones above & below. Above: dark streaking on blackish crown, undefined medial stripe; indistinct face pattern; dark-centered feathers on dorsum with buff edges. Below: buff with heavy dusky streaks, belly whitish. Independent 20 d post-fledging.

Baird's Sparrow (*Centronyx bairdii*)

Adult. (AZ) 15–20.3 g, 12 cm (4.7 in.). Diet: insects & seeds. Nest: cup on ground in depression in grass. **Nestling.** MC: pale carmine. GF: bright naples-yellow. Bill: pinkish-gray upper, lower pale pink. Skin: bright pink, turns pale orange. Down: pale smoke-gray on 6 tracts. LF: pale pink. G&D: no mass info, egg 2.2 g; d1 PinVis; d2–3: PinEm; d3–5: eyes open; d4–5: ventral PinUn, light colored; d7: PinUn; d9: dark "eye spot" evident (unique to Baird's). **Fledgling.** Fledges: 8–11 d, flies 13 d. Plumage above: scaly appearance; back black with pale buff edges, black rump

& rect edged ochreous-buff, scapulars with buff edges, prim & sec covs tipped buff. Below: more heavily streaked than Ad.

Henslow's Sparrow (*Centronyx henslowii*)

Adult. 10.6–15 g, 12.7–14 cm (5–5.5 in.). Diet: mostly insects, some seeds. Nest: cup near ground in grasses. **Nestling.** Down: smoke-gray to buffy or brownish-gray. G&D: d0: 1.5 g, d5–6: PinUn, d6: 9.8 g. **Fledgling.** Fledges: 9–10 d. Plumage like Ad but facial markings weak (lateral throat stripe sometimes absent). Above: crown with two broad lateral blackish stripes (dark spots or streaks); dull yellow lores, auriculars brownish to gray; back & scapulars with pale fringes creating scalloped effect; wings light fuscous or olive-brown with brownish to cinnamon barring. Below: buffy, no dark streaks, more olive on lower throat.

Savannah Sparrow (*Passerculus sandwichensis*)   Plate 33

Geographically variable especially with amount and color of streaking. **Adult.** 20 g (avg, varies by ssp); (CA) 15–17.8 g; (MB, Canada) M 20.6 g, F 19.5 g; 14 cm (5.5 in.). Diet: insects & spiders (spring/summer), seeds & fruits all seasons. Nest: cup. **Nestling.** MC: pink. GF: cream or yellow. Bill: yellow. Skin: yellowish-orange. Down: sparse, long, mouse-gray or dull grayish-brown to black (d5). G&D (NL, Canada): d0: 1.9–3 g; d1: 3.6; d2: 5.8; d3: 7.5, prim PinEm; d4: 10.1; d4–5: eyes open, PinEm legs, back, belly; d7: can tell sex; d8: 18 g; d9: 17.7. Voc: faint lisping note. **Fledgling.** 15.4 g, fledges 7–9 d, cannot fly, first flight 8–14 d. Plumage: two color morphs predominately gray or brown, variable amount of yellow; crown flecked dark brown, white & chestnut; eyebrow stripe indistinct or absent, little to no yellow; broad chestnut edges on sec, wing covs & scapulars. Below: buffy or yellowish (paler belly), breast dark streaked. Tail: short & slightly forked. LF: dusky to bright pinkish. Independent 15 d post-fledging.

Song Sparrow (*Melospiza melodia*)   Plate 33

The number of subspecies, depending on how they are classified, varies from 24 to 38 and is among the highest of N.A. birds. High degree of variability in all nestling & fledgling features from mouth color to plumage. **Adult.** (CA, varies by ssp) 18.6–21.6 g; (PA) M 18–22.6 g, F 17.1–23.4 g; 16 cm (6.25 in.). Diet: omnivorous, shifts seasonally, insects (spring/summer). Nest: open cup. **Nestling.** MC: red (ruby), salmon, or pink. GF:

white to yellow, brighter on inside at rictus. Bill: H orange-gray, some brown at tip, pointed. Skin variable: yellow or light orange with pink; (n. CA) *M. m. gouldii*: bright orange with pink. Down: variable, black, sepia-brown, light or dark gray on dorsum. Face: bare. LF: relatively short legs; H pink to yellow-orange. G&D (see Table C-11, Appendix C): d0: 1.5 g (egg mass 1.8–2.85). Voc: quiet *chirp*; d7–9: scream, new begging & location call. **Fledgling.** 17.8 g, fledges 9–12 d, flies well 16–17 d. Plumage like Ad but less distinct facial stripes. Above: wings dark, variable dull to rusty edging on sec; olive-brown tail. Below: dull white, broad malar streaks, throat unstreaked, streaked undersides (paler than Ad). LF variable: grayish-brownish, some with soles dull to bright yellow. Independent 22–32 d post-fledging. *Sim sp Swamp Sparrow and Lincoln's Sparrow: look at soles; Song Sparrow has large, stouter bill, dark crown; Swamp rect (more pointed) & sec more rusty with redder edging; Lincoln's has indistinct gray supercilium. Song learning: must learn song from Ad, sensitive period between 5 & 10 wks (Mulligan 1966), "singing" appears 13–15 d, "crystalized" song 120 d.*

Lincoln's Sparrow (*Melospiza lincolnii*)

Adult. (PA) 14.3–19.4 g, 14.6 cm (5.75 in.). Diet: insects (spring/summer), small seeds (winter). Nest: woven cup concealed on ground or elevated. **Nestling.** MC: red to bright red. GF: H whitish, then bright yellow d2. Bill: upper dusky, lower yellowish. Skin: orange-pink or dark reddish-buff. Down: grayish-black or dark blackish-gray. LF: dark. G&D: d0: 1.4–1.9 g, d2: PinEm; d4–6: eyes open, d6–7: contour PinUn. **Fledgling.** 14.7 g, fledges 10–11 d, cannot fly, hops, flaps wings, flies 6 d post-fledging. Plumage much like Ad but crown stripe & markings indistinct, mottled nape, rusty edges on wing covs. Below: throat & belly white. Tail: short. *Sim sp: see "Song Sparrow."*

Swamp Sparrow (*Melospiza georgiana*)

Adult. (PA) 13.1–19.2 g; 14.6 cm (5.75 in.). Diet: mainly arthropods (spring/summer); seeds, fruits & aquatic invertebrates (winter). Nest: cup of grasses over or near water. **Nestling.** MC: H pink, N bright or reddish-orange. GF: pale yellowish, turns more yellow. Skin: pink. Down: sepia-brown to blackish-brown, all tracts (some white on ventral?). G&D: d0: 1.37–1.55 g, d4–5: eyes open; d5: PinUn spinal tract; d7: PinUn femoral and ventral; d8: PinUn capital. **Fledgling.** Fledges 9–11 d, flies d14, fed until 15 d post-fledging. Bill: narrow, dark. Plumage sim to Ad DefB.

Above: crown tinged olive, median crown stripe pale, olive-gray super-ciliary line, blackish postocular stripe. Black wings & tail edged chest-nut, wing covs tipped rusty, pointed rect. Below: olivaceous wood-brown (may or may not have streaks), brownish peppering on flanks & belly. LF: buff, turn dark. *Sim sp: see "Song Sparrow."*

Canyon Towhee (*Melozone fusca*)

Adult. (AZ) 36.6–52.5 g, (TX) M 46.5 g, F 44.6 g; 23 cm (9 in.). Diet: seeds primarily, few insects. Nest: open cup in tree or shrub. **Nestling.** GF: yellow. Down: brown. LF: very long. G&D: unknown. **Fledgling.** Flies ~9 d. Plumage above: mostly dull gray-brown with wing covs edged pale fulvous, crown finely streaked cinnamon. Below: dull white with fine dark & dusky streaking, lacks buffy throat & rufous crown of Ad, crissum ochre to orange-rufous.

Abert's Towhee (*Melozone aberti*)

Adult. (AZ) M 40–54 g, F 39.5–51 g; 24 cm (9.5 in.). Diet: insects primar-ily (all seasons) & seeds. Nest: cup. **Nestling.** MC: red. GF: creamy (lined yellow). Bill: pink. Skin: pinkish. Down: grayish-brown. G&D: d0: 3.6 g, d6–7: 14–19. **Fledgling.** 32.8 g (70% of Ad weight). Fledges 11–13 d on foot. Plumage above: drab, rump & uppertail covs tinged rusty. Below: sides of chin with dusky malar marks, black spots across breast; rust on flanks, lower belly & legs. Independent 4–5 wks post-fledging.

California Towhee (*Melozone crissalis*)   Photos 53-55

Adult. 37–67 g, 23 cm (9 in.). Diet: seeds (51%), grains (28%), insects (15%), fruit (5%); young fed mostly grasshoppers & caterpillars. Nest: cup. **Nestling.** MC: deep pink to red. GF: cream to pale yellow, thin, not prominent. Bill: pointed. Skin: pink or dark gray. Down: long, brown-gray, darker by d5. Gray down on head along superciliary line longer like a "crown." LF: very long, pale brownish. G&D: d1–4: 10–14 g; d3–5: PinVis; d5: 17g (length 21–25 cm, no tail), eyes open; d5–7: 20–33 g; d6–8 PinUn, 8 d cowers; d10–13: 25–35 g, rect 2 cm long. Voc: high-pitched, repeated, like crickets, then single sharp, high-pitched *chink*. Bobs & weaves while begging. **Fledgling.** 35–45 g (50 g good release weight), fledges 6–11 d (varies by ssp), runs, hops, flies poorly. Plumage above: grayish-brown (cinnamon tinged), brown crown, pale buff wing-bars. Below: streaked, rich cinnamon crissum. Tail only 2 cm long.

Rufous-crowned Sparrow (*Aimophila ruficeps*)

Adult. M 16–23.3 g, F 15–20.3 g; 15.2 cm (6 in.). Diet: insects (spring & summer), seeds (winter). Nest: cup, on ground. **Nestling.** Skin: orange. Down: sparse, black. G&D: unknown. **Fledgling.** Fledges 9 d, cannot fly. Plumage like Ad but more buffy, dull brown (streaked dark brown), dusky-brown streaking on breast & flanks; brownish malar stripe not distinct.

Green-tailed Towhee (*Piplio chlorurus*)

Adult. (CA) M 23.5–39.4 g, F 22.5–33.2 g; (AZ) 21.5–37.3 g; 18.4 cm (7.25 in.). Diet: seeds (small insects, some fruit). Nest: cup. **Nestling.** MC: bright red. GF: yellow. Skin: light pinkish-tan, turns bluish-gray by 12 hrs. Down: sparse gray above. G&D: d0: 3–4.1 g; d6: PinEm; d6–8: eyes open; d8 prim PinEm; d10: FF. **Fledgling.** Fledges 11–14 d, cannot fly, can run & climb. Bill: light slate. Plumage above: light brownish-olive (or rufous-brown), broad streaking all over, olive to pinkish-buff wingbars. Below: dull or buffy-white.

Spotted Towhee (*Pipilo maculatus*)   Plate 33

Mainly ground dwelling. **Adult.** (CA) M 36–43.8 g, F 34.3–41.4 g; (AZ) 30.6–45.5 g, 21.6 cm (8.5 in.). Diet: plant matter (76%), insects (24%), some fruit. Nest: cup on ground in litter or elevated in vegetation. **Nestling.** MC: pink to red or scarlet-orange. GF: pale yellow. Bill: pinkish-tan. Skin: bright pink, fades d4. Down: sparse, bluish-black or grayish. LF: very long. G&D: mass may be sim to Eastern Towhee based on Ad mass; d0: PinVis most tracts; d1–2: prim PinEm; d3: rem PinEm; d5–7: eyes open; d7: nearly all PinUn; d8: tail (~5mm) PinUn, down wisps only. **Fledgling.** Fledges 9–11 d, fed by Ad for 30 d. GF: cream. Plumage above: dorsum variable yellowish-brown to buffy-brown (newly fledged in n. CA appear darker grayish all over) & broadly streaked (less on crown), buffy "spots" on wings & covs more distinct in interior versus coastal subsp. Below: lighter & streaked dusky-brown to blackish, appears mottled. Tail: stub at first; white tail corners on outermost rect (varies by ssp). LF: long, grayish-brown or light brown. Dependent 3–4 wks post-fledging. *Song learning: learns father's song, then after establishing own territory the following spring, learns songs from neighboring M.*

Eastern Towhee (*Pipilo erythrophthalmus*)   Plate 33

Northern populations migratory, southern populations resident. Mainly ground dwelling. **Adult.** (PA) M 34–46.6 g, F 34.8–44.8 g; 21.6 cm (8.5 in.). Diet: omnivorous; arthropods (winter), mostly insects (spring/summer) + seeds & fruits (late summer/fall). Nest: open cup on or above ground. **Nestling.** MC: red, fades to pink. GF: bright yellow, fades to pale yellow. Bill: N pinkish-buff. Skin: pinkish. Down: sparse, drab grayish, all dorsal tracts. LF: very long (Ad length by d8). G&D (Barbour 1950): d0: 3.2 g; d1: 6.5 (41.3 mm long), PinVis all tracts except ventral; d2: 9.8 g, ventral PinVis, PinEm body, prim & sec; d3: 12.7 g, spinal PinEm; d4: 15.2 g (70 mm), cowers in nest; d4–6: eyes open; d5: 17.7 g; d5–6: rect PinUn; d6–7: 21.2–22.5 g (75–85 mm), PinUn all dorsal & ventral tracts; d8: 86 mm long, no traces of down. **Fledgling.** 24.7 g, fledges 10–11 d. Plumage above: cinnamon-brown, heavily streaked, black feathers unspotted. Below: buffy, heavily streaked; darker & distinct in M, less distinct in F, indistinct wingbars. Tail: stub at first, then rounded, M with brownish-black to black rect (F browner); tail corners tipped white on r3–6 or r4–6. LF: pinkish-buff. Iris: light brown, turns reddish-brown (by winter). Dependent 3–4 wks post-fledging. *Song learning: nonmigratory birds learn similarly as resident Spotted Towhees.*

YELLOW-BREASTED CHAT

(Family: Icteriidae)

Only one species represents this family; it raises its young throughout the US (and parts of Mexico).

Yellow-breasted Chat (*Icteria virens*)   Plate 34

Adult. Medium-size songbird. (PA) M 22.3–25.8 g, F 23.7–27.7 g; 19 cm (7.5 in.). Diet: insects & berries (about equal); young fed arthropods. Nest: cup, dense shrubs. Bill: medium-long, thick, strongly decurved culmen, nares not covered. Wings: 9 (visible) prim, rounded. Tail: long, rounded. LF: smooth, rounded in front. **Nestling.** MC: pink-red with yellowish areas. GF: yellow or cream & thin. Bill: large, grayish. Skin: pink or dull yellow. Down: none. LF variations: blue, slate, olive-brown, black. G&D: little info, (IN) d0: < 5 g, d3: all PinVis. **Fledgling.** 16.3 g, fledges 7–10 d. Plumage above: grayish-brown tinged olive, head olive, narrow eyebrow; wings & tail olive-brown or olive-green. Below: paler olive-gray, white chin & throat (tinged yellow); olive-gray band on breast, brownish flanks;

yellow appears with pre-basic molt on natal grounds before migration. LF (variable): blue, slate, or olive-brown.

BLACKBIRDS AND RELATIVES

(Family: Icteridae)

Adult. Medium to large songbirds, females generally smaller than males and less colorful. Diet: mainly insects (spring/summer), fruits and grains other seasons, and nectar (orioles). Nests: bulky cup on ground, in aquatic vegetation, or in trees, some in colonies. Orioles weave basket-shaped nests. Cowbirds do not build a nest but parasitize nests of other species. Eggs: long elliptical to short oval, white to blue, with strong markings. Bill: medium to medium-long (orioles range short to long, pointed, slightly hooked), seldom longer than head; somewhat conical, strong, straight, slender, sharply pointed (acute); culmen sometimes slightly decurved, elevated toward base and extending far back to part the feathers of the forehead, commissure somewhat angulated; rictal bristles lacking. Plumage: many are iridescent black, some have bright red or yellow markings. Orioles are mostly yellow or orange with black and white markings. Wings: 9 (visible) primaries, variable in length and shape, usually long and pointed. Tail: variable in length and shape, short to medium (always more than half as long as wings, never conspicuously longer, except in some orioles and grackles). Legs/feet: very strong, tarsi usually equal to, or slightly longer than, middle toe with nail; black. **Young.** Mouth: pale pink at hatch, then ranges deep pink to cherry-red. Gape flanges: relatively thin, may be swollen only at commissure (blackbirds), mainly white or yellow. Bill: relatively long and pointed, some with prominent operculum. Skin: varies in ranges of pinks, salmon, reds, yellows, oranges. Down: scattered or abundant on head, back, wings, thighs; variable shades of white, buffy or tawny, and grays (light gray to blackish-gray). Leg colors in orioles may be bluish. Fledge: 8–17 days, varies by species. Care: one or both parents.

Yellow-headed Blackbird (*Xanthocephalus xanthocephalus*)   Plate 34

Adult. (OR) M 72.5–85.5 (or more) g, F 42.4–56 g; 24 cm (9.5 in.). Diet: aquatic insects spring/summer, cultivated grains & weed seeds fall/winter. Nest: cup. **Nestling.** MC: bright red & pink. GF: yellowish-white. Bill: N brownish with white egg tooth. Skin: H yellowish-pink to salmon-pink, then gray-blue. Down: tawny or buff. LF: H reddish-pink. G&D: d0: 2.6–4.8 g (3.7 avg); d10: M 45–55, F 30–40; d3: eyes open; d8–9: FF.

Fledgling. 45 g (avg), fledges 9–12 d, climbs around vegetation until able to fly (~20 d). Plumage: M have tawny backs (F darker) & white wingbars (F buff). LF: buff-tan with buff nails.

Bobolink (*Dolichonyx oryzivorus*)   Plate 34

Adult. (NY) M 33.9 g, F 29.2 g, 17.8 cm (7 in.). Nest: cup, always on ground, hay fields, grasslands. **Nestling.** MC: pinkish. GF: cream. Skin: pink. Down: sparse, buff mostly on capital & dorsal. G&D: d0: 1.7–3.96 g; d3: wing PinVis; d4: wing PinEm; d4–6: eyes open; d6: PinUn capital, ventral (yellow) & wings; d8: appears FF; d10: 22.5 g (avg). **Fledgling.** Fledges 10–14 d, cannot fly, can run d7. Plumage above: dull brownish-black with buff edges & grayish-white tips; head buff with 2 lateral dark crown stripes (median crown stripe buff). Below: rich buff, faint flecks on sides of throat & breast center.

Eastern Meadowlark (*Sturnella magna*)

Differentiating Eastern Meadowlarks from Western Meadowlarks can present problems, even when in hand. Recently fledged individuals may be impossible to separate (Pyle 1997b). **Adult.** M 104–131 g, F 86–94.2 g; (FL) 102 g; 21–24 cm (8.3–9.5 in.). Diet: insects, seeds in winter (+ some fruits). Nest: cup of grasses (with roof or tunnel), ground. **Nestling.** MC: pinkish or red. GF: bright yellow or ivory. Skin: pinkish-orange or orange-red. Down: pale or "pearl-gray," thickest on back. LF: pinkish-buff. G&D: little info, d0: ~5 g; d3: PinVis; d4: eyes open; d6: PinEm; d10: FF. **Fledgling.** ~59 g, fledges 10–12 d, runs well, takes short flights, sustained flight by d21. Bill: pinkish-buff. Plumage above: feathers have whitish or buffy margins, stripes on head less defined, malar mark mostly white, white on outer tail feathers. Below dull yellow, dusky spots on breast. *Sim sp: Western Meadowlark.*

Western Meadowlark (*Sturnella neglecta*)   Plate 34

Adult. (Great Plains, US) M 112 g, F 89.4 g; 24 cm (9.5 in.). Diet: insects (+ seeds); young fed insects primarily. Nest: cup in dense vegetation. **Nestling.** MC: pinkish or red. GF: bright yellow or ivory. Skin: pinkish-orange or orange-red. Down: abundant pale gray especially on dorsal tract, above eyes & occiput. G&D: d1: 7 g; d3: PinVis; d4: eye slits; d6: PinUn (yellow feathers show on chest & sides); d10–12: FF; d8: 71 g. **Fledgling.** 40 g, fledges 10–12 d, cannot fly. Nostrils: ovate, may have prominent operculum. Bill & LF: pinkish-buff. Plumage sim to Ad, but paler yellow below with dusky spots or streaks on breast (black "V" not evident), yellow malar mark; more white on outer tail feathers than Eastern Meadowlark. Dependent 2 wks post-fledging. *Sim sp: see "Eastern Meadowlark."*

Orchard Oriole (*Icterus spurius*)   Plate 34

Adult. (NE) 16–28 g, 20.3 cm (8 in.). Diet: insects (primarily spring/summer), mostly fruit, some nectar (summer), berries (fall). Nest: rounded cup, suspended, fork of outer branch. **Nestling.** MC: reddish. GF: whitish, thin (not prominent). LF: long, pinkish tan. Down: pale gray to buff. G&D: unknown. **Fledgling.** Fledges 11–14 d. Bill: long, pointed, tan, dark culmen. Plumage like Ad F; olive-green above & more yellow below, white wingbars on brownish wings. LF: bluish-gray.

Hooded Oriole (*Icterus cucullatus*)   Photo 56

Adult. (AZ) 20.6–33 g, 20.3 cm (8 in.). Diet: insects, nectar, fruit. Nest: cup-shaped woven with small opening at 1 side or top, often sewn to underside of palm frond. **Nestling.** MC: pink, then cherry-red. GF: white, thin, thicker & turned down sharply at corners. Bill: long, pointed, narrow; nares large, exposed. Down: grayish-brown. LF: grayish-tan with brown nails. G&D: little info, H: unknown, N: 15 g (avg CA), d4: eyes open. Voc: grating, throaty notes, usually 3–4 continuous (similar to Ad "chatter"). **Fledgling.** < 17 g (one bird in Nevada County, CA); 23.2 g (avg Napa County, CA). Fledges 14–16 d, partly flighted. Plumage like Ad F DefB, but duller, wingbars buffy. Above: olive-brown crown & back, no black on face. Below: olive-yellow, yellow throat (no black), whitish abdomen. Dependent ~14 d post-fledging.

Bullock's Oriole (*Icterus bullockii*)   Photo 57

Adult. (OK) 32.4–42.5 g, 23 cm (9 in.). Diet: insects, nectar, fruit. Nest: pendulum. **Nestling.** MC: deep pink, red, or orange. GF: cream to yellow; not prominent. Bill: long, pointed, narrow. Skin: pale pinkish-apricot. Down: fluffy white. G&D: little info; d7: PinEm; d10–12: 23 g, FF (yellow, gray, white underparts). **Fledgling.** 23.8 g (avg CA), fledges 12–14 d, partly flighted. Plumage sim to Ad F but duller. Above: grayish olive-yellow or pale grayish-brown, yellow wingbars & edges of wing feathers. Below: yellow or light olive breast, throat pale (M may show some black in throat). Dependent ~14 d post-fledging.

Baltimore Oriole (*Icterus galbula*)   Plate 34, Photo 58

Adult. (PA) M 30.5–39 g, F 28–35.2 g; 17–19 cm (6.7–7.5 in.). Diet: arthropods (caterpillars, insects, spiders), more fruits & nectar (winter). Nest: pensile pouch, usually near tip of tree branch. **Nestling.** MC: reddish or orange-pink. GF: white or yellow, thin (not prominent). Bill: long, pointed, narrow. Skin: pink to orange-pink, bare around eye. Down: long, thick, whitish, or light buffy-yellow, thickest on spinal (37 neossoptiles) and coronal (16). G&D: little info, d0: 2–2.63 g. Voc: chatter of notes, either on same note or rising & falling. **Fledgling.** 34 g, fledges 11–14 d. Bill: pinkish-buff to light tan. Plumage above (including rump) olive-brown with yellow-orange tinge, grayish-brown wing covs, buff wingbars, brownish-gray rem & rect. Below: yellow to orange throat, pale grayish-olive to yellow-orange breast to belly. LF: olive-gray or bluish-gray.

Scott's Oriole (*Icterus parisorum*)

Adult. (AZ) 32–41 g, 23 cm (9 in.). Diet: insects + fruit & nectar. Nest: semi-pensile basket (like Hooded Oriole) often in yuccas. **Nestling.** MC: rose-pink or pink. GF: cream or grayish, thin (not prominent). Bill: H apricot, N long, pointed. Skin: pink-apricot. Down: pale gray to buff. G&D: d0: 2 g, d7: wing PinUn, eyes open. **Fledgling.** 26–28 g, fledges 9–15 d, flies poorly. Plumage sim to Ad F but paler with fewer markings. Above: olive-gray, wingbars indistinct yellowish. Below: yellowish, light citrine-yellow throat.

Red-winged Blackbird (*Agelaius phoeniceus*)   Plate 34

Highly polygynous, sexually dimorphic in color & size. **Adult.** (PA) M 55.4–72 g, F 34.5–42.7 g; varies in CA. 15.2–25 cm (6–9.8 in.). Diet: insects (& some seeds). Nest: open cup in marsh, upland, or agricultural habitats; in colonies. **Nestling.** MC: pink or red. GF: thin, white, cream, or yellow. Bill: long, pointed, yellowish near tip. Skin: salmon-pink to orange-red (or even scarlet) to pinkish-red (some with yellowish tint). Down: scant buffy, grayish, or white on capital, dorsal (lower half), humeral, femoral, crural & abdomen. Feathered N have bald head (around eye, cheek & throat). LF: long legs, nails white, dark as fledgling. G&D (see Table C-12, Appendix C): d0: 3.8. **Fledgling.** Close to Ad mass, fledges 10–14 d, d10 climbs around reed to reed, may not be FF. GF: reduced, pinkish & downcurved. Bill: dark grayish-brownish. Plumage sim to Ad F DefB except bald face. Above: overall sepia with buffy tips, wing edges buffy-brown, white downy tufts on crown over yellow-green eye orbital & dorsal areas. Below: variable, streaking may appear (1) dark on buffy-yellow background or (2) buff-yellow or white on dark background.

Tricolored Blackbird (*Agelaius tricolor*)

Adult. (CA) M 60–79 g, F 46–54.5 g; 17.8–24 cm (7–9.4 in.). Diet: insects & grains (opportunistic foragers). Nest: cup. **Nestling.** MC: H pink, then red. GF: cream, thin (not prominent). Bill: N long, pointed. Skin: orange to pinkish-red. Down: dark grayish (darker than Red-winged Blackbird). G&D: d0: 2–3 g; d1: PinVis; d5: PinEm; d5–6: eyes open; d7–8: prim & sec PinUn; d10: head FF. **Fledgling.** Fledges 11–14 d, remains in vegetation until about 4 d, then flies to trees. Bill: paler than Ad. Plumage like Ad F DefB but paler, gray & buff. Lores are still bare at d22. Very vocal.

Bronzed Cowbird (*Molothrus aeneus*)

Brood parasite of many species. **Adult.** (TX) M 69 g, F 57 g; 20.3 cm (8 in.). Diet: seeds & arthropods. Nest: does not build nest. **Nestling.** MC: red or reddish. GF: cream or white, thick. Bill: dusky-yellow. Skin: orange-pink (becomes brownish), greenish-blue around eyes. Down: mouse-gray. LF: dusky-yellow, light yellow claws. G&D: little known. d0: 4.2–5 g; d4: wing PinEm; d5: rect PinEm, eyes open. **Fledgling.** Fledges 10–12 d. Plumage: dull sooty-black (no gloss); below paler, dusky streaking.

Brown-headed Cowbird (*Molothrus ater*)   Photos 59-60

Brood parasite of many species. **Adult.** (PA) M 43–57.3 g, F 32.3–42.3 g; (CA, *M. a. artemisiae*) M 47.5 g, F 37.6 g; (CA, *M. a. obscurus*) M 40.2 g, F 32 g; 19 cm (7.5 in.). Diet: seeds (75%) & arthropods (25%). Nest: does not build one, lays egg in nests of host sp. **Nestling.** MC: variable range of pink to cherry-red. GF: thin, color variable by geography & genetic factors: either white or yellow (in some or all populations of *M. a. obscurus*). Bill: pointed but more like finches than most blackbirds. Skin: light pinkish-tan & light rose. Down: buff-yellow to olive-gray. Head shape: "flattened" forehead from tip of bill to top of crown. LF: N long, pinkish-tan. G&D: d0: 1.8–3.1 g; d1: 4.5; d2: 7–8, eyes open; d3: 11–13 g; d3–5: PinEm; d5: 15 g; d7–8: 20–25; d8–9: may be FF with wing PinUn; d9: 25.5 g; d10–12: 29–32. Voc: continuous, high-pitched vibrating sound. Begging: rapid wing shivers. **Fledgling.** Fledges 10–11 d (range

8–13 d). Plumage like Ad F, but more streaked below, bald face & very short tail. Above: streaked & mottled appearance, sooty grayish-brown or hair-brown & whitish, wing covs tipped buffy or whitish (juv F paler).

Rusty Blackbird (*Euphagus carolinus*)

Adult. (PA) M 45.9–80.4 g, F 47–76.5 g; 23 cm (9 in.). Diet: omnivorous, invertebrates all year. Nest: bulky cup usually near water. **Nestling.** MC: presumably red. GF: yellow. Bill: N grayish. Down: long, thin, fuscous. G&D: unknown, d5–9: eyes open, d10: FF, down still present. **Fledgling.** Fledges 11 d, cannot fly. Plumage: slate washed olive-brown on dorsum & throat, tail darker (may have metallic greenish). Iris: dark.

Brewer's Blackbird (*Euphagus cyanocephalus*)   Photos 61-62

Adult. (OR) M 60–73 g, F 51–67 g; 23 cm (9 in.). Diet: insects (+ seeds & fruit). Nest: cup, on ground or in tree, some near water, often colonial. **Nestling.** MC: red or bright pink, becomes tinged grayish d12. GF: white, thin, not prominent. Bill: grayish near tip; turns black d12; becomes long & very pointed; nares raised. Skin: bright pink & yellowish. Down: nearly naked at first, then plentiful. Color varies geographically: blackish-gray or pinkish-buff to grayish-white. LF: long, pinkish; d0–1: nails white; d5: medium gray proximally, pale gray or white distally; d10–11: black. Iris: dark brown. G&D (see Table C-13, Appendix C): d0: 3.7 g. Voc: H: *peeping*, then (N) raucous, repeated call (like a rusty hinge). Begging posture: tail held approximately parallel to ground. **Fledgling.** 46 g, fledges 12–16 d (9 d if disturbed), flight fully developed 23 d, joins flocks. Plumage like Ad F DefB. M above vinaceous-brown, fuscous crown; below brownish mouse-gray. F lighter, more rufescent-brown. Iris: grayish-white (M only) d12–21.

Common Grackle (*Quiscalus quiscula*)   Plate 34

Adult. (IL) M 91–142 g, F 74–124 g; 28–34 cm (11–13.4 in.). Diet: H fed spiders; older N beetles & corn. Nest: cup, typically high in conifer. **Nestling.** MC: bright pink or red. GF: whitish to yellowish, mod flared. Bill: large, long, dark, very wide gape; nares are "bumps." Skin: pinkish-apricot, N bare yellowish-apricot face. Down: thick sepia-brown above, sparse pale gray on belly. LF: very long, sepia-brown, nails pale yellowish. Iris: brown. G&D: d0: 5.6–6.7 g (or 3.5–7 from another source); d1: 10–11.8; d2: alar PinEm, eye slits; d3: 17.6 g; d5–6: 25–32, wing PinUn;

d7: 36.7 g, eyes fully open; d9: 43 g; d10–13: contour PinUn, smoke-gray to black; d12: 50.3 g; d14: 63, one-half of face covered with feathers; d15: 69 g; d17–18: 75–80; d20: 88; d21: 94. Voc: barky, loud, scratchy, or growly. **Fledgling.** 63–68 g (avg, M > F), fledges 10–17 d (avg 12 d). Face still bare. Plumage overall: F paler than M; dull brown with pale brown edges. Below: paler than upperparts, obscure barring & streaking. Tail short, dark, some purple & greenish iridescence. LF: dark. Care: fed by Ad several wks.

Boat-tailed Grackle (*Quiscalus major*)

Adult. (SC) M 154–239 g, F 93–147 g; M 36.7–39.9 cm, F 25.6–29.5 (10–15.7 in.). Diet: omnivorous; variety of invertebrates (frogs, turtles, lizards, arthropods), grain, fruit & human foods. Nest: cup, in colony. **Nestling.** MC: bright red. Down: sparse, gray or cream. G&D: d0: 6.64 g (M > F), d2: wing PinEm, d3: body PinEm, d3–5: eyes open, d13: M avg 32.1 g (twice that of F). **Fledgling.** 78 g, fledges ~14 d. Plumage: sepia with dusky streaking. Iris: brown. *Range sympatric with Common Grackle along coast of s.e. US.*

Great-tailed Grackle (*Quiscalus mexicanus*)

Adult. M 190–253 g, F 102–137 g; 38–45.7 cm (15–18 in.). Diet: insects & other animal matter spring/summer, plant matter fall/winter. Nest: cup, often near water. **Nestling.** MC: pink (salmon), then cherry-red. GF: white, thin. Bill: pale, turns dark, long. Skin: pale salmon, then dark apricot. Down: long, sparse, brown. G&D: d0: 6.2 g; d3–5: eyes open; d14: FF; near-fledge: 80–120 g. **Fledgling.** 100 g, fledges 12 d (earliest) or later, cannot fly. Plumage: above dark brown, bare around face & eye, dark wings; below buffy-brown. Tail shorter than Ad.

WARBLERS, NEW WORLD

(Family: Parulidae)

Of all passerines, New World Warblers ("wood-warblers"), sometimes confused with vireos and kinglets, may offer the most variety within the passerine families. Characteristics of the genera and an understanding of breeding biology can narrow down species. Look closely at flight feathers, particularly "patterns to the rectrices and colors of the edges of primaries, secondaries, and primary coverts" (Pyle et al. 2015). Useful features are pointed rectrices and the color of the toe pads. Species in the genera

Setophaga are brightly colored with conspicuous white or yellow patches in the tail, usually wingbars, and a short, rounded bill. Fledgling identification is difficult because the juvenile plumage is ephemeral; many molt out of this plumage while in the nest or shortly after fledging (Pyle et al. 2015). This molt before the first fall involves all the feathers of the body and wing coverts; feathers of wings, tail, and primary coverts are retained (Chapman 1917). **Adults.** Most are small songbirds. Diet: mainly insects by gleaning leaves, twigs, and bark; some forage like creepers; some feed on nectar and small fruits during migration and in winter. Nest: cup or domed; ~60% on ground or in low vegetation, ~40% in trees or cavities. Eggs: subelliptical, whitish, creamy, light blue, or pale olive; speckled or stippled with reddish-brown. Bill: variable, but usually short, usually slender, straight, some compressed, acute, never hooked. Plumage: rictal bristles are sometimes conspicuous or may be lacking; males usually brightly colored yellows or olive with black, gray, blue, white, brown, orange, or red. Wings: 9 (visible) primaries, length and shape variable, mostly long and somewhat pointed. Tail: generally shorter than wings, varies from square to slightly rounded. LF: tarsi usually less than twice as long as middle toe without nail, posteriorly ridged (not rounded). **Young.** Mouth: highly variable; most in red range, some in orange range, a few in yellow range. Down: most hatch with down, a few hatch naked then develop down by day 2. Fledge 8–12 days, leave by foot or fly poorly at first. Plumage of newly fledged (for a few weeks): dull olive, brown, buff, and grays with streaking, spotting, or mottling; wings not fully grown, tail stumpy. Biparental care; young may be dependent several weeks. *Sources: Billerman et al. 2020; Chapman 1917; Curson et al. 1994; Pyle 1997b; Pyle et al. 2015.*

Ovenbird (*Seiurus aurocapilla*) Plate 35

Adult. (PA) 15.4–21.4 g, 15.2 cm (6 in.). Diet: invertebrates in leaf litter, few seeds. Nest: domed, on ground. **Nestling.** MC: pink or red. GF: H cream, yellow d4. Bill: large, brown. Skin: pinkish-tan. Down: gray (darker on head) or pale sepia-brown. G&D: d0: 1.5–2.2 g; d1: 3.3; d3: 7.0, wing PinEm; d4: 9.0 g, eyes open; d5–6: 11–12.8 g; d7–8: 13.6–14, PinUn. **Fledgling.** Fledges 7–10 d. Plumage: above cinnamon-brown streaked black on scapulars & back, olive-brown edges on prim, pale cinnamon-ochre or rusty wingbars, faint lateral crown stripes, indistinct eye ring. Below: breast & sides pale cinnamon streaked narrowly olive-brown or black, belly whitish. Tail: green edging, no spots. LF: pink.

Worm-eating Warbler (*Helmitheros vermivorum*)

Adult. (s. IN) 11.5–15.5 g, 11–12.7 cm (4.3–5 in.). Diet: arthropods, spiders, slugs, plus caterpillars (fed to young). Nest: cup on ground, in small hollow. Bill: long, spikelike. **Nestling.** MC: pinkish. GF: buff-yellow. Skin: pinkish-red. Down: hatches naked; d2 develops buffy to dark grayish-brown. LF: H pale, buffy tinged. G&D: unknown. **Fledgling.** Fledges 10–11 d, 8.5 d (avg in CT). Plumage (June–July): sim to Ad DefB, buffy-cinnamon above but head stripes duskier & indistinct, indistinct cinnamon wingbars.

Louisiana Waterthrush (*Parkesia motacilla*)

Adult. (PA) 16.8–23 g, 15.2 cm (6 in.). Diet: aquatic invertebrates. Nest: cup in cavity, hollow, or under log, near stream. **Nestling.** MC: bright red or red. GF: yellowish or yellow. Skin: H red, then lighter. Down: long, dark gray. G&D (NY, birds in GA slightly less): d0: 2.2 g; d1: 4, PinVis; d2: alar PinEm; d3: 8.5 g, caudal PinEm, eyes open; d5: 13.3 g, body PinEm; d7–8: 16.9 g, d9–10: 17. **Fledgling.** 17 g, fledges d10. Plumage above: sooty olive-brown, prominent whitish supercilium, dark brown postocular stripe, distinct buffy or rusty wingbars. Below: dull whitish with olive-brown or black streaking.

Northern Waterthrush (*Parkesia noveboracensis*)

Occurs in Alberta boreal forests. **Adult.** (PA) 14.5–18.6 g, 15.2 cm (6 in.). Diet: insects, spiders, snails, other invertebrates. Nest: cup in cavity near water (swamps). **Nestling.** MC: red. GF: yellow. Down: dark gray or dark olive-brown. G&D: d0: 1.9 g; d1: 3, PinVis; d3: 5.6 g; d4: 7.2, alar PinEm, eye slits; d5: 9.7 g; d6: 14, PinUn wings & back; d7–9: FF, d8: rect PinEm ~3 mm; d9: 13.5 g. **Fledgling.** Fledges d9–10. Bill: large, dark. Plumage sim to Ad DefB but buffy or yellowish supercilium, dark eyeline, mottled appearance, narrow dusky-olive or rusty wingbars. Below: buffy-white or sulphur-yellow heavily streaked black, less defined. Tail: no spots.

Blue-winged Warbler (*Vermivora cyanoptera*)

Adult. (WV) 7.5–10.3 g, 11–12 cm (4.3–4.75 in.). Diet: arthropods (especially crickets, grasshoppers) & spiders. Nest: basket-like cup. **Nestling.** MC: pink. GF: cream or yellow. Skin: pinkish. Down: thin, sparse, gray or yellowish-gray on capital, humeral & middle of back. G&D: d0:

unknown; d1: 1.5 g; d4: 4.9; d4–5: eyes open; d6: 7.3 g; d9: 9.2. **Fledgling.** Fledges d9. LF: bright pink. Bill: sharp. Plumage above: olive-brown tinged yellow, dusky lores, indistinct eyeline, yellowish-white wingbars. Below: pale ochre-yellow, olive-brown throat. Tail bobbing.

Black-and-white Warbler (*Mniotilta varia*)   Plate 35

The only N.A. warbler with "black-and-white wings" (Pyle et al. 2015). **Adult.** (PA) 8.8–12.9 g, 11–12.7 cm (4.3–5 in.). Diet: variety of arthropods; bark probing. Nest: cup in cavity or depression near ground, favors swampy areas. "Nuthatch-like" vertical creeping behavior. Elongated hind claw. **Nestling.** MC: H yellowish to pinkish-orange. GF: pale yellow. Bill: long, pointed. Skin: pink to rosy pink. Down: dark gray or brown. G&D: unknown. **Fledgling.** Fledges 8–12 d, flies poorly. Bill & LF: pinkish-buff; soles are pale brownish-yellow. Plumage above: dull blackish-brown & buffy-white; head boldly striped; wings & tail blackish edged grayish, buff-white wingbars. Below: dull white with brownish streaks, buffy-white belly. Wings: bright white outer edge of longest tertial.

Prothonotary Warbler (*Protonotaria citrea*)   Plate 35

Adult. 13.3 g, (OK) 14–14.8 g; 13.3 cm (5.25 in.). Diet: insects mainly, some mollusks, some seeds, fruit & nectar. Nest: usually in cavities in swampy woodlands. Bill: large. Tail: short. **Nestling.** MC: orange-red. GF: pale yellow. Bill: H tannish; N dark on upper, d5 long, very pointed, raised nares. Skin: orangish-pink or orange-red. Down: thick, grayish-brownish. LF: orangish-pink. G&D: d0: 1.8–2.1 g; d3: eye slits, alar & leg PinEm; d5: rem PinUn; d6: olive-green PinUn; d9: FF. Voc: *tschip* d6. **Fledgling.** 11.5–12.7 g, fledges d10–11, can run d9. Plumage above: head & mantle dusky olive-green or brownish-olive, olive wingbars. Below: dusky-grayish tinged yellow or brown, paler belly.

Swainson's Warbler (*Limnothlypis swainsonii*)

Adult. (FL) 14.3–20.4 g, 12.7–14 cm (5–5.5 in.). Nest: cup in low, dense vegetation. Bill: long, heavy. **Nestling.** MC: reddish. GF: cream. Bill: H pale pinkish-buff; N long, sharp, pointed. Skin: dark pink. Down: dark brownish. Head: flat forehead. G&D: no mass info; d4: eye slits; wing PinEm; d8: wing cov PinUn; d9: most wing PinUn, head FF. **Fledgling.** LF: pink. Plumage: cinnamon-brown to olive-brown above, supercilium lacking or indistinct, buffy wingbars. Below: off-white.

Tennessee Warbler (*Leiothlypis peregrina*)

Adult. (ON, Canada) 8–11.8 g, 12 cm (4.75 in.). Diet: invertebrates (spruce budworms), some fruit & nectar in winter. Nest: cup, concealed on ground, often in moss. **Nestling.** Bill: large, pinkish-buff. Down: no info. G&D: little info. (ON, Canada) d0: 1.34 g. **Fledgling.** 8–11 g, fledges 11–12 d. Plumage above: dull olive-gray (tinged greenish), wings dark with green edges, eyeline dusky, pale yellow supercilium, broad white wing-bars. Below: dull yellow-buff, dusky olive-yellow throat & breast. Tail: may or may not have white patch on outermost rect. LF: pinkish-buff.

Orange-crowned Warbler (*Leiothlypis celata*)   Plate 35

Adult. (CA) 7.3–11.6 g, (AK) 7.4–10.8 g, 12.7 cm (5 in.). Diet: insects (mostly), fall/winter takes small fruits (berries), sap, nectar, suet; few seeds. Nest: cup in rock crevice or on ground. **Nestling.** MC: crimson or red. GF: yellow. Bill: H pink, then horn. Skin: pinkish. Down: sparse, grayish-white tinged brown. LF: pink. G&D: d0: 1.3–1.6 g; d1: 2.3, alar & caudal PinVis; d2: PinVis remaining tracts; d3: 4.2 g; d5: 6.0, PinUn; d6–7: eyes open; d7: 7.8 g; d9–10: 9.6–10.5. **Fledgling.** Fledges 11–13 d. Plumage above (variable by ssp, Foster 1967): dull olive-brown, olive-greenish, or russet-olive on rump & uppertail covs, olive-brown tail & wings, faint buffy-yellow wingbars. Below: dull brownish-buff breast, pale yellow sides, grayish undercoat, white or pale yellow belly. Tail: a stub.

Nashville Warbler (*Leiothlypis ruficapilla*)

Adult. (PA) M 7.2–9.6 g, F 7.1–9.1 g; 11.4–12 cm (4.5–4.75 in.). Diet: insects. Nest: open cup under cover often near water. **Nestling.** MC: red. GF: yellow. Bill: H pinkish. Down: sparse, sepia-brown. LF: pinkish-buff. G&D: d0: 0.75–1.75 g; d5: eyes open; d8: 8–8.5 g; d9: 11. **Fledgling.** Fledges 8–11 d. Bill: pinkish-buff. Plumage above: dusky olive-green, brownish-gray head & neck, buffy-yellow wingbars, white eye ring. Below: white chin & throat, breast & flanks brownish-yellow, belly pale yellow. Tail bobbing. LF: soles dull yellow.

Virginia's Warbler (*Leiothlypis virginiae*)

Adult. (AZ) 6–10.5 g, 12 cm (4.75 in.). Diet: may be entirely insectivorous. Nest: open cup on ground. **Nestling.** Skin: pink. Down: gray-white.

G&D: unknown, egg mass: 1.2 g (AZ); d2: PinVis, eye slits; d7: PinUn tips. **Fledgling.** Fledges 10–14 d. Above: grayish-brown, no rufous in crown, dull yellow rump, buff wingbars, complete white eye ring. Below: pale gray-brown, lacks yellow, dull whitish belly, saffron (orange-yellow) uppertail and undertail covs. Independent 2 wks post-fledging.

MacGillivray's Warbler (*Geothlypis tolmiei*)

Adult. (AZ) 8.6–12.6 g, 12.7 cm (5 in.). Diet: insects. Nest: cup in riparian habitat in low shrubs. **Nestling.** MC: orange. GF: yellow. Down: not described, maybe none, d2 down appears on crown, dorsal & alar, color "possibly gray." LF: pale pinkish-buff. G&D: little info. d4–5: eyes open. **Fledgling.** Fledges 8–10 d. Plumage like Ad F but duller. Above: dark brownish-olive, pale cinnamon wingbars; crown olive-brown, lores grayish or brownish, eye arcs duller, indistinct. Below: brownish-olive washed yellowish, whitish-gray breast & throat. *Hybridizes with Mourning Warbler where sympatric.*

Mourning Warbler (*Geothlypis philadelphia*)

Adult. (PA) M 9.9–13.6 g, F 9.7–13 g; 10–15.2 cm (4–6 in.). Diet: primarily insects. Nest: cup on or near ground. **Nestling.** MC: red. GF: yellow. Down: sparse, dark gray. G&D: unknown. **Fledgling.** Fledges 8–9 d, cannot fly, not FF until 3 wks post-fledging. LF: bright pinkish. Plumage above: dark olive-brown; wing edges & rect olive-green, cinnamon wingbar; crown brownish olive-green, partial eye ring indistinct or lacking (becomes complete by fall). Below: tawny-olive, grayish-white or buffy throat; blackish basally. Tail: green, pointed, no markings. *Hybridizes with MacGillivray's Warbler.*

Kentucky Warbler (*Geothlypis formosa*)

Adult. (PA) 12.1–16.9 g, 12.7 cm (5 in.). Diet: insects. Nest: cup on or near ground in small shrub. **Nestling.** MC: orange. GF: yellow. Bill: H blackish, turns yellow by d6–8. Skin: pinkish-orange. Down: olive-brown or gray tufts. LF: pale pinkish-tan. G&D: d0: 2.5–3.5 g, d1: 4–4.5, d5: 10.3, eyes open; d8–9: 11.7–12.5 g. **Fledgling.** 12 g, fledges 8–10 d. Plumage duller than Ad, olive-brown or brownish above, brownish or rusty wingbars, yellowish spectacles. Below: yellowish.

Common Yellowthroat (*Geothlypis trichas*)　　　　　　　Plate 35

Most widespread N.A. warbler, behaves like a wren; considerable geographic variation. **Adult.** (PA) M 8.1–12.2 g, F 7.8–11.5 g, (w. US) M 9.5–10.7 g; 11–12.7 cm (4.3–5 in.). Diet: insects & spiders. Nest: cup on/near ground, well concealed. **Nestling.** MC: deep red, reddish-pink or reddish-orange. GF: pale yellow, or bright yellow, rounded. Bill: N light tan. Skin: dark orange. Down: grayish or mouse colored on capital, lower back & humeral. LF: pinkish-tan with yellow nails. G&D: d0: 0.90–1.14 g; d3: alar PinEm; d4–5: eyes open; d6: fear reaction; d6–7 all PinUn; d8 FF; wood-brown chin & throat. **Fledgling.** 10.3 g, fledges 9–10 d. Bill & LF: pinkish-buff. Plumage above: brownish olive-green (or brown), cinnamon-buff wingbars, pale buff eye ring. Below: chin, throat & upper breast may or may not show yellow after the pre-formative molt. Tail: bright olive-green, pointed rectrices. S. California ssp: grayish cinnamon-brown above, rusty-buff wingbars; brownish-white below.

Hooded Warbler (*Setophaga citrina*)

Adult. (PA) M 9.6–12.5 g, F 8.6–12.8 g; 12.7 cm (5 in.). Diet: insects (+ spiders in spring/summer). Nest: cup near ground in shrub. **Nestling.** MC: dark pink. GF: pale yellow. Skin: pinkish-orange. Down: pale sepia-brown to gray. LF: pinkish-buff. G&D: d0: 1.9 g, d5: 7; d5–7: eyes open. **Fledgling.** 8.5–9.5 g, fledges 8–9 d. Plumage: yellowish-brown or pale brownish-olive above, buff or brownish wingbars. Below: dull buff, breast light yellow mottling, straw-yellow belly. Tail flashing shows white outer rect. Independent 4–5 wks post-fledging.

American Redstart (*Setophaga ruticilla*)　　　　　　　Plate 35

Adult. (PA) M 6.5–10.2 g, F 6.7–10 g; 11–12.7 cm (4.3–5 in.). Diet: insects (+ some berries & fruits late summer). Nest: open cup on branch. **Nestling.** MC: red or reddish-orange. GF: yellow. Down: hair-brown. G&D: d0–2: mass unknown; d3: wing PinEm, eye slits; d4: 5 g; d6: wing PinUn; d7: 7.5 g; d8: FF, shows yellow in tail; molt out of juv plumage before fledging. **Fledgling.** Fledges 9 d. Bill: sharp. Plumage: brownish-gray or olive-brown above, paler crown, whitish-yellow or pale buff wingbars. Below: whitish; M darker than Ad. **Juvenile.** Sexually dimorphic: base of M tail with orange or salmon; F tail yellowish (base).

Kirtland's Warbler (*Setophaga kirtlandii*)

Endangered status. **Adult.** (MI) M 12.3–14.6 g, F 12.4–16 g; 13.7–14.2 g, 14.6 cm (5.75 in.). **Nestling.** MC: yellowish (throat pink), turns orange. GF: yellow. Bill: H pinkish-brown, tip dark. LF: pinkish. Down: brown, longer on head. G&D: d0: 1.3 g; d1: 2.6; d3: 5.1, PinVis, Prim PinEm; d5: 9.4 g, eyes open; d7: PinUn; d8: 12.7 g, appears FF; d9–10: 13g. Voc (d8): *ti-ti-ti-ti-ti* when parents bring food. **Fledgling.** Fledges 9 d, short flights, sits quietly for long periods. Plumage: gray-brown above, pale buff wing-bars. Below: buff, heavy markings on breast. Tail: 1 cm (12–13 d), 2 cm (15 d), 5 cm (23 d); white spots in outer rect.

Cerulean Warbler (*Setophaga cerulea*)

Adult. (PA) M 9.3 g, F 8.8 g; 11.4 cm (4.5 in.). Diet: insects. Nest: tiny cup on lateral limb of deciduous tree. **Nestling.** Bill: H pinkish-buff. Down: light mouse-gray to light drab. LF: H pinkish-buff. G&D: unknown. **Fledgling.** Fledges: 10–11 d. Plumage above: brownish-gray (M bluer than F, F yellower than M), grayish-white median crown stripe, pale super-cilium, drab-gray postocular stripe, dark wings edged greenish, buffy wingbars. Below: all white. Iris: dark brown.

Northern Parula (*Setophaga americana*)   Plate 35

Adult. (PA) M 7.2–8.4 g, F 6.5–8.8 g, (FL) 7–10.2 g; 10.8 cm (4.25 in.). Diet: insects & spiders; berries & seeds in winter. Nest: cup in hang-ing mass of moss or other epiphytic growth. **Nestling.** MC: presum-ably red or reddish-orange. GF: yellow. Bill: H pinkish to buff. Skin: yellowish-pink. Down: long, white, silky, on dorsal tracts, turns smoke-gray. LF: pinkish-buff. G&D: little info. **Fledgling.** Fledges 10–11 d, FF, cannot fly. Bill: sharp, pointed, dark upper, yellowish lower. Plumage (May–Aug): brownish-gray tinged green, pale eye arcs, faint dark eye-line; wings & tail edged blue-green, broad white wingbars. Below: pale grayish-white, pale yellow throat, yellow patches on breast sides; tail spots.

Magnolia Warbler (*Setophaga magnolia*)

Unique tail pattern, white inner band with bold black tip. **Adult.** (PA) M 6.9–10.3 g, F 6.6–9.7 g; 11–12.7 cm (4.3–5 in.). Diet: arthropods, mainly caterpillars. Nest: flimsy cup in dense coniferous vegetation. **Nestling.**

GF: yellow. Skin: orange-red. Down: sparse, black or sepia-brown. G&D: no mass info; d2: PinVis; d3–4: eyes open, wing PinEm ~6 mm; d5: all PinEm; d6: PinUn; d8: FF. Voc: *zee zee* d9–10. **Fledgling.** Fledges d10. LF (soles): yellow (juv). Plumage above: dark sepia-brown, indistinctly streaked black, paler crown & rump; wings & tail black with grayish-brown edges, narrow buffy wingbars. Below: pale buff, dusky buffy-olive breast, wide blackish streaks on belly & flanks, white undertail covs. Tail pattern (unique): white oval-shaped spots on all except central pair of rect.

Bay-breasted Warbler (*Setophaga castanea*)

Adult. (PA) 10.2–13.8 g; 12.7–14 cm (5–5.5 in.). Diet: insects & spiders. Nest: cup in conifers. **Nestling.** MC: red. Down: brown. G&D: unknown, d7: wing flapping, d10: FF. **Fledgling.** Fledges: d10–11. LF: little or no dull yellow on soles. Bill: pinkish-buff, turns dark. Plumage above: dull olive-gray faintly streaked black. Auriculars: pale gray-buff, mottled darker brown. Lores: dusky. Wings & covs chestnut-brown with greenish edging, buffy wingbars; brownish-black tail. Below: pale buffy-white; throat, breast, sides spotted dusky. *Sim sp: Yellow-rumped Warbler (Myrtle) and Bay-breasted Warblers.*

Blackburnian Warbler (*Setophaga fusca*)

Adult. M 10 g, F 9.5 g; 11–12.7 cm (4.3–5 in.). Diet: insects & spiders (+ some fruit in winter). Nest: cup shape in dense vegetation. **Nestling.** MC: yellow. GF: yellow. Down: sepia-brown or dark brown. G&D: unknown. **Fledgling.** Bill & LF: pinkish-buff. Plumage above: dull gray-brown with indistinct clove-brown streaking, buffy wingbars, buffy supercilium (eye to nape), gray-brown auricular patch. Below: white, throat & breast suffused with buff & brownish, dull sepia spotting on breast & flanks.

Yellow Warbler (*Setophaga petechia*)   Plate 35

Widespread breeding range, strong geographic variation, especially in overall brightness. **Adult.** (PA) M 8.5–11.5 g, F 7.3–11 g; 12–12.7 cm (4.7–5 in.). Diet: insects & other arthropods (some fruit). Nest: cup. **Nestling.** MC: pink, red, or reddish-orange. GF: pale yellow then bright yellow. Bill: N brownish-gray, pointed. Skin: reddish or yellow-orange. Down: sparse, light smoke, mouse-gray, or cream. G&D: d0: 1–1.2 g;

d1–2: 2.2–3.2, PinVis all tracts; d3: 4.8 g; d3–5: eyes open; d4–5: 6.1–7.5 g; d6–7: 8.5–9. **Fledgling.** Fledges 9–12 d. Bill: pale base. LF: pinkish-buff. Plumage duller than Ad. Above: brownish-olive, back & rump streaked dusky. Below: pale grayish-olive to buffy-white. Wings & tail: dusky to clove-brown, edged bright lemon-yellow, buffy-yellow wingbars. Tail pattern: unique yellow patches.

Chestnut-sided Warbler (*Setophaga pensylvanica*)

Populations declining. **Adult.** (PA) M 8.4–10.9 g, F 7.9–10.4 g; 12.7 cm (5 in.). Diet: insects, spiders (+ some seeds & fruit). Nest: cup in fork of tree. **Nestling.** MC: H yellow, brighter d1, then gradually changes to bright red by fledge. GF: pale cream or yellow. Skin: yellowish-tan or cinnamon. Down: sparse, short, light gray or dark gray, all dorsal areas. G&D: d0: 0.97 g (1 chick incubator-hatched); d3–4: eyes open; d8: FF. **Fledgling.** Fledges d9–10 (or 10–12, another source). Bill: stout, slate above, dull white lower. Plumage above: dark umber-brown with dull black indistinct streaking on back, buffy-yellow wingbars, white patches on outer 2–3 pairs of rect; white eye ring. Below: gray throat, white belly & crissum.

Black-throated Blue Warbler (*Setophaga caerulescens*)

Adult. (FL) M 8.4–12.4 g, F 8.8–12 g; 12.7 cm (5 in.). Diet: insects, spiders & other arthropods. Nest: cup in fork of low shrub. **Nestling.** MC: H ochre, becomes pink by d7. GF: pale yellow. Skin: pale yellowish-pink, turns darker by d1. Down: dark gray on capital, humeral, femoral, small patch middorsal. LF: N pale yellow-pink. Iris blackish-brown. G&D: d0: ?; d1: 1.3–1.6 g; d4: 5–5.5, prim PinUn, eyes open; d6: 7.5–8.2 g. **Fledgling.** 9–10 g, fledges d8–9, flies weakly. Bill: stout. Plumage above: M brownish olive-green; buff supercilium (nearly absent in F), blackish lores & auriculars (except F); wings & tail blackish with bluish-gray or greenish edging (F is browner with duller greenish edges, no bluish), white patch at bases of prim (reduced or lacking in F), outer rect with subterminal white blotches. Below: whitish-buff or dusky-yellowish, brownish-olive mottling on breast.

Pine Warbler (*Setophaga pinus*)   Plate 35

Adult. (MN) 9.4–15 g, (FL) 10.5–17 g, (NC) 7–16 g; 12.7–14 cm (5–5.5 in.). Diet: mostly arthropods, some fruits & seeds. Nest: cup in pine tree. **Nestling.** Little known, nest inaccessible. GF: whitish. Down: grayish-brown. **Fledgling.** Fledges d10–12. Plumage above: warm grayish-brown, buff wingbars, buff broken eye ring; below buff mottled with yellowish-olive, throat & flanks gray, belly white. Prejuvenile molt completed before fledging.

Yellow-rumped Warbler (*Setophaga coronata*)   Plate 35

Possibly the most widespread N.A. warbler. **Adult.** (CA) 10–16 g, (PA) M 10–14.7 g, F 9.8–13.6 g; 12.7 cm (5 in.). Diet: insects (& berries in winter). Nest: cup shape on branch of conifer (usually). **Nestling.** MC: pink to reddish. GF: white or yellow. Down: sepia to dark brown. Bill & LF: dusky-buff. G&D: little published info. Nestling (Grass Valley, CA): 9 g (on intake), gained 1 g/d for 6 d (15 g); d6–7: PinEm, d8: 13 g. **Fledgling.** Fledges 10–14 d, flies poorly, tail barely grown, still downy (grayish). Plumage above (May–Aug): buff-brown streaked dusky; bare face; rem & rect dark with grayish or brown edges; buff or white wingbars; no yellow on rump but quickly appears after fledging. Below: grayish-white, some lightly washed yellow, some streaked dusky-brown. Juv Audubon & Myrtle ssp similar until fall, when yellow rump & side patches appear. Juv Audubon: breast more solid color (white streaked black), pale throat, white on outer 4–5 pairs of rect; Myrtle juv: breast white, heavily streaked, pale throat extends around back of auriculars, white on outermost 2–3 pairs of rect. May have yellow soles. *Myrtle fledglings have sim plumage as Blackpoll & Bay-breasted Warbler fledglings. Hybrids: Myrtle & Audubon's ssp.*

Yellow-throated Warbler (*Setophaga dominica*)

Adult. (PA) M 9.3–10.8 g, F 8.8–10.5 g; 12.7–14 cm (5–5.5 in.). Diet: arthropods. Nest: cup, high in canopy. **Nestling.** Little known. **Fledgling.** Fledges 10 d. Plumage above: dull olive-brown, dark brown streaking or mottling; dark through auriculars & lores, buff-white supercilium & spot below eye. Below: white finely streaked clove-brown. Wings & tail: dull brownish-black with plumbeous-gray edges; buff-white wingbars. White patches on outer 2–3 rect.

Prairie Warbler (*Setophaga discolor*)

Adult. (FL) M 6–10 g, F 5.7–10.8 g; 11.5 cm (4.5 in.). Diet: insects & spiders (+ some fruit). Nest: cup in fork of tree or shrub. **Nestling.** MC: red (or yellow). GF: light yellow. Bill: H buffy-yellow. Skin: light brownish-red or yellowish-orange. Down: medium gray (darker on head), all regions including 4 downs on abdominal. G&D (Nolan 1978): d0: 0.8–1.7 g; d1–2: 1.2–2.6; d3: 3.6, darkened tracts; d3–6: eyes open; d4: 4.7 g; d5: 5.7; d7: 6; d9: 6.3. Voc: begging *churr*. **Fledgling.** 5–5.7 g, fledges 8–11 d. LF: buff-yellow. Plumage above: olive grayish-brown, blackish wing covs, buff-yellow wingbars; faint grayish line over eye. Below: lighter, streaked olive. Voc: some "singing" at 40 d post-hatching. Tail bobbing.

Grace's Warbler (*Setophaga graciae*)

Adult. (AZ) M 7.5–9 g, F 7–9.8 g; 12 cm (4.7 in.). Diet: arthropods. Nest: cup, well concealed. **Nestling.** Down: grayish. G&D: unknown. **Fledgling.** Plumage like Ad but less black streaking overall, no black on lores & no yellow; F less streaked & more brownish than Fl M. Above: drab gray to olive-brown streaked dusky; wing covs blackish, white wingbars. Below: buffy-white, breast spotted/streaked dusky, buff-white belly & crissum.

Black-throated Gray Warbler (*Setophaga nigrescens*)

Adult. (OR & CA) 6–9.8 g; 11–12.7 cm (4.3–5 in.). Diet: insects. Nest: cup on horizontal branch of tree. **Nestling.** MC: dull red. GF: pale yellow. Down: drab-gray. LF: tawny, tinged rufous. G&D: little info. **Fledgling.** Plumage sim to Ad F but dull brownish-gray, darker crown, broad whitish supercilium. Below: dull gray, breast lightly streaked dull brown, belly whitish.

Townsend's Warbler (*Setophaga townsendi*)

Adult. (CA) 7.3–10.7 g; 12 cm (4.7 in.). Diet: insects. Nest: cup in coniferous tree. **Nestling.** MC: pink. GF, bill, LF: yellow. Skin: tannish to brownish-black. Down: dark gray or brown above. G&D: d0: unknown; d3–6: eyes open; d6–7: 8.4–9.2 g; FF. **Fledgling.** Fledges d9–11, weak fluttering flight. Plumage like Ad F but streaking indistinct on back. Above: olive-brown, broad buffy eyebrow & malar stripes; crown gray olive-brown, dark wings & tail (white patch on r4), white wingbars. Below: chin yellow, flanks & breast sides grayish or brownish, belly light streaked dusky. LF: pinkish-tan. *Hybridizes with Black-throated Green & Hermit Warblers where sympatric.*

Black-throated Green Warbler (*Setophaga virens*)

Adult. (PA) M 8–10 g, F 7.4–9.5 g; 11–12 cm (4.3–4.7 in.). Diet: insects & caterpillars, some berries (migration). Nest: cup, usually in conifer. **Nestling.** MC: pink. GF: yellow. Skin: burnt-orange d3. Down: dark brown or burnt umber d3. G&D: d1: 1.7–1.9 g, d3–5: eyes open, d4: alar PinEm, d9: 8.5 g. **Fledgling.** Fledges 8–10 d, flies weakly. Plumage above: grayish olive-brown obscurely spotted/streaked slaty; brownish-white eyeline. Below: whitish streaked dusky. *Hybrids: see "Townsend's Warbler."*

Canada Warbler (*Cardellina canadensis*)

Adult. Mass F < M: (PA) 8.5–11.7 g, (AL) 8.2–15.5 g; 12–15.2 cm (4.7–6 in.). Diet: insects (flying) & spiders. Nest: on ground, cup in thickets, with dense cover. **Nestling.** MC: pale yellow to yellow. GF: pinkish-buff, turns dusky. Skin: yellow-tan. Down: sepia-brown or medium gray. G&D: d0–5: mass unknown; d5–6: alar PinEm; d6: 9.6 g. **Fledgling.** Fledges 8.1 d (avg), weak flight. Bill & LF: bright yellow. Plumage similar to Ad but duller & more variable, eye ring paler. Above: brownish, gray wings edged pale, indistinct (or absent) buffy wingbars. Below: dull yellow washed brown with indistinct breast streaking. Tail: unmarked.

Wilson's Warbler (*Cardellina pusilla*)   Plate 35

Adult. (CA) 5.4–9 g, (PA) 6.5–8 g; 10–12 cm (4–4.7 in.). Diet: insects (+ some berries). Nest: cup. **Nestling.** MC: pinkish-orange? GF: yellow. Bill: buffy-yellow, short & wide, rictal bristles. Skin: pink. Down: long, sepia-brown, brownish-gray, or olive-brown. LF: pinkish. G&D: little info, egg (AK): 1.65–1.78 g; d0: 1–1.2 g (based on egg mass); d2: PinVis; d3: dorsal PinEm; d4: prim PinEm; d5–6: eyes open; d8: FF, dull brown with dull yellow underparts. Voc: quiet until d6. **Fledgling.** Fledges 9–11 d. Plumage above: light or sepia-brown, can appear mottled, olive crown, wings & tail dull olive-brown with olive-green edges, buff wingbars, yellow lores. Below: yellowish-buff, throat & flanks brownish. LF (soles): dull yellow (juv).

CARDINALS AND RELATIVES

(Family: Cardinalidae)
   Groups in this family: cardinals, tanagers, grosbeaks, buntings, Dickcissel. **Adult.** Mostly medium-size songbirds. Diet: insects mainly + fruit.

Nest: open cup or saucer usually near end of horizontal branch. Eggs: subelliptical to short elliptical; pale blue or pale green; markings vary from speckled to boldly blotched with gray or brown. Bill: groups variable; cardinals and grosbeaks are large billed; buntings are small billed. Plumage: many are brightly and boldly patterned with reds, greens, blues, and yellows. Wings: 9 (visible) primaries. Legs/feet: scutellate, tarsi variable in length, hallux nail equals nail length of middle toe (except tanagers). **Young.** Mouth: variations from yellow to orange to red. Bill: starts out light, turns darker with age. Skin colors: varying shades and combinations of pink, red, yellow, and orange; many newly fledged have bare faces except crown. Down: varies from white to shades of gray. Fledge ~9–13 days. Nestling & fledgling plumage is colorful: nestling M may be distinguished in some sp.

Tanagers: 17.8–20.3 cm (7–8 in.). Bill: short to medium, most as long as head; stout, somewhat conical, culmen decurves toward tip, commissure not angulated, slightly hooked with notch on upper mandible near tip and tomium toothed near middle (variable by ssp); inconspicuous rictal bristles. Plumage: adult male more or less red, female olive-green. Fledgling plumage drab, varying dull yellow to olive. Some juveniles confused with juvenile orioles but oriole beaks are sharply pointed and wings are darker than body. Wings moderately long and somewhat pointed. Tail shorter than wings; may be squared to slightly rounded, sometimes slightly notched. LF: tarsi longer than middle toe with nail.

Cardinals, grosbeaks, buntings, Dickcissel: 14–22 cm (5.5–8.75 in.). Bill: short; bulky; conical; culmen slightly decurved; commissure sharply angulated; lower mandibular tomia flat, not rolled inward; nasal fossa oval. Plumage: rictal bristles usually present; males often brightly colored. LF: tarsi length varies; length of hallux nail equal to middle toenail. Wings and tail: variable in length and shape.

Hepatic Tanager (*Piranga flava*)

Adult. 38 g, 7.6–9 cm (3–3.5 in.). **Nestling.** Unknown. **Fledgling.** No mass or fledge info. Plumage variable by ssp. Above: dull yellow with dusky-brownish wash, heavily streaked or barred olive-brown or fuscous, grayish-tan rump, buff-yellow wingbars. Below: yellow-amber or straw-yellow & heavily streaked, crissum yellowish-tawny.

Summer Tanager (*Piranga rubra*)   Plate 36

Adult. (FL) 25.8–33.6 g, 17 cm (6.7 in.). Diet: insects, especially bees & wasps (+ fruit in winter). Nest: cup. Bill length (nares to tip): *P. r. cooperi* 13.2–16.8 mm, *P. r. rubra* 12–14.7 mm (Pyle 1997b) **Nestling.** MC: yellow to bright yellow. GF: yellow. Bill: gray. Skin: pink to pinkish-red. Down: plentiful buffy-gray or gray-brown. G&D (3 chicks, KS): d0: 4.2 g; d1: 5.2; d2: 8.7 (30 mm long); d3: rem PinF 11.5 mm; d5: 17.3 g (44 mm), many PinUn; d5–6: eyes open; d6: 17 g (45 mm), can hop; d7 (3 chicks left nest): 17.3 g (46 mm); d10: 18.2 g (50 mm). Voc: soft begging calls, loud two-syllabled *zhurri*. **Fledgling.** 17–18 g, fledges 7–12 d, flies poorly. Bill: olive with yellow tip. Plumage may be variable, mostly yellow with brownish wash. Above: crown & hind neck dull olive-buff with dusky-gray streaks, olive-brown back with less distinct streaking, face bare at first, bright yellow eye ring (incomplete), grayish down still present, wings & tail olivaceous, yellowish-buff wingbars. Below: white or dull yellow, breast with heavy dusky-gray streaks, yellowish or orange-buff crissum. Immature M (as Ad F) is yellow & spotted or splotched with red & orangish-red patches. Alarm call: *sgwee e err* (repeated several times); hunger call loud two-syllable *zhurri*. Dependent ~3 wks post-fledging. *Sim sp: Scarlet Tanager with smaller bill (notch not as evident) & underparts that are more yellow.*

Scarlet Tanager (*Piranga olivacea*)   Plate 36

Adult. (PA) 23.8–33.5 g; 16–17 cm (6.3–6.7 in.). Diet: insects (spring/ summer), fruit & earthworms. Nest: open cup on tree branch, flimsy. Bill: 10.5–12 mm (length, nares to tip) (Pyle 1997b) & notch is evident. **Nestling.** MC: bright orange-yellow, orange (d3), or orange-red. GF: yellow or bright orange-yellow. Bill: pinkish-buff, d3 yellowish (tip dark). Skin: orange, d4 tan. Down: grayish-white or silvery-gray, very thick mostly on spinal (35 neossoptiles), abdominal (12). LF: pinkish olive-gray, pinkish dull metallic blue (d6). G&D: d0: 2.7–2.97 g; d1: 4.7; d2: 6.9, wing PinVis; d3: 9 g, wing PinEm; d3–6: eyes open; d4: 11.6 g; d4–5: dark olive PinUn above, yellowish on ventral; d5–6: 14–15.4 g, prim & sec PinUn, yellowish tail PinUn; d6–7: yellow & white ventral PinUn; d7: 16.5 g, greenish PinUn on dorsum; d9: 19–22 g. Voc: soft, high *peep*, fl *veer*. **Fledgling.** 19–20.5 g, d8–9 walks on branches, flutter flight (fledges) 9–15 d. Bill: shorter than other tanagers. Plumage above: dark dull olive (F) or olive-green (M), wings & tail darker, pale wingbars, dusky streaking throughout, pale yellow incomplete eye ring. Below: yellowish with olive-brown streaking, bright yellow crissum, turns rich cinnamon-buff rump. Iris: grayish to grayish-brown. *Sim sp: see "Summer Tanager."*

Western Tanager (*Piranga ludoviciana*)   Photo 63

Adult. (CA) 22.5–34.5 g; 16–19 cm (6.3–7.5 in.). Diet: insects (+ fruit). Nest: cup. Bill: size similar to Scarlet Tanager (Pyle 1997b). **Nestling.** MC: red-orange. GF: H bright yellow, thin, down-curved. Bill: long, distinctive. Skin: cinnamon-rufous. Down: white to pale gray. G&D: little info; d7: eyes open; d10: buff-yellow wingbars. **Fledgling.** 11–15 g avg (about ¾ Ad size), fledges 11–15 d. GF: pale yellow. Bill: olive-brown upper; lower dull yellow, light orange, or buff. Eyes: large. Plumage: like Ad F but washed dusky. Above: brownish-olive to grayish-olive, 2 downy tufts above eyes, brown wings edged olive-yellow to dull white, yellow wingbars. Below: dull buffy or yellowish-white with indistinct brownish-gray streaking, pale lemon crissum. First-year young in fall: throat & rump become bright yellow. Voc: soft *chi-wee* notes.

Northern Cardinal (*Cardinalis cardinalis*)   Plate 36, Photo 64

Sexually dichromatic. Both M & F sing, and songs are sexually dimorphic. **Adult.** (PA) M 37–52.2 g, F 36.2–47.5 g; 21.3–23 cm (8.4–9 in.). Diet: plant foods 71%, animal food 29% (annually). Nest: cup, wedged in fork, concealed. **Nestling.** MC: deep pink or reddish-orange (edged yellow). GF: H whitish & thin; N creamy & thicker. Bill: H pointed, pinkish upper; d4: gray tip, N gray to black, large; shape of open mouth rounded; nares large & raised, placement more toward beak tip. Skin: apricot or orange (Ad has black skin), darker d5. Down: medium gray to white (+ blackish-gray), sparse & long on dorsal areas. G&D: d0: 3.5 g, d1: 6.3, primary PinEm; d2: 9.5 g; d3 alar PinEm; d4: 16.6 g, PinEm dorsum, wing PinUn & longer; d5: crown PinUn; fear response strong; d6: 23.1 g, PinF longer all over, wing PinUn, eye slits; d7: 23 g, some contour, more wing & rect PinUn; d8: 25.8 g, crown PinUn; d9: FF. LF: medium brown. Voc: begging call loud by d5. Chicks do not move or gape until parents arrive. **Fledgling.** 27 g, fledges 7–13 d (usually 9–10 d), flies well 19 d. Bill: orange by winter. Plumage like Ad F DefB but flight feathers reddish & sepia. M may show some red on breast & flanks, is bare faced, small crest, tail very short. *Sim sp: see "Pyrrhuloxia." Song learning: both M & F must learn songs in the 1st year of life, F stops learning at 70 d posthatching, M continues to learn until 7 mos old (Yamaguchi 1998).*

Pyrrhuloxia (*Cardinalis sinuatus*)

Adult. 35 g, 20.3 cm (8 in.). Diet: seeds, insects, fruits. Nest: open cup, dense brush or tree. Bill: yellow, decurved upper mandible, "parrot-like." **Nestling.** MC: bright red. Skin: grayish brown. Down: drab-gray. Bill: bright yellow. G&D: no info except d8–9: FF. **Fledgling.** Unknown, except cannot fly or flies weakly at first. Bill: dark gray. Similar to Ad F with little to no red except maybe in crest, dorsal bases of prim & outer sec. Inconspicuous wingbars. *Sim sp: Northern Cardinal; distinguish by size & bill features.*

Rose-breasted Grosbeak (*Pheucticus ludovicianus*)   Plate 36

Adult. (PA) 34.4–50.6 g; 17.8–21.3 cm (7–8.4 in.). Diet: insects & vegetable matter (wild fruit & seeds). Nest: open cup in tree, shrub, or vine. **Nestling.** MC: reddish-orange. GF: yellow, thin. Bill: very large, pale pink-horn with gray tip. Skin: yellowish-orange. Down: long, silky, thick, grayish-white. Voc: high-pitched whistle followed by scratchy *eek*. G&D: d0: 2.8–3.3 g (5 cm long); d5: wing PinEm; d7: 24.9 g, breast & tail PinEm; d8–9: 26–27.5 g (10 cm), d12: almost FF. **Fledgling.** Fledges 10–12 d. Plumage above: dark brown (F) or black (M), bold buff or cinnamon wingbars, underwing covs (at wrist) pinkish-apricot (M) or yellowish-orange (F); bare around eyes. Below: white with buff wash on breast (M) or buff with olive-brown streaks (F). LF: grayish-bluish, white nails. Dependent 3 wks post-fledging. *Sim sp (on Great Plains): sympatric & hybridize with Black-headed Grosbeak.*

Black-headed Grosbeak
(*Pheucticus melanocephalus*)   Plate 36, Photo 65

Adult. (CA) M 35–46 g, F 37–48.8 g, (NM) M 46 g, F 48.2; 17.8–19 cm (7–7.5 in.). Diet: insects (+ seeds & fruit), in CA important consumer of destructive oak moth; young fed pale green mash until d4, then insects. Nest: cup. Female sings. **Nestling.** MC: red to reddish-apricot, seed grinding plate on upper palate. GF: yellow or ivory-white. Bill: H apricot-orange, darkens d1–4. Gape very large. Skin: apricot-orange on body, turns paler d1–4. Down: grayish-white (long d3), all tracts including ventral. LF: apricot-orange, paler by d5. G&D: d0: 3 g; d3: 6; d4: 10; d8: eyes open; d10–12: ~20 g. Voc: faint, high-pitched *peeps*; N makes continuous *whee-you* or *whee-urr*, builds to loud begging when parent arrives (Weston 1947). Swallows food without closing mouth. **Fledgling.**

Fledges 10–14 d, cannot fly, d15 flies short distance, follows Ads incessantly begging. Plumage: sim to Ad F DefB. Above: back streaked dark brown & buff, brown head with buffy crown & eye stripe, wings & tail with indistinct buffy spots, white wingbars. Below: variable from bright to drab-buff (M are generally brighter than F), upper breast & flanks lightly streaked. Voc: very high, loud, single "pip," sings basic song before migrating. Dependent: weeks(?) post-fledging. *Sim sp: see "Rose-breasted Grosbeak."*

Blue Grosbeak (*Passerina caerulea*) Plate 36

Adult. M 22–40.5 g, F 23–32.5 g; 15.2–16 cm (6–6.3 in.). Diet: insects, snails, seeds. Nest: cup, may incorporate snakeskin. **Nestling.** MC: orange. GF: yellow. Bill: lower mandible gray, brownish-pink upper. Skin: pinkish. Down: sparse, brownish mouse-gray. LF: brownish-pink. G&D: unknown. Based on 1 sample nest, d6: eyes open, wing PinUn (buffy & sepia); d8: rem & rect PinUn. **Fledgling.** Fledges 9–10 d. Bare face. Plumage: dark brown above, buff below, buffy or buff-yellow wingbars, sepia tail (M may have blue wash or edging).

Lazuli Bunting (*Passerina amoena*) Photo 66

Adult. (CA) M 13–19.5 g, F 12.7–16.9 g; 12.7–15.2 cm (5–6 in.). Diet: insects, seeds, fruits. Nest: cup. **Nestling**. MC: orange. GF: yellow. Bill: yellow. Skin: pinkish-apricot. Down: light gray, crown light vinaceous-buff. G&D: d0: 1.3 g, d10: 12.3, wingbars show. **Fledgling.** 12.6 g, fledges 9–13 d, flies poorly. Bill: light brown (paler lower). Plumage above: light gray-brown feathers with pale edges (may have mottled appearance), tan wingbars. Below dull cream-buff with faintly streaked (rust) breast (F paler, streaking less heavy than for M). Independent 14 d post-fledging. *Hybridizes with Indigo Bunting.*

Indigo Bunting (*Passerina cyanea*) Plate 36

Adult. (MI) M 12.5–17.5 g, F 11.9–18.5 g; 11.4–12.7 cm (4.5–5 in.). Diet: insects & spiders, berries (spring/summer), small seeds, berries, buds & insects (winter). Nest: cup in bushes concealed under canopy. **Nestling.** MC: orange (or deep pink) or red. GF: pale yellow. Bill: H orangish. Skin: (light) pinkish-orange. Down: sparse, light or mouse-gray. G&D: d0: 2.1 g; d1: 3.3; d2: 4.5; d4: PinVis; d4–5: 9.1 g, eyes open; d5: dorsal PinEm; d6: PinUn. **Fledgling.** 9–12 g (14 g avg CA), fledges 9–12 d. LF:

bluish-gray. Plumage above: brownish, pale buff wingbars, bare face. Below: throat & upper breast streaked. M sometimes with bluish cast to rump & rect edges (F have no bluish). *Hybridizes with Lazuli Bunting.*

Painted Bunting (*Passerina ciris*)

Adult. (FL) 13–19 g, 12–14 cm (4.7–5.5 in.). Young fed caterpillars, grasshoppers, small beetles (larvae). **Nestling.** GF: yellow. Down: scant & light drab. G&D: d0: 2 g, d3: eyes open. **Fledgling.** 10–11 g, fledges 8–10 d. Bill: gape & base of lower mandible orangish. Plumage sim to Ad F but light brownish-olive (not green) & upperwing feathers lighter, crown buff-brown. Below: buff to light cream-buff.

Dickcissel (*Spiza americana*)   Plate 36

Adult. M 28.5 g, F 25.2 g; 14–16 cm (5.5–6.25 in.). Diet: arthropods (spring/summer) + seeds other seasons. Young fed arthropods. Nest: cup near ground. **Nestling.** GF: bright yellow to orange-yellow. Bill: brown. Skin: orangish to bright pink. Down: white, all dorsal tracts, turns gray. LF: pinkish-buff. G&D: d0: 2.8 g; d1: 4.7; d2: 7.2, wing PinEm; d3: 10.3 g; d5: 13, eyes open; d7–8: 18–18.5 g; d10: 19, all PinUn; d12: 22 g; d18: 26.6. **Fledgling.** Fledges 7–10 d, cannot fly until 11–12 d. Plumage above: crown sepia, supercilium & malar stripe orange-buff to ochre-buff, mantle & back fuscous edged cinnamon-buff, cream wingbars, bare face. Below: pale buff to off-white, no streaking.

PLATES

PLATE 1

GEESE, DUCKS, AND OTHER WATERFOWL

Canada Goose
a. Gosling. Distinctive round crown patch.

Wood Duck
b. Duckling. Eyeline does not continue to bill. Feet boldly patterned, usually edged yellow.

Mallard
c. Duckling. Eyeline continues through the eye to bill. Orange patches on feet.

Hooded Merganser
d. Duckling. Darker than Common Merganser. Reddish-brown nail on bill tip.

Common Merganser
e. Duckling. Bolder markings than Hooded Merganser. White stripe below eye.

b.

c.

a.

d.

e.

273

PLATE 2

QUAIL, NEW WORLD

Distinguishable features of chicks are small size and the dorsal pattern of down.

Mountain Quail

a. Down pattern similar to Northern Bobwhite with 3 buffy stripes along back.

Northern Bobwhite

b. Two buff-yellow stripes down crown and back.

California Quail

c. Just hatched. Egg tooth. Buff background, two black stripes on back bordered by three buff-yellow stripes.
d. Juvenile. Plumage varies geographically.

Gambel's Quail

e. Three yellowish-buff stripes along back.

TURKEY, GROUSE, AND RELATIVES

Most distinguishing features of young are cryptic dorsal pattern of down and features of the feet in some groups.

Wild Turkey

f. Small, dark markings on head, large dark splotches on back.
g. 1–2 weeks old.

Ruffed Grouse

h. Older chick. Dark blotch behind eye.

Spruce Grouse

i. Older chick. Markings surrounding eye.

Greater Sage-Grouse

j. Large dark blotches on sides of head.

Ring-necked Pheasant

k. Black patch near base of bill. Black patches on upper wings and rump.

a.

b.

e.

f.

c.

d.

g.

h.

k.

i.

j.

PLATE 3

GREBES

Western Grebe and Clark's Grebe younger chicks difficult to distinguish.

Pied-billed Grebe
a. 1st downy stage

Horned Grebe
b. 1st downy stage

Red-necked Grebe
c. 1st downy stage

Eared Grebe
d. 1st downy stage

Western Grebe
e. 1st downy stage
f. Juvenile

Clark's Grebe
g. Juvenile. Generally paler gray than Western Grebe.

LOONS

Taxonomically, loons follow the Laridae family but are shown here with grebes as young chicks of both families ride on the backs of adults.

Red-throated Loon
h. 2nd downy stage

Pacific Loon
i. 2nd downy stage

Common Loon
j. Chick (riding on back of adult). 2nd downy stage.

a.

b.

e.

c.

d.

g.

f.

h.

i.

j.

PLATE 4

PIGEONS AND DOVES

Most distinguishing features of squabs are the bill and "hairlike" stringy down.

Rock Pigeon
a. Hatchling

Band-tailed Pigeon
b. Nestling

CUCKOOS

Greater Roadrunner
c. Hatchling, 14 grams

Yellow-billed Cuckoo
d. Days 1–2, 14–15 grams

Black-billed Cuckoo
e. Days 1–2, ~15 grams

a.

b.

c.

d.

e.

PLATE 5

NIGHTJARS AND RELATIVES

Lesser Nighthawk

a. Nestling

Common Nighthawk

b. Days 1–2, 10–12 grams
c. Fledgling. Days 15–16. 66 grams. See also Photo 4.

Common Poorwill

d. Nestling
e. Fledgling

Chuck-will's-widow

f. Days 1–3

Eastern Whip-poor-will

g. Day 7

SWIFTS

Chimney Swift

h. Hatchling, 1–1.5 grams
i. Nestling

Vaux's Swift

j. Nestling

HUMMINGBIRDS

Anna's Hummingbird

k. Day 7

a.

b.

c.

d.

e.

f.

g.

h.

i.

j.

k.

PLATE 6

RAILS, GALLINULES, AND COOTS

Virginia Rail
a. 6–8 grams

Sora
b. 5–7.5 grams

Common Gallinule
c. 10–13 grams
d. Wing claw

American Coot
e. 19–22 grams

STILTS AND AVOCETS

Black-necked Stilt
f. Dorsal view
g. Egg
h. 14 grams

American Avocet
i. 17–21 grams

CRANES

Sandhill Crane
j. Chick

OYSTERCATCHERS

American Oystercatcher
k. 37 grams

PLOVERS

Killdeer
l. 15 grams

a.

b.

d.

c.

e.

f.

g.

h.

i.

j.

k.

l.

PLATE 7

SANDPIPERS AND RELATIVES

American Woodcock

a. Day 1
b. Dorsal view

Wilson's Snipe

c. Chick

Spotted Sandpiper

d. Older chick

GULLS AND TERNS

Herring Gull

e. Chick

Common Tern

f. Day 7

Forster's Tern

g. Chick

a.

b.

c.

d.

e.

f.

g.

PLATE 8

HERONS AND RELATIVES

American Bittern
a. Nestling

Least Bittern
b. Nestling

Snowy Egret
c. Nestling

Cattle Egret
d. Nestling

Yellow-crowned Night-Heron
e. Nestling

a.

b.

c.

d.

e.

PLATE 9

VULTURES, NEW WORLD

Turkey Vulture
a. Nestling

Black Vulture
b. Nestling

OSPREY

Osprey
c. Nestling

FALCONS

American Kestrel
d. Days 4–5
Nestling: see Photo 21

Merlin
e. Nestling

Peregrine Falcon
f. Nestling

a.

b.

c.

d.

e.

f.

PLATE 10

HAWKS, EAGLES, AND KITES

Northern Harrier
a. Nestling

Sharp-shinned Hawk
b. Days 9–10

Cooper's Hawk
c. Days 9–12, 147 grams
d. Days 14–18, 238 grams
Near-fledgling: see Photo 10

Red-tailed Hawk
e. Nestling
f. Older nestling

BARN OWLS

Barn Owl
g. Hatchling
h. Nestling

a.

e.

b.

f.

c.

d.

g.

h.

PLATE 11

OWLS

Western Screech-Owl

a. Near-fledgling

Eastern Screech-Owl

b. Hatchling
c. Fledgling

Great Horned Owl

d. Hatchling
Nestling and fledgling: see Photos 12 and 13

Burrowing Owl

e. Hatchling
f. Days 5–6

Spotted Owl

g. Hatchling
h. Near-fledgling

Barred Owl

i. Fledgling
Nestling: see Photo 16

Northern Saw-whet Owl

j. Fledgling

a.

b.

c.

d.

e.

f.

g.

h.

i.

j.

PLATE 12

KINGFISHERS

Belted Kingfisher

a. Nestling

WOODPECKERS

Red-headed Woodpecker

b. Hatchling, 8 grams
c. Fledgling
Nestling: see Photo 17

Acorn Woodpecker

d. Hatchling, 4–5 grams
e. Day 7
f. Day 10
Days 18–20: see Photo 18
g. Near-fledgling, 52 grams

Red-bellied Woodpecker

h. Hatchling head, egg tooth, and knobs
i. Fledgling

a.

b.

c.

d.

e.

f.

g.

h.

i.

PLATE 13

Yellow-bellied Sapsucker

a. Fledgling

Red-naped Sapsucker

b. Fledgling

Black-backed Woodpecker

c. Fledgling

Downy Woodpecker

d. Days 1–2, 5–8 grams
e. Day 7, 18–23 grams
f. Days 10–12, 25–27 grams
g. California fledgling (female, no red), 26 grams. Dark bars on tail. White tuft at base of bill. Eastern birds have more white spotting on wings and wing coverts.

Hairy Woodpecker

h. Fledgling (eastern). Outer tail feathers all white.

Northern Flicker

i. Hatchling, 6 grams
Nestling: see Photo 20

a.

b.

c.

d.

e.

f.

g.

h.

i.

PLATE 14

TYRANT FLYCATCHERS

Cassin's Kingbird

a. Hatchling
b. Fledgling

Western Kingbird

c. Hatchling
d. Fledgling. Darker upperparts than Cassin's.

Eastern Kingbird

e. Day 12
f. Fledgling

Scissor-tailed Flycatcher

g. Nestling
h. Fledgling

PLATE 15

TYRANT FLYCATCHERS (*continued*)

Ash-throated Flycatcher

a. Nestling. White undersides.
b. Fledgling. Crest feathers, rust on tail.

Great Crested Flycatcher

c. Nestling. Dark skin, yellow undersides.
d. Fledgling. Rust wingbars and tail.

Say's Phoebe

e. Fledgling. Cinnamon wingbars and underparts.

Black Phoebe

f. Hatchling
g. Fledgling. Buff wingbars, contrasting plumage.

Eastern Phoebe

h. Day 6
i. Day 16, fledgling. Rusty wingbars.

a.

b.

c.

d.

e.

f.

g.

h.

i.

PLATE 16

TYRANT FLYCATCHERS (*continued*)

Western Wood-Pewee

a. Fledgling. Eye ring, yellow underparts.

Eastern Wood-Pewee

b. Fledgling. Buff wingbars.

Willow Flycatcher

c. Hatchling. Bright yellow gape flanges.
d. Fledgling. Olive eye ring, buff wingbars.

Least Flycatcher

e. Fledgling. Buff wingbars.

Dusky Flycatcher

f. Hatchling
g. Fledgling

Pacific-slope Flycatcher

h. Hatchling
i. Nestling
j. Fledgling. Buff to cinnamon wingbars.

a.

b.

c.

d.

e.

f.

g.

h.

i.

j.

PLATE 17

SHRIKES

Loggerhead Shrike

a. Nestling. Short, very sparse down.
b. Fledgling. Dark mask, white wingbars, fine barring on underparts.

VIREOS

Black-capped Vireo

c. Fledgling. Distinct white wingbars.

White-eyed Vireo

d. Hatchling. No down.
e. Fledgling. Pale yellow spectacles, brownish streak through eye.

Bell's Vireo

f. Hatchling. No down.
g. Fledgling. Distinct wingbars, white underparts.

a.

b.

c.

d.

e.

f.

g.

PLATE 18

VIREOS (*continued*)

Hutton's Vireo
a. Nestling
b. Fledgling. Rounded head, broken eye ring.

Yellow-throated Vireo
c. Fledgling. Yellow eye arcs.

Plumbeous Vireo
d. Fledgling. Buff-yellow wingbars.

Warbling Vireo
e. Fledgling. White supercilium, grayish eyeline, pale eye ring.

Red-eyed Vireo
f. Hatchling
g. Day 7
h. Fledgling. Dark wings edged olive-green.

a.

b.

c.

d.

e.

f.

g.

h.

PLATE 19

CROWS AND RELATIVES

Steller's Jay

Hatchling and Nestling: see Photos 22–24
a. Fledgling. See also Photo 25.

Blue Jay

b. Hatchling
c. Nestling
d. Fledgling

California Scrub-Jay

e. Egg
f. Hatchling
g. Day 3. Grayish-white appearing in ventral tracts.
h. Days 8-10. See also Photo 26.
i. Fledgling. White supercilium.

a.

b.

c.

d.

e.

f.

g.

h.

i.

PLATE 20

CROWS AND RELATIVES (*continued*)

Black-billed Magpie

a. Hatchling, 18 grams. Re-created likeness from MVZ fluid specimen.
b. Fledgling. White upperwing coverts.

American Crow

Hatchling: see Photo 28
c. Nestling
d. Fledgling
e. Head. Bristles barely cover nares, pinkish bare areas.

Common Raven

f. Hatchling. Few tufts of long, stringy down.

a.

b.

c.

e.

d.

f.

PLATE 21

LARKS

Horned Lark (streaked subspecies of Pacific Northwest, *E. a. strigata*)

a. Near-fledgling. Mottled appearance.

SWALLOWS

Bank Swallow

b. Fledgling. Slender notched tail.

Tree Swallow

c. Hatchling, 1.3–1.8 grams. See also Photo 29.
d. Days 4–6, 9–12 grams
e. Days 10–11
f. Days 12–14, 22–24 grams
g. Fledgling. Wing tips reach tail tip.

Violet-green Swallow

h. Hatchling
i. Nestling
j. Fledgling. Wing tips project beyond tail tip.

a.

b.

c.

d.

e.

f.

g.

h.

i.

j.

PLATE 22

SWALLOWS (*continued*)

Northern Rough-winged Swallow

a. Hatchling
b. Day 4. Dark tip on upper mandible.
c. Days 7–10
d. Fledgling. Square tail.

Purple Martin

e. Hatchling
f. Days 6–8
g. Fledgling

Barn Swallow

h. Hatchling, 2–4 grams
i. Day 10, ~20 grams
j. Fledgling. Tail not as forked yet, shorter than adult tail.

Cliff Swallow

k. Hatchling
l. Days 7–9
m. Fledgling. Tail short, square-edged appearance, with spotting.

Cave Swallow

n. Fledgling. Whiter throat, rump patch.

a.

b.

c.

d.

e.

f.

g.

h.

i.

j.

k.

l.

m.

n.

PLATE 23

CHICKADEES

Carolina Chickadee

a. Hatchling
b. Day 7
c. Day 10
d. Fledgling

Black-capped Chickadee

e. Hatchling
f. Nestling
g. Fledgling

Mountain Chickadee

h. Day 6
i. Day 9

Chestnut-backed Chickadee

j. Hatchling

Boreal Chickadee

k. Hatchling. Likeness re-created from 100-year-old fluid museum specimen.

a.

c.

b.

d.

e.

f.

i.

g.

h.

j.

k.

PLATE 24

TITMICE

Oak Titmouse

a. Hatchling
b. Nestling
c. Near-fledgling
d. Fledgling

Tufted Titmouse

e. Hatchling
f. Fledgling

PENDULINE TITS

Verdin

g. Hatchling (d0). See also Photo 30.
h. Fledgling

LONG-TAILED TITS

Bushtit

i. Hatchling. No down.
j. Days 6–7
k. Fledgling

a.

c.

b.

d.

f.

e.

g.

h.

j.

i.

k.

PLATE 25

NUTHATCHES

Red-breasted Nuthatch

a. Nestling, 4.5 grams
b. Fledgling

White-breasted Nuthatch

c. Hatchling
d. Day 11
e. Fledgling

Pygmy Nuthatch

f. Beak. Dorsal view, arrowhead shape.
g. Hatchling. Prominent gape flanges.
h. Fledgling

Brown-headed Nuthatch

i. Fledgling

TREE CREEPERS

Brown Creeper

j. Hatchling
k. Day 14
Fledgling: see Photo 31

a.

b.

c.

d.

e.

f.

g.

h.

i.

j.

k.

PLATE 26

WRENS

Canyon Wren

a. Hatchling
b. Fledgling

House Wren

c. Hatchling
d. Nestling
e. Fledgling

Marsh Wren

f. Fledgling

Carolina Wren

g. Hatchling
h. Nestling
i. Fledgling

a.

b.

c.

e.

d.

f.

g.

h.

i.

PLATE 27

WRENS (*continued*)

Bewick's Wren
a. Hatchling
b. Day 10
c. Fledgling

Cactus Wren
d. Hatchling
e. Fledgling

GNATCATCHERS

Blue-gray Gnatcatcher
f. Fledgling

DIPPERS

American Dipper
g. Fledgling

KINGLETS

Golden-crowned Kinglet
h. Fledgling

Ruby-crowned Kinglet
i. Fledgling

SYVIID WARBLERS

Wrentit
j. Day 7
k. Fledgling

a.

b.

c.

d.

e.

g.

f.

h.

i.

j.

k.

PLATE 28

THRUSHES

Eastern Bluebird

a. Hatchling
b. Days 12–14

Western Bluebird: all California birds

Growth progression: see Photos 32–35
c. Hatchling, 3.3 grams
d. Days 7–8, 18 grams
e. Days 7–8
f. Days 12–14
g. Days 20–21, fledgling

Mountain Bluebird

h. Fledgling

Townsend's Solitaire

i. Fledgling

a.

b.

c.

d.

e.

f.

h.

g.

i.

PLATE 29

THRUSHES (*continued*)

Hermit Thrush
a. Hatchling
b. Fledgling

Wood Thrush
c. Fledgling

American Robin
d. Nest and eggs
e. Hatchling
f. Days 5–6, 36 grams, 9 cm
g. Day 6
h. Fledgling. See also Fig. G8, Illustrated Glossary.

a.

b.

c.

d.

e.

f.

g.

h.

PLATE 30

MOCKINGBIRDS AND THRASHERS

Gray Catbird

a. Hatchling
b. Nestling
c. Fledgling

Brown Thrasher

d. Nestling
e. Fledgling

Northern Mockingbird

f. Hatchling and egg
g. Days 2–3
h. Nestling. See also Photo 36.
i. Fledgling

a.

b.

c.

d.

e.

f.

g.

h.

i.

PLATE 31

STARLINGS

European Starling
a. Day 3
b. Day 7
c. Fledgling

WAXWINGS

Cedar Waxwing
d. Hatchling
e. Day 10. See also Photo 37
f. Fledgling

SILKY-FLYCATCHERS

Phainopepla
g. Nestling
h. Fledgling

SPARROWS, OLD WORLD

House Sparrow
See also Photos 38–40
i. Egg
j. Hatchling
k. Hatchling. Gaping.
l. Day 6
m. Fledgling
n. Adult male
o. Adult female

a.

b.

c.

d.

e.

f.

g.

h.

i.

j.

k.

l.

m.

n.

o.

PLATE 32

FINCHES AND RELATIVES

Evening Grosbeak
a. Fledgling

Red Crossbill
b. Fledgling

Pine Siskin
c. Fledgling, ~10 grams

American Goldfinch
d. Hatchling. Gaping.
e. Nestling
f. Fledgling

SPARROWS, NEW WORLD

Grasshopper Sparrow
g. Fledgling

Chipping Sparrow
h. Hatchling
i. Nestling
j. Fledgling

a.

b.

c.

d.

e.

f.

g.

h.

i.

j.

PLATE 33

SPARROWS, NEW WORLD (*continued*)

Vesper Sparrow

a. Fledgling

Savannah Sparrow

b. Fledgling

Song Sparrow

c. Hatchling
d. Nestling
e. Fledgling

Spotted Towhee

f. Near-fledgling

Eastern Towhee

g. Nestling
h. Fledgling

a.

b.

c.

d.

e.

f.

g.

h.

PLATE 34

YELLOW-BREASTED CHAT

a. Juvenile. 1st fall.

BLACKBIRDS AND RELATIVES

Yellow-headed Blackbird
b. Fledgling

Bobolink
c. Fledgling

Western Meadowlark
d. Fledgling

Orchard Oriole
e. Fledgling

Baltimore Oriole
f. Nestling
Fledgling: see Photo 58

Red-winged Blackbird
g. Hatchling
h. Fledgling

Common Grackle
i. Hatchling
j. Day 6
k. Fledgling

a.

b.

c.

d.

e.

f.

g.

h.

i.

j.

k.

PLATE 35

WARBLERS, NEW WORLD

Ovenbird
a. Fledgling

Black-and-white Warbler
b. Fledgling

Prothonotary Warbler
c. Fledgling

Orange-crowned Warbler
d. Fledgling

Common Yellowthroat
e. Fledgling

American Redstart
f. Fledgling

Northern Parula
g. Fledgling
h. Tail underside

Yellow Warbler
i. Fledgling

Pine Warbler
j. Fledgling

Yellow-rumped Warbler
k. Fledgling, newly fledged, grayish, "fluffy"
l. Fledgling, darker, sleeker, longer tail

Wilson's Warbler
m. Fledgling

a.

b.

c.

d.

e.

f.

g.

h.

i.

j.

k.

l.

m.

PLATE 36

CARDINALS AND RELATIVES

Summer Tanager
a. Fledgling

Scarlet Tanager
b. Fledgling
c. Fledgling. After molt.

Northern Cardinal
d. Nestling, days 2–3. See also Photo 64 (nestling gaping).
e. Fledgling

Rose-breasted Grosbeak
f. Fledgling

Black-headed Grosbeak
g. Days 3–4, 8 grams
Near-fledgling: see Photo 65

Blue Grosbeak
h. Fledgling

Indigo Bunting
i. Fledgling

Dickcissel
j. Fledgling

a.

c.

b.

g.

d.

e.

f.

h.

i.

j.

PHOTO GALLERY

1. Eurasian Collared-Dove hatchling | 2. Mourning Dove, day 2 | 3. Mourning Dove, days 10–15 | 4. Common Nighthawk fledgling | 5. Green Heron nestling | 6. Black-crowned Night-Heron nestling, 111 grams | 7. Black-crowned Night-Heron nestling | 8. Black-crowned Night-Heron brancher

Vonda Lee Morton

Bobbie Hefner

Vonda Lee Morton

Linda Adams

Linda Adams

Annette Purther

9

10

11

12

13

14

9. Mississippi Kite nestling | 10. Cooper's Hawk, day 28, 340 grams |
11. Red-shouldered Hawk nestlings | 12. Great Horned Owl nestling |
13. Great Horned Owl juvenile | 14. Northern Pygmy-Owl nestling

15. Northern Pygmy-Owls, 61 and 67 grams | 16. Barred Owl, day 21 |
17. Red-headed Woodpecker, 51 grams | 18. Acorn Woodpeckers, day 18 |
19. Red-breasted Sapsucker fledgling | 20. Northern Flicker, days 19–20 |
21. American Kestrel nestling

22. Steller's Jay hatchling, day 3, 18 grams | 23. Steller's Jay nestling, day 7, 42 grams | 24. Steller's Jay nestling day 10, 62 grams | 25. Steller's Jay fledgling, day 18, 100 grams (with photo of adult) | 26. California Scrub-Jay nestlings, rose-pink gape | 27. Yellow-billed Magpie nestling, MVZ fluid specimen | 28. American Crow hatchlings

29

Evelien De Greef

30

Linda Adams

31

Linda Adams

32

Linda Adams

33

Linda Adams

34

Linda Adams

35

Linda Adams

36

Vonda Lee Morton

37

Connie Black

29. Tree Swallow hatchlings, three siblings | 30. Verdin nestling, MVZ fluid specimen | 31. Brown Creeper fledgling | 32. Western Bluebird hatchlings, days 0–1, 2–4 grams | 33. Western Bluebird hatchlings, days 3–4, 7.3–10.5 grams | 34. Western Bluebird nestlings, days 7–8, 17.8–19.5 grams | 35. Western Bluebird nestlings, days 9–10, 22.5–24.4 grams | 36. Northern Mockingbird nestling, bright yellow gape | 37. Cedar Waxwing nestling, cherry-red gape

38   *Linda Adams*

39   *Linda Adams*

40   *Linda Adams*

41   *Linda Adams*

42   *Linda Adams*

43   *Linda Adams*

44   *Linda Adams*

45   *Linda Adams*

46   *Linda Adams*

38. House Sparrow hatchling, day 2, 5.5 grams | 39. House Sparrow nestling, day 7, 16.5 grams | 40. House Sparrow nestling, day 11, 24 grams | 41. House Finch, just hatched with egg, 1.1–2.3 grams | 42. House Finch, day 2, pink gape | 43. House Finch, day 3, 5–6 grams | 44. House Finch siblings, day 7, 12.5 and 15.7 grams | 45. House Finch, day 7, 14 grams | 46. House Finch, just fledged

47 *Linda Adams*

48 *Linda Adams*

49 *Linda Adams*

50 *Ashton Klutz*

51 *Linda Adams*

52 *Janice Barbary*

53 *Linda Adams*

54 *Linda Adams*

55 *Linda Adams*

47. Lesser Goldfinch hatchling | 48. Lesser Goldfinch nestling | 49. Lesser Goldfinch fledgling | 50. Lark Sparrow, just fledged | 51. Lark Sparrow, older fledgling | 52. Dark-eyed Junco nestling | 53. California Towhee hatchling, day 3, 7–8 grams | 54. California Towhee nestling, day 6, 15 grams | 55. California Towhee nestling, day 12, 29 grams

56. Hooded Oriole fledgling | 57. Bullock's Oriole fledgling | 58. Baltimore Oriole fledgling | 59. Brown-headed Cowbird nestling | 60. Brown-headed Cowbird nestling, salmon-pink gape | 61. Brewer's Blackbird nestling, red gape | 62. Brewer's Blackbird near-fledgling | 63. Western Tanager nestling, 17 grams | 64. Northern Cardinal nestling, red gape | 65. Black-headed Grosbeak near-fledgling | 66. Lazuli Bunting fledgling, female

APPENDIX A

COLOR CHART AND TERMINOLOGY

COLOR VARIATION AND INTERPRETATION

It may seem that colors on baby birds should be simpler to assess than on adults because there are fewer colors to choose from. However, having fewer colors makes it more difficult to distinguish between species. As explained in Chapter 2, there is variation in the coloration of soft tissues, the integument, and associated structures such as feathers that are affected by diet, health, and other factors related to where individuals live. Plumage colors are richest in newly fledged birds, but gradually, as delicate feather edges become worn and plumage fades, identification becomes more challenging.

The descriptions in published literature often disagree owing in part to regional differences and because of the absence of a single standard color reference. Some descriptions in published literature were from observations during the 1800s when the only references available were unsuitable for naturalists. In one of the first color-naming systems for birds, *A Nomenclature of Colors for Naturalists*, Dr. Robert Ridgway (1886, 9) states, "Undoubtedly one of the chief desiderata of naturalists, both professional and amateur, is a means of identifying the various shades of colors named in descriptions, and of being able to determine exactly what name to apply to a particular tint which it is desired to designate in an original description." Ridgway produced an updated version in 1912, *Color Standards and Color Nomenclature*, that is still being used today. Both of these volumes are available online to download. Another system commonly used by birders is the Munsell soil-color charts, because they cover all the browns and grays. Newer systems may be difficult to find, expensive to purchase, and quite technical. Currently, hundreds of color names exist, and dozens of color order systems have been created in the last two hundred years.

Disagreement in determining color in nature also results from human interpretation and ambiguity. We can measure a gram, but color is more illusive. It can be skewed by lighting, shading, translucency, and transparency. What is red to one person may be red-orange or a shade of brown to another. Some colors are ambiguous or vague, such as tan, horn, and flesh. Some artists' renderings of young birds, especially those where information on downy feathers was not available, were obtained by deducing *probable* colors from faded material in museum collections or from wisps of down remaining on juvenile birds. Historical descriptions of color may be described differently today because of current technology. For example, the natal down color of a hatchling Tree Swallow was described by Wetherbee (1957) as "white or pallid mouse-gray." Pallid, by current definition, means *abnormally* pale or lacking intensity of color, suggesting poor health. However, the Tree Swallow is one of the best-studied birds in North America, and many photos are available from numerous nest box studies. By enlarging photos on an Apple computer monitor to 24 × 13 inches, the down appears very vibrant and shimmery in colors of whitish and pale brownish or gray. Color descriptions in the species accounts were verified whenever possible by referencing multiple sources, including photographs and live and preserved specimens. Where colors could not be verified and only published descriptions were available, all descriptions (even if they contradict) have been included. Most important, when comparing the color of your bird with descriptions, illustrations, or photos, some latitude should be allowed for variation based on the source and date of the description and interpretation by the examiner.

NAMES OF COLORS AND COLOR TERMINOLOGY

Names of colors in this guide have been defined or described in the table below. Most of the color names are from Ridgway 1886 and 1912, as they often coincide with watercolor paint colors I used to paint the birds in this guide. *Hue* is a color or shade and is the dominant or principal color. *Intensity* refers to the brightness of a color. A *tint* is any hue with white added, which makes the color paler but does not change the original color. A *tone* is produced when gray is added to a color, resulting in a duller or more neutral intensity. *Tinge* (or tinged) means a slight shade, stain, or "wash" over another color. Modifiers of colors such as pale, drab, dull, light, dark, or medium may indicate lightness, darkness, or brightness. Some modifiers may also be a color, such as drab. The suffix "-ish" at the end of a color refers to a tinge and allows for more variability. For example, "grayish-brown" is having a gray tinge on brown feathers.

Table A-1. Colors found in birds

Main color	Variations and description	
Apricot	Light yellowish-orange	
Black	Technically a "shade"; variations: slate-black, blackish-slate.	
Blue	Cerulean blue	Fine light blue. The name of a paint color.
	Cobalt blue	Bright deep blue (lighter than Prussian blue). The name of a paint color.
	Indigo blue	Dark dull blue. The name of a paint color.
	Phthalocyanine blue	Bright greenish-blue. The name of a paint color.
	Prussian blue	Intense, rich blue; darker and more greenish than ultramarine and cobalt. The name of a paint color.
	Smalt blue	Deep purplish-blue. The name of a paint color.
	Ultramarine blue	Deep pure blue. The name of a paint color.
Brown	Burnt sienna	Rich reddish-brown. The name of a paint color.
	Burnt umber	Deep rich brown, more reddish than sepia. The name of a paint color.
	Chestnut	Rich, dark reddish-brown with slightly purplish cast.
	Cinnamon	Light reddish-brown (or yellowish-brown).
	Clove-brown	Lamp-black + cadmium orange.
	Fawn	Light, warm brown (burnt umber + white).
	Fulvous	Brownish-yellow or yellowish-brown tint, like tanned leather; tawny.
	Fuscous	Dark brown.
	Hair-brown	Clear, somewhat grayish tint of brown (similar to the brown hair of humans).
	Mouse-brown	Dull light brown.
	Raw sienna	Bright yellowish-brown. The name of a paint color.
	Raw umber	Light yellowish-brown. Umber in natural form. The name of a paint color.
	Russet	Dark brown with reddish-orange tinge.
	Sepia	See "Sepia" below.
	Umber	Natural brown or reddish-brown earth pigment. See "Raw Umber" and "Burnt Umber."
	Walnut-brown	Deep warm brown, like heartwood of the black walnut.
	Wood-brown	Light brown, like some varieties of wood.
	Vandyke brown	Rich deep brown, similar to burnt umber but less reddish. The name of a paint color.

Table A-1. Colors found in birds (*continued*)

Main color	Variations and description	
Buff	Pale yellow-brown, of undyed or clean chamois leather. Variations: buff-cream, buffy, buffy-white, buffy-yellow, buffy-gray, Tilleul-buff.	
Coral	A shade of orange. Various tones of coral from coral-oranges to coral-pinks.	
Cream	Light pinkish-yellow, like the oily, yellowish part of milk.	
	Creamy-buff	Paint mix: yellow ochre + white
Crimson	See "Red."	
Drab	Dull, light brown, such as drab-brown. Also used as an adjective to describe dullness, such as drab brown.	
	Drab-gray	Black + white + burnt umber.
Dusky	Dark color of more or less indefinite or neutral tint.	
Ferruginous	Rust-red or the color of rusted iron.	
Flesh or flesh-color	Ambiguous and may be considered offensive, with definitions ranging from buff to salmon. Original meaning: pinkish color, "like that observable in the cheeks of a person of fair complexion" (Ridgway 1886, 79). Wherever possible, "flesh" has been replaced with a more appropriate color.	
Fuscous	See "Brown."	
Gray (or grey)	When gray is mixed with a color, it is defined as a *tone* and of various shades depending on the amount of black or white. Variations: gray-brown, white-gray, blue-gray, bluish-gray, brown(ish)-gray, smoke(y)-gray, sooty-gray, slate-gray, charcoal-gray.	
	Cinereous	Ash-gray, a clear bluish-gray, lighter than plumbeous.
	Mouse-gray	Beige-gray, the color of a house mouse (black + white + sepia).
	Pearl-gray	Pearl = pale tint of off-white or cream.
	Plumbeous	Deep bluish dull gray, like tarnished lead.
	Slate (gray)	Dark or blackish-gray, less bluish in tint than plumbeous or lead.
Green	A mixture of blue and yellow.	
	Olive or olivaceous	Greenish-brown or muddy green, like that of olives. Variations: olive-brown, olive-buff, olive-gray, olive-green, olive-yellow.
	Sage-green	Dull grayish-green, like leaves of the herb.
Ochre	See "Yellow."	
Olive	See "Green."	
Orange	Deep reddish-yellow, like the rind of an orange (between red and yellow on the light spectrum). Derived from pigments found in plants and animals. Variations: orange-yellow, orange-red, orange-buff.	
	Burnt-orange	Medium dark (reddish) orange.
Pallid	Not a color; an adjective indicating pale, faded, or lacking color.	

Table A-1. (*continued*)

Main color	Variations and description	
Pink	Pale rose-red with a large range of variations.	
	Pink(ish)-buff	Light pinkish-brown.
	Geranium-pink	Lighter than geranium-red.
	Rose-pink	Pure purplish-pink color.
Plumbeous	See "Gray"	
Purple	Red + blue.	
Red	One of the three primary colors. Red is derived from pigments, melanins, and porphyrins. Variations: red-orange, orange-red, copper-red, cherry-red, ruby-red, watermelon-red, brick-red, pink (see "Pink"), reddish-brown.	
	Carmine	Very pure and intense crimson, or deep red.
	Coral-red	Light, dull vermilion.
	Crimson	Rich, deep reddish-purple. Crimson color comes from bodies of scale insects that are eaten by birds.
	Ferruginous	Rust-colored, reddish-brown.
	Geranium-red	The purest possible red color.
	Rose-red	The purest possible purplish-red color.
	Ruddy	Reddish rosy-crimson, closer to red than to rose.
	Rufous or rufescent	Brownish-red or reddish-brown.
	Russet	See "Brown."
	Salmon	Intermediate between flesh-color and orange, like the flesh of the fish. Ranges: pale pinkish-orange to light pink.
	Scarlet	Brilliant red with a tint of orange. Lighter and less rosy than carmine, richer and purer than vermilion.
	Vermilion	Fine red with a bit of orange, lighter and less rosy than carmine, not as rich as scarlet.
	Vinaceous	Brownish-pink or delicate brownish-purple; soft, delicate wine-colored pink or purple. Resembles red wine or grapes. Variations: vinaceous-buff, vinaceous-pink, vinaceous-rufous.
Ruddy	See "Red."	
Rufous	See "Red."	
Salmon	See "Red."	
Scarlet	See "Red."	
Sepia	Black-brown, a very deep dark color, found abundantly in wild birds and used extensively in the paintings.	
Tan	Pale tone of brown.	
Tawny	Of tanned leather, an orange-brown or yellowish-brown color.	
Tilleul	Pale yellowish-green. Variation: tilleul-buff.	

Table A-1. Colors found in birds (*continued*)

Main color	Variations and description	
Vermilion	See "Red."	
Vinaceous	See "Red."	
Violet	Purple color of blue and red.	
White	Technically a "shade" defined as "light" rather than a color. Adding white to a color "lightens" a color, making it less intense. Variations: snow, creamy-white, buffy-white.	
Yellow	Variations: Lemon-yellow, yellow-orange, orange-yellow, buff(y)-yellow, citrine-yellow, Naples-yellow, fulvous (see "Brown").	
	Ochre (ochreous)	Ridgway 1886: A brownish-orange or intense buff.
	Primrose-yellow	Like the flower, varies pale to bright yellow.
	Straw	Very light impure yellow, like cured straw.
	Sulphur-yellow	Very pale pure yellow, less orange than lemon-yellow.

IDENTIFICATION WORKSHEET

Family or Group _____ **Species** _____
    Natal Type: altricial _____ precocial _____ Weight (g) _____
    Where bird was found: _____

Use this worksheet along with the guide while you are learning what clues to look for on your bird. You will also need both a good field guide to reference adults and range maps.

Questions to Ask the Finder (before you receive the bird):

1. TOES and FEATHERS. See Tables 2.6, 3.1, and 3.3. If not immediately obvious, see "Foot Type" and "Age Assessment," below.

2. ADULT BIRDS. Present? Are they the parents? Are they feeding or protecting?

3. HABITAT and NEST. See Tables 4.1 and 4.2. Was bird found near water, in a field, below a tree, in a parking lot, near a building, on the road, etc.? Is the nest open-cup, near a tree cavity, made of mud, "sock-like," or like a woven basket? If the empty nest was dislodged, have the finder bring it.

4. EGGS. If any eggs or bits of shell were found (have the finder bring), note the shape, size, color, and any markings.

5. BODY SIZE. Compare to a common bird (small like a goldfinch, medium like a robin, large like a crow or hawk) or compare to an

inanimate object, for example, dime (~0.5 in., 1.8 cm); quarter
(~1 in., 2.4 cm); golf ball (~1.7 in., 4.3 cm); tennis ball (~2.5 in., 6.5
cm); baseball (~3 in., 7.5 cm); football, etc.

Examining the Bird

1. WEIGHT and SIZE OF BIRD (in hand). Weigh bird (to the nearest
   tenth of a gram if small). Record above.

2. FOOT TYPE (arrangement of the toes and webbing). See Table 2.6
   and Fig. G5 a–h (Illustrated Glossary). How many toes are pointing
   forward and backward? Is there webbing, and if so, how many toes
   are involved? Are the nails long and acute or chicken-like?

3. MOUTH. Mouth color is useful and most reliable in passerine birds
   (Table 2.1) and hummingbirds. In most other species, mouth color is
   not a reliable or useful feature for ID. Note the gape flange color and
   thickness (thick, thin, not evident, reduced); if more like "knobs" on
   the lower mandible (and zygodactyl feet), consider woodpeckers. Is
   the gape (open) large, small, wide, or angulated, and are there any
   markings or unusual structures inside the mouth on the tongue or
   upper palate?

4. BEAK SIZE, SHAPE, and FEATURES. See Tables 2.2 and 2.3 and
   Fig. G3 (Illustrated Glossary). Size: *relative to head*, is the beak
   shorter (short), same as (medium), longer (long), tiny, small, thin,
   big (like grosbeak)? Is the culmen short, medium or long, decurved or straight? What is the beak shape: conical, circular, acute,
   compressed, depressed, wide, flat, or stout? Does it have a "tooth"
   (notch), a hook at the tip, or a bump on top at the base?

5. AGE ASSESSMENT based on down, juvenile feathers, and behavior.
   See Tables 3.2 and 3.3 to determine a stage of life.

6. FAMILY or GROUP. From features you have noted, go to the species
   accounts in the guide and find a family or species that fits. For passerine birds, mouth color will eliminate almost half of your choices.

7. DOWN COLOR and APPEARANCE. For altricial species, if the bird
   has no down, see Table 2.9. If downy, use Table 2.11. For precocial
   species of newly hatched chicks, note the color of down, conspicuous
   dorsal pattern, beak shape, and details of the tarsus and toes.

8. FEATHER ATTRIBUTES. These would be the juvenile feathers.
   Colors. See Table A-1, Chart of colors (Appendix A). Is the bird
   uniformly light or dark or strikingly two-toned (dark above, light
   below)? What colors are evident, and where are the colors?

Remember that juvenile males may resemble the adult female.

Patterns or features that stand out. Note stripes, streaks, spots, speckling, or mottling on the chest, belly, or dorsal areas. Does plumage resemble the bark of a tree (see owls or nightjars)? Is the coloring disruptive?

Head. Describe the shape, conspicuous facial pattern (auricular or cheek patches, "mustache stripe") and color, eyelines (above eye, through eye), eye rings or spectacles (complete or incomplete). Head feathers may be crested. Is the crown striped? Is the face bald (especially around the eyes)?

Upperparts (dorsal surface). Starting at the crown and working down to the tail, include top of wings. What is the overall color? Is there a rump patch or crissum?

Underparts (ventral surface). Begin at the chin, work down to underside of tail (undertail coverts). Include side, flank, and under surface of the wing. Is the underside striped, spotted, mottled? For example, the young of thrushes have conspicuous spots on throat and chest, and many sparrows have mottling or streaks on their undersides. Are the undertail coverts a different color?

Wings. Are they rounded or sharply pointed? Are they very long relative to the body? Still in sheaths? Wingbars: one or two, color, bold, defined, indistinct? Spots or barring on the wings and back?

Tail. See Table 2.10 for shapes and characteristics. How long are the tail feathers relative to the body? Are the feathers still in sheaths? Is the tail forked, notched, square tipped, pointed, or long and graduated? Are outer feathers all white, partly white, with patches, spots, or patterns? Are there stripes?

9. TARSUS, TOES, NAILS, and SCALES. See Tables 2.5, 2.6, and 2.7 and Fig. G5 (Illustrated Glossary). Note colors. Are toes small, weak, or strong, and what is the length relative to the leg? Describe scales on the feet.

10. SKIN COLOR. See Table 2.4. Color is variable, changes daily, and is subject to age and health.

11. EYES. Are the eyelids closed, partly open (slits), or fully open? In general, for passerines, eyelids are closed the first 3–4 days post-hatching, partly open days 3–5, and fully open days 5–7 or longer for bigger birds. Eyelids of some altricial species are open at hatch (Table 3.1).

12. BEHAVIOR (for assessing age and possible family group). Nestling: Does it weave or shiver while begging? Does it evacuate in the nest or over the nest? What kind of begging behavior does it display? Does it flatten out or show a fear response when disturbed? Is it stretching, preening, and exercising wings? Can it perch on the nest edge or a branch?

13. VOCALIZATIONS (described in species accounts). Are sounds weak or strong? What kind of sounds: chirp, cheep, buzz, churr, peep, squawk, whistle, staccato trill, whee-you (continuous), like a rusty hinge (raucous, repeated), vibrating, wheezy quality, high pitched, one note, several notes? Does the bird call incessantly?

GROWTH AND DEVELOPMENT TABLES

Key to abbreviations: see Introduction to the Species Accounts.

Table C-1. Growth and development of Yellow-billed Cuckoo
(*Coccyzus americanus*)

Age in days	Development
0 = hatch d	8–9 g, gains 4.9 g/d (avg) 1st 4 days. Lifts head. Gapes.
1	PinF short, sparse. Eye slits. Stands, flaps wings. Strong grip. Begging: slight wing shivering (more rapid with development).
2	14–15 g. Perches.
3	18–19 g. Defecates over nest. Eyes open one at a time.
5	27–30 g. Wings and tail continue to grow. Fully covered, long grayish quills (4–5 cm long). Very alert. Eyes fully open. Threat response: hunches down, motionless.
6–7	26–29 g. Appears FF, grayish.
7–9	Fledges, 32–38 g, cool weather may delay. Can run. Flies short distances.

Sources: Preble 1957; Parkes 1984; Franzreb and Laymon 1993.

Table C-2. Growth and development of Black-billed Cuckoo (*Coccyzus erythropthalmus*)

Age in days	Development
0 = hatch d	7.5–9 g (7 chicks), 6–7 cm; gains 4.7 g/d 1st 4 days.
1	Strong grip. Begging: slight wing shivering (more rapid with development). Voc: begging like "buzzing of a bee."
2	Eye slits.
3	Neossoptiles < 3 mm long. Ventral PinF barely visible.
4	Eyes open. Follows moving objects. Stretches.
5	Defecates over nest edge. Aware of sounds outside of nest. Feather tips may still have downy tips.
6	Feathers resemble porcupine quills (17–21 mm long). Much preening. Response to threat: loud *bark*.
6–7	28.5 g. May fledge. Cannot fly. Can jump and climb onto branch from ground.
7	Preens. PinUn within 6–12 hrs (except head and neck). Fluffy appearance.
21–24	First flight.

Source: Spencer 1943.

Table C-3. Growth and Development of Downy Woodpecker (*Picoides pubescens*)

Age in days	Development (Mass [avg] for wild raised listed first; mass for captive raised in parenthesis. Length is from wild raised.)
0 = hatch d	3.5 (2–3) g. Gapes. Stretches neck. Voc: *pips* and *rasps*.
2	6 (5–8) g; ~54 mm long. PinVis posterior ventral tract, greater prim and sec covs of most developed N. Holds head erect several seconds.
4	9 (14–17.5) g. Eyes and ears begin to open. White PinF outer rect. Feather tracts and nails darkened. Stands on hocks.
6–7	12 (17.7–23) g; ~93 mm long. PinEm most tracts, including wings and tail.
8	18 (21.2–24.2) g. Eyes and ears fully open. M has red crown.
8–11	Most PinUn. Begins to thermoregulate.
12	24 (25–27) g. Knobs at rictus reduced.
12–14	Capital (anterior) PinUn. Can cling. Hunches down when light at nest entrance changes.
14	27 (25.5) g; ~125 mm long. Peak body mass reached. d15 climbs cavity walls. Short flights.
16	142 mm long. Mass decreases. Mostly FF except for rect. Flies well. Climbs. Tries to escape. Hides under vegetation. Voc: adult-like *whinney* call.
18–21	Fledges, 25–26 g.

Sources: Mass and length for wild raised: Hadow 1976. Mass for captive raised, from Nevada County, CA (Nancy Barbachano).

Table C-4. Growth and development of American Crow
(*Corvus brachyrhynchos hesperis*)

Age in days	Weight (g)	Development (Age, weights, and developmental stages *highly* variable by ssp, regions, individuals, sex, and availability of resources)
Egg	16.5–17	42 × 29 mm.
0 = hatch	10–13	15.6 g (SK, Canada). Size of head (tip of bill to back of head): 27 mm (adult head 85 mm). Gapes. Voc: weak sounds.
1–2	13–23	First downy-like feathers appear on crown, humeral, alar, and spinal tracts.
3	23.4–26.4	Feathers continue to grow.
4–5	33–46	Prim PinEm, tail ~1 mm. d5: eyes still closed or barely slits. Yolk sac: 10 mm × 8 mm. Voc: variety of sounds.
6–7	53–66	Body PinVis. Bill hardens.
8–10	81–120	Period of rapid growth begins (continues to d18). Skin: darkened. Eyes opening. PinEm under "downy" coat on dorsal and ventral tracts.
11–15	135–210 (Emlen 1942)	Brooding by adult tapers. Eyes fully open. Wings: longest prim sheaths (nos. 5 and 6) 16 mm; d15 unsheathed tips appear (6–7 mm). d12–14: yolk sac gone. d14–17: fear response begins.
16–18	225–250	d18: 250 g. Voice lower in pitch. Fear reaction: crouching.
21		Middle primaries (nos. 5–7) extend ~25 mm beyond sheath.
19–25	255–300	Crown and body FF: fine, short, fluffy feathers that continue growing. Wings and tail in blood-feathers. Middle prim (nos. 5–7) extend ~50 mm beyond sheath. Still gapes. Begging: high-pitched *whine*.
25–30	300+ (avg)	Can thermoregulate. Wings: d25: prim > than 50 mm beyond sheaths; d29: middle primaries (nos. 5–7) extend ~75–100 mm beyond sheath. d28: tarsus adult size. More aware of surroundings, watches movements. May attempt to escape.
Fledges 28–38	300–370 (variable)	Older nestlings (+30 d) begin moving in and out of nest. GF reduced. Plumage: abdomen usually covered (still bare in some). d33: middle prim extends ~100 mm beyond sheath. d38: middle prim extends ~125 mm beyond sheath. Still fed by parents.
44		Middle prim extends ~152 mm beyond sheath.

Sources: Emlen 1942; Parmalee 1952; and author's rehabilitation records.

Table C-5. Growth and development of Tree Swallow (*Tachycineta bicolor*)

Age in days	Development
Egg	1.2–2.6 g.
0 = hatch d	1.3–1.8g (2.3 AL, Canada), 1.6–1.7 g (ON, Canada). Gains 2.14 g/d (NY). Wing chord 6 mm, length 12.1 mm. Uncoordinated. Raises head to gape. Waves tiny wings.
2–3	PinVis. Wing PinEm. Active. Voc: begging calls.
4–6	9–12 g. Prim and rect PinEm.
5	Eyes begin to open (slits).
6–7	12–15 g; dorsal (wings and flanks mainly) contour PinUn. Down may have disappeared in some. Voc: loud and persistent *peeping*.
7–8	8th prim PinEm.
8–9	~18 g. May be able to thermoregulate (depending on ambient temperature and other variables).
10	Eyes fully open. Stretches wings. Begins to preen. GF shrinking.
10–12	Wing and tail PinUn. Downy remnants may or may not be present.
12–14	22–24 g, the most it will weigh (loses weight before fledging). Wing PinUn ~50%. Creeps around. More fearful (crouches). May leave nest early; will not stay if replaced.
16	FF, flight feathers dark gray.
18–25	Fledges, 20–21 g. Flies well. Wing chord ~78 mm.

Table C-6. Growth and development of Western Bluebird
(*Sialia mexicana*)

Age in days	Development[a]
Egg	1.81–2.09 g, pale blue, occasionally white, unmarked.
0 = hatch	Mass (CA): 2.1-3 g; (OR): 2.9 g. Weak, can barely raise head to gape. Voc: faint *peeping* sounds.
1–4	Mass (CA) d1: 3.6–3.9 g (4.4 cm); d2: 6.1 (5 cm); d3–4: 7.3–10.5. Mass (OR) d2: 7.3 g, d3: 10 g. PinVis, areas along feather tracts darkening.
5	Eyes begin to open. Skin ruddier. Bill and legs darkening. PinEm wings and tail. Vocalizes strongly when parents arrive.
6–7	PinEm on capital, spinal, and rear portion of ventral tracts. Throat and chest feathers fill in gradually.
7–8	CA: 17.8–19.5 g (6.7–7.1 cm). Eyes begin to open. d8: grasp is stronger; crouches in nest.
8–11	d10–11: 90% of mass achieved. Eyes fully open. PinUn (d8). Bill-snap when parent alarm call is given (d10). Freezes or startles when disturbed (d11).
9–12	CA: d9–10: 22.5–24.4 g (7.5 cm). Daily energy requirement for wild nestlings ~15.5 g insects/day/individual.
12–14	CA: d12–13: 25.2–27.4 g. Feathers of wings and tail of M show smalt or cobalt blue; of F are dull gray-blue or brownish. Legs still unable to support full weight, unsteady.
15–16	Becomes restless. Peers out cavity entrance. Freezes when nest approached.
17–21	Almost FF (d18). d21: LF dark gray, little down evident (disappears on crown last). Very vocal but freezes if disturbed.
~21 days (avg)	Fledges ~28 g avg. Ranges: 22-23 d (CA), 18–25 d (AZ) and 16–23 d (OR). FF, some bare areas under wings and cloaca, most down gone. Wings short. Tail very short.

[a] Mass and length of CA birds are given as a range between the smallest nestling and largest nestling from one clutch of 6 that hatched in a 24-hour period from one nest box study, Nevada County, CA, May 2017. Data from OR and AZ is from a larger sampling (Billerman et al. 2020).

Table C-7. Growth and development of American Robin (*Turdus migratorius*)

Age in days	Development*
0 = hatch	(2) 4.1-6.7 g, 5-6.2 cm
1	(1) 8.9 g, (2) 5.4-11.9 g, 6.1-7.3 cm.
2–3	(1) d2: >14.3 g, (2) d2-3: 8.4-25.2 g. More down covers body. Mouth color: richer and darker. Gape very large. Beak: beginning to darken, thrush-like shape.
3	(1) 21.3 g, (2) 12.2-25.2 g. Feather tracts: dark on dorsal, wing, thigh, and top of head. PinEm wings.
4–5	(1) 26.6 g, (2) 17.9-39.5 g, 8.9-11.5 cm (varies). Beak size can vary considerably d4–7. Eyes begin to open (slits). PinVis all tracts, bird appears darker and bluish. Tail PinEm. PinF longer on wings. Measurements (1 N, CA): mass 24.9 g, length 8 cm, beak culmen 12 mm, tarsus 12 mm, hallux with nail 12 mm, middle toe with nail 17 mm, wing chord 28 mm.
5–6	(1) 32.2 g, (2) 23.7-45.9 g. PinUn at tips all over (including ventral tract), dorsal tract longer. Some upperwing covs rust colored. Beak: strongly thrush-like. GF: swollen at corners.
6–7	(1) 40.1 g, (2) 32.5-58.8 g. Eyes: fully open, very large, dark (typical of thrush family).
7	(1) 47 g, (2) 36.2-58.8 g. Wing covs grown in.
8–9	(1) 52 g, (2) 42-61.4 g. Feathers have broken out all over with some down poking through on crown, humeral areas, and lower dorsal region. Parent stops brooding (siblings keep each other warm). Food: whole worms and large insects.
9–10	(1) 55 g, (2) 43.4-63.2 g. Appears FF. Beak: darkened and longer. GF: beginning to reduce in size (varies), somewhat down-curved. May leave nest early if alarmed, hide in vegetation.
10–11	(1) 56 g, (2) 49.0-61.1 g, 13-15 cm. FF but still partly in sheaths (especially wings and tail). Wings rounded and shorter or as long as body. Tail very short.
13–16	Fledge 56 g (avg) (range 50–62 g, fledglings received at rehab centers). Cannot fly well. Wings and tail not full-length.

*Source of weights and lengths: (1) Hamilton 1935; (2) Howell 1942.

Table C-8. Growth and development of European Starling (*Sturnus vulgaris*)

Age in days	Development (c = culmen, t = tarsus)
Egg	Size (cm): 29.46 × 21.4 (e. US), 29.75 × 21.37 (Mid-w. US), 29.35 × 20.93 (w. US)
0 = hatch	5–6.4 g. Nearly naked. c = 8 mm, t = 9 mm. Gape in response to disturbance.
2	Begins to move around.
3	PinVis dorsal, scapular, and femoral tracts. In dorsal tract, pigmentation widens into a diamond shape in center of the back, then forms an inverted Y-shape about the uropygial gland. PinVis prim and sec.
4	~8.2 cm long (neck stretched). Crawls around. Uses wings and legs to propel.
6–7	PinUn on some prim, sec, ventral, and dorsal tract toward tail. Eye slits. (Note: growth of 1st sec is used as an indicator of age in starlings, emerges ~6.5–7 d, and on d9 begins to unsheathe at distal tip of the sheath.)
8	Eyes fully open. Ventral PinUn. Parent ceases brooding.
10	Most wing and tail PinUn. Ventral feathers are becoming brown.
11	71 g (d11–12). c = 17 mm.
12	Shows fear, attempts to escape if disturbed. t = 29 mm.
13	Able to thermoregulate.
14	Most capital PinUn. Very active, stretches and beats wings.
15–21	78 g (d17). d18: c = 21 mm, t = 33 mm. Nearly FF. Some can fly by d19.
21–23	Fledges, 71 g. Flies well. Some retain egg tooth. Down may still be attached to contour feathers, especially head and rear regions. d22: c = 22 mm.

Source: Kessel 1957.

Table C-9. Growth and development of House Sparrow (*Passer domesticus*)

Age in days	Development
0 = hatch	1.5–3.0 g; length 40 mm. Hatch naked, no down. Dorsal tracts appear (6–10 hrs post-hatching).
2	4.5–5.5 g; 48.6 mm. PinVis dorsum, Prim PinEm. Can crawl, turn head. Voc: faint *cheep*.
3	6–8.5 g; 56.1 mm. GF enlarged, color varies. Eyes begin to open (d3–5). Legs and wings longer.
4	9.5–11.5 g; 63.8 mm. Egg tooth nearly gone. Ear openings evident. Head and dorsal tracts darker. Prim and some sec 1–2 mm.
5	11.5 g; 67.6 mm. PinEm all over. Eyes open.
6	14 g; 73 mm. PinUn dorsal and ventral. Prim, sec, and tail PinF longer. Sec cov PinUn, tipped brown.
7	16.0–17.0 g; 82.5 mm. GF reducing. Wing prim ~6.6 mm, prim covs ~4 mm. Brownish head feathers resemble spikes in rows; back brown, neck darker. Ventral PinEm buffy. Rect PinUn, 5.3 mm. Legs darker, scales evident.
8–10	19–23 g; 91 mm. More brownish overall. PinUn capital and spinal. Bill and nails darker. d10: prim ~20 mm, rect 11 mm. Fear response (d8), cowers, backs away, chest lower than rear; holds tail up (like wrens); may leave nest if disturbed. Voc: shrill *cheep* if alarmed or handled.
11–12	23–24.5 g; 103–107 mm. GF thinner. Beak: dark "ring" evident. Can thermoregulate. FF, sleek appearance, supercilium, wingbars and back stripes evident. Can perch well and flies if needed. May fledge. Voc: loud *cheep*.
13	23–25.5 g; 115 mm. Perches at front of nest. May leave.
14–16	Fledges, 25–27 g. d14: 117 mm, d15: 123 mm, d17: 126 mm. GF much reduced. d15: prim 44.6 mm, rect 30.7 mm.

Source: Wild clutches, Nevada County, CA.

Table C-10. Growth and development of House Finch (*Haemorhous mexicanus*)

Age in days	Development (c = culmen, t = tarsus, w = wing chord)
Pre-hatch	Eggs laid: 1/day, 2–7 in clutch; pale blue (sometimes white), speckling variable. Asynchronous (may hatch over 4–5 d), time of day varies, most overnight or early morning. Development: slower compared with other passerines.
0 = hatch	Mass: 1.1–2.3 g (35 mm long). c 4 mm, t 4 mm, w 7 mm. Raises head. Gapes.
1	2.3–3.9 g (38 mm). Skin darkening. Crawls. Rises up on legs.
2	3.5–4.2 g (40–44 mm).
3	4.7–6.2 g (42–48 mm), c 5 mm, t 6–8 mm, w 11–12 mm. Alar PinEm. Eye slits d3–5. Ear dent appears in skin.
4	6.7–7.5 g (49 mm), c 6 mm, t 8–10 mm, w 13–14 mm.
5	6.4–7.7 g (51–58 mm), c 6 mm, t 9–12 mm, w 18–19 mm. Wing PinF 3 mm, back and breast PinEm. Ear: small orifice opens.
6	10.5–12 g (54–65 mm), c 6–7 mm, t 10–13 mm, w 22–24 mm. Wing PinF 9.5 mm. Tail PinF 6 mm. Eyes fully open. Back and breast PinUn (9.5 mm). Tail: middle rectrix 3 mm, outermost rectrix 4 mm, longest wing feather at wrist 8 mm. Very active, strong grip. Fear response: lies flat until parents arrive with food.
7	12.5–14.2 g (56–65 cm), c 6–7 mm, t 14–15 mm, w 28 mm. Tail: PinUn (tips). Eyes fully open; may not hold open.
8	13.8–15.7 g (60–68 mm, 75–85 with neck fully stretched), c 6–7 mm, t 15 mm, p9 22–23 mm. Tail 8–10 mm (unsheathing at tips). Wing PinUn (tips).
9	15.4–16.1 g (70 mm, 75 stretched), c 7 mm, t 15–17 mm, p9 25 mm. Tail: 9–12 mm (inner shortest to outermost longest). Dorsal PinUn.
10	16.9–17.7 g (70–75 mm), c 7 mm, t 15–17 mm, p9 28–31 mm. Tail: 15–19 mm. Prim tips extend 11 mm past sheath. Head and dorsal PinUn. Perches on nest rim and fans wings. Fear response strong; fledges prematurely if disturbed. If renesting is attempted, remaining nestlings may jump if oldest nestling emits alarm call.
11	17.5–19.1 g (74–88 mm), c 7–8 mm, t 16 mm, p9 32 mm. Tail 21–22 mm. Can perch, strong grip, could fly if attempted. Some down left on head, sparse in other areas, most of wings and tail out of sheaths; ear coverts first emerge.
12	19.5 g (largest chick in clutch), 88 mm, c 8 mm, t 16 mm, p9 35 mm. Appears FF with remnants of down. Capable of short flight if disturbed.
14–16	May fledge (11–19 d), 18–19 g. If nest becomes too small for five fledglings, one or two nestlings lay on top of others. Feedings by parent(s) much less frequent. Oldest (usually most developed) fledges first.

Source: Wild clutches, Nevada County, CA.

Table C-11. Growth and development of Song Sparrow (*Melospiza melodia gouldii*)

Age in days	Development
0 = hatch	1.5 g (egg mass: 1.8–2.85 g).
1	1.73–2.83 g (32.4–40.7 mm), alar PinVis.
2	2.65–4.13 g, dorsal PinVis, ventral PinVis (look like spots).
3	4.3–6.6 g, PinEm capital, humeral, alar, and femoral. Eyes open (d3–6).
4	6.15–9.14 g (44.5–51.5 mm), ears open, humeral PinEm.
5	8–11.6 g, some contour PinUn, ventral PinUn ~1 mm. Stands d5–6, stretches.
6	9.3–14.3 g.
7–8	12.16–15.44 g. Appears FF; prim PinUn, capital PinF 2 mm, wing and tail PinUn, breast and belly may show short, thin streaks on dull white.
9–12	15 g (d9). May fledge.
11–15	18–19 g. Adult size and shows sexual dimorphism by 8 wks.

Source: Jonsomjit et al. 2007.

Table C-12. Growth and development of Red-winged Blackbird (*Agelaius phoeniceus*)

Age in days	Mass of males (g)	Mass of females (g)	Length of males (mm)	Length of females (mm)	Growth and development
0	3.9	3.8	48.0	47.0	
1	6.1	6.0	52.5	53.5	
2	9.5	9.0	62.5	60.5	Prim PinEm.
3	13.4	12.7	69.5	68.5	50% of adult tarsus length attained in both sexes.
4	18.9	16.6	79.5	77.5	
5	24.2	21.1	88.0	84.5	d6: eyes open.
6	28.4	24.3	94.5	90.0	Prim PinUn (tips).
7	32.5	26.1	101.0	96.0	
8	34.9	27.3	105.5	102.0	50% of adult length and wing length attained in F nestlings.
9	37.7	28.5	112.0	107.5	d9–10: FF. 50% of adult length attained in M nestlings.
10	38.4	28.9	116.5	111.5	May climb out of nest onto vegetation. 100% of adult tarsus length attained in both sexes. 50% of adult wing length attained in M nestlings.
11			122.5	115.0	d11–14: leaves nest.
12			126.5		

Source: Mean weight (grams) and mean total length in male and female nestlings from Holcomb and Twiest (1968).

Table C-13. Growth and development of Brewer's Blackbird
(*Euphagus cyanocephalus*)

Age in days	Development
0 = hatch	Mass: 3.7 g (about 6% of Ad F). Gapes.
1	5 g. PinVis all tracts except capital.
2	6.5 g. Prim PinEm.
2–3	Eyes: slits. 1st prim PinEm.
3	10 g. Body PinEm.
3–5	PinEm all areas. Tarsus and hallux 50% fully grown. Grasps nest lining with toes.
4	15 g. Central rectrix PinEm.
5	20 g. Sheaths all tracts.
6	26 g. Prim PinUn (tips). Briefly grasps with toes.
7	30 g. 1st prim 7.4 mm long. Wing growth peaks at 9.2 mm.
8	35 g. Eyes fully open. More alert.
9	38 g. May fledge prematurely.
10–11	40 g. Avg N mass 70% of adult F. Almost all PinUn. Tarsus and hallux adult size, nearly black. Bill wiping, preening, stretching. Voc (d10–12): *tutz-utz-utz* repeated.
12	Some down attached to feather tips. Head scratching.
13	46 g. Fledges (d12–16) usually by hopping, walking, or flutter-fly short distance.
14	Exploratory pecking (hand-raised chicks).
18	Begins bathing.
21–28	Catches insects. Picks up seeds. d23: hand-raised can land on perch and maintain balance.
39	Wild and hand-raised completely self-feeding.

Source: Balph 1975.

SPECIES COMPARISON TABLES

Key to abbreviations: see Introduction to Species Accounts.

Table D-1. Plumage comparisons of fledgling and juvenile North American hummingbirds

Species	Overall	Head and throat	Wings and tail
Allen's	Fl M similar to adult F, but crown and back edged buff or cinnamon-buff. Below: grayish, buffy-rufous flanks (black base).	Head: orange eyebrow. Throat: M heavily speckled dusky-bronze to greenish-bronze. F like juv M but appears finely streaked without orange-red feathers.	Wings: dark sepia. Tail: pointed tips (white on outer rects), outer rects narrow, central rects with rufous along shaft.
Anna's	Above: (iridescent) green or golden-green edged brownish-buff, rufous not extensive. Below: pale gray, green mottling on sides.	M: dusky-greenish feathers on crown and head are scalloped (have rounded edges). M throat: light with dusky (or drab) markings in several vertical rows, with one to several pinkish-red iridescent feathers. F: as M but no red anywhere, fewer dusky markings. White spot behind eye may appear 21–28 d.	Wings: p10 usually broader and blunter. Tail: no rufous; M notched, outer rects relatively narrow and pointed, black subterminal band, white tip; F outer rects broad, no notch, double-rounded, variable tip pattern (usually with terminal grayish-white pointing centrally into black of r4 and r5, more so on M).

Table D-1. (continued)

Species	Overall	Head and throat	Wings and tail
Black-chinned	Plumage like adult F but edged buff. Above: bronzy-green. Below: grayish-buff.	Grayer crown and auriculars. Throat: M has whitish, heavy dusky streaking, may have 1–2 black or iridescent gorget feathers post-fledging. F whitish without black or iridescent violet feathers, few dusky markings.	Wings: M p6 inner web indistinctly attenuate at the tip, p10 relatively narrow and curved; F p6 relatively broad at tip, p10 avgs broader and blunter. Tail: no rufous; M outer rects narrow, tapered, white tips; F r2 often tipped white, outer rects relatively broad.
Broad-tailed	M like adult F but duller green (washed grayish) above and edged whitish-buff to cinnamon (appears scaled). Below: flanks pale rufous. F like juv M except for throat and tail.	Throat: M neatly spotted, whitish-gray with dark green vertical markings; F markings duller and smaller than juv M.	Tail: M and F outer rects with white tips (wider on juv F), M with rufous at base mainly on outer rect, F lacks rufous.
Calliope	M like adult F, bronzy-green above. Fledgling F more bronzish above with dull brownish or grayish-buff narrow margins. Below: flanks deep buffy.	M usually with moderately heavy to heavy bronze-green markings and one to many metallic reddish or violet feathers. F without iridescent reddish or violet feathers and with moderate amount of dull bronze markings.	Wings: F p10 avg broader and blunter. Tail: short, square, little (or no) rufous; M r5 often < r1, central rects green with rufous on sides at base. F r5 usually < r1, central rects green but little or no rufous on sides at base, r3–r5 with white, the white tip on r5 typically larger.
Costa's	Plumage: like adult F. Above: grayish (less green or bonze than other sp), grayish-buff edging. Below: flanks whitish-buff. Bill (adult): slightly curved.	F: crown and head without iridescent violet feathers, usually without dusky markings. Throat: lightly marked dusky-greenish; M may have 1–2 iridescent violet feathers; F with few spots, no violet.	Wings: p10 usually broader and blunter. Tail: short, no rufous; M squared, outer rects narrow and tapered, white tip usually with a point of black extending from subterminal band; F graduated, r2 often tipped white, outer rects relatively broad, usually with an extensive terminal white patch pointing centrally into the black subterminal band.

Table D-1. (*continued*)

Species	Overall	Head and throat	Wings and tail
Ruby-throated	Above: bright green, grayish-brown edging. Below: flanks green and buff or cinnamon.	Green forehead with more contrasting pattern. Throat: M heavily spotted or streaked grayish to dusky markings, may have red iridescent markings (amount varies). F whitish, no dusky or red.	Wings: M p10 narrow and curved; F p6 relatively broad and less attenuate at tip, p10 averages broader and blunter. Tail: no rufous; M forked (moderately deep), outer rects long and narrow, tapered, tipped white; F double-rounded with shallow notch, outer rects broad.
Rufous	Similar to adult F except crown with cinnamon edging. Below: buffy-rufous flanks (black base).	Throat: M spotted greenish-bronze with orange-red (iridescent) feathers. F dull bronze markings, may have few orange-red feathers.	Wings: dark sepia. Tail: rufous on uppertail covs, more on F.

Note: All juveniles have bills that are soft, short, and grooved (corrugated or striated) on upper mandible.

Table D-2. Comparison of six genera of Tyrant Flycatchers

Genus	Common adult features	Hatchling down	Mouth interior	Gape flanges
All	Bill: wide, flat, tapers to point.			
Contopus	Generally charcoal-gray with wing-bars. Long wings; aerial hawking. Bill: broad and long. Large pewees have longer bill than Wood-Pewees.	Whitish or gray	Yellow to yellow-orange	Yellow
Empidonax	Some are indistinguishable except by vocalizations and nest structure. Bill: relatively short; lower mandible yellowish. Have eye rings and contrasting patterns on underside.	White to mouse-gray	Orange or bright orange to reddish-orange	Yellow to bright yellow
Myiarchus	Cavity (or box) nesters. Bill relatively larger. Ash-throated has longer bill and hook.	Dark gray	Yellow, orange throat	White
Pyrocephalus	Small, brightly colored (vermilion) M; F and juv grayish-brown above and streaked below. Bill: marginally broad tapering to finer point.	Light gray or creamy-brown	Bright yellow-orange	Yellow or bright yellow
Sayornis	Tail-flicking. Bill: relatively slimmer, smaller, hooked at tip.	Light to medium gray	Bright yellow-orange to reddish-orange	Creamy-yellow to yellow
Tyrannus	Long, pointed wings; aerial hawking. Most aggressive, will attack a large predatory bird.	White to buffy-white or dusky-brown	Bright yellow to orange	Yellow

Table D-3. Comparison of nestling and juvenile American, Lesser, and Lawrence's Goldfinches

American Goldfinch *Carduelis tristis*	Lesser Goldfinch *Carduelis psaltria*	Lawrence's Goldfinch *Carduelis lawrencei*
Adult: 13 g, 11.5–14 cm (4.5–5.5 in.).	Adult: 9.5 g, 10–11.5 cm (4–4.5 in.).	Adult: 11.5 g, 10–11.5 cm (4–4.5 in.).
d0: 0.94–1.52 g.	d0: < 1 g.	d0: no info, probably ~1 g.
Mouth: pinkish red.	Mouth: pink, then orange or orange-red.	Mouth: red.
Gape flanges: pale creamy-yellow, pinkish corners.	Gape flanges: pale creamy-yellow, pinkish corners.	Gape flanges: color not noted.
Down: pale grayish, head and body.	Down: drab gray to buff.	Down: color not noted, well developed on spinal tract.
Fledgling: cinnamon buff rump and wing markings. Acquires adult plumage after first breeding season.	Fledgling: all greenish with some faint streaks on breast.	Fledgling: grayish-brown with traces of yellow, green tinge on back; black on forehead, lores, and chin. White spots on center of inner webs of tail feathers.

Table D-4. Weight comparison of American Robin and Northern Mockingbird

Age in days	American Robin Mass (g)	Northern Mockingbird Mass (g)
Adult	56–112 (length 25.4 cm)	39–57 (length 21–25.5 cm)
Egg	6.3	3.5–4.6
0 = hatch day	4.1–6.7	2.6–3.5
1	8.9	4
2-3	14.3	8
3	21.3	13.3
4	26.6	17
5	32	21
6	40	26
7	47	No data
8	52	33.5
9	55.2	35
10-11	54.9	38
12	54.8 (drop in weight is normal before fledging)	36.3 (avg fledge mass, d12–14)
13–16	56	No data
17–18	> 56	40–41

NOTES

Introduction

1 A "baby bird" in this guide, although not a scientific term, encompasses all the stages of altricial young (hatchling, nestling, fledgling, nidifuge, and chick) and recently hatched precocial chicks.

Chapter 1. The Importance of Correct Identification

1 *Conspecific* refers to belonging to the same species.
2 "Airplane wing" refers to when the wing feathers are twisted so they point out laterally.
3 *Heterospecific* refers to belonging to different biological species.
4 In wildlife rehabilitation, *hacking* is a process that involves the use of a *hack box* (an artificial nest) from which the birds will eventually be released.
5 An obligate brood parasite is a species that does not build its own nest, lays eggs in nests of other species, and does not rear its own young.
6 Partners in Flight is a network of over 150 organizations throughout the Western Hemisphere working to reverse population declines of birds. See partnersinflight.org.

Chapter 2. Anatomy

1 A trophic category is based on the kinds of food an organism consumes. For example, a granivore eats primarily plant matter (such as seeds), an insectivore eats mainly insects, and an omnivore consumes both plant and animal matter.
2 Noun: the gape (or gapes) of a bird is the interior of the mouth. Verb: to gape is to open the mouth. A bird with an open mouth is gaping.

Chapter 4. The Process of Identification

1 *Sympatric species* refers to two related species or populations existing in the same geographic area.

LITERATURE CITED

Alcock, J. 1979. Animal behavior: An evolutionary approach. 2nd ed. Sunderland, MA: Sinauer Associates.

Baicich, P.J., and C.J.O. Harrison. 2005. Nests, eggs, and nestlings of North American birds. 2nd ed. Princeton, NJ: Princeton University Press.

Balph, M.H. 1975. Development of young Brewer's Blackbirds. Wilson Bull. 87(2): 207–230.

Banks, R.C. 1959. Development of nestling White-crowned Sparrows in central coast California. Condor 61:96–109.

Baptista, L.F., and K.L. Schuchmann. 1990. Song learning in the Anna's Hummingbird (*Calypte anna*). Ethology 84:15–26.

Barbour, R.W. 1950. Growth and feather development of towhee nestlings. Am. Midl. Nat. 44 (3): 742–748.

Bateson, P.P.G. 1966. The characteristics and context of imprinting. Biol. Rev. 41:177–220.

Bateson, P.P.G. 1978. Early experience and sexual preference. *In* Biological Determinants of Sexual Behavior, ed. J.B. Hutchinson, 29–53. London: John Wiley.

Bateson, P.P.G. 1979. How do sensitive periods arise and what are they for? Anim. Behav. 27(2): 470–486.

Bateson, P.P.G. 1981. Control of sensitivity to the environment during development. *In* Behavioural development, ed. K. Immelmann, G.W. Barlow, I. Petrinovich, and M. Main, 432–453. Cambridge: Cambridge University Press.

Bateson, P.P.G. 1990. Is imprinting such a special case? Phil. Trans. R. Soc. Lond. B. 329:125–131.

Bateson, P.P.G. 2017. Imprinting and attachment. *In* Behaviour, development and evolution, 19–26. Cambridge: Open Book Publishers. http://www.jstor .org/stable/j.ctt1sq5tz0.5.

Beason, R.C., and E.C. Franks. 1973. Development of young Horned Larks. Auk 90:359–363.

Bent, A.C. 1940. Life histories of North American cuckoos, goatsuckers, hummingbirds, and their allies. United States National Museum, Bulletin 176.

Bent, A.C. 1961. Life histories of North American birds of prey. Pts, 1 and 2. New York: Dover Publications.

Billerman, S.M., B.K. Keeney, P.G. Rodewald, and T.S. Schulenberg, eds. 2020. Birds of the World. Cornell Laboratory of Ornithology, Ithaca, NY.

Bischof, H.J. 1979. A model of imprinting evolved from neurophysiological concepts. Z. Tierpsychol. 51:126–139.

Bolhuis, J. J., G.J. de Vos, and J.P. Kruijt. 1990. Filial imprinting and associative learning. Q. J. Exp. Psychol. 42B(3): 313–329.

Bortolotti, G.R., J.E. Smits, and D.M. Bird. 2002. Iris colour of American Kestrels varies with age, sex, and exposure to PCBs. Physiol. Biochem. Zool. 76(1): 99–104.

Botelho, J.F., D. Smith-Paredes, and A.O. Vargas. 2015. Altriciality and the evolution of toe orientation in birds. Evol. Biol. DOI 10.1007/s11692-015 -9334-7.

Boulton, R. 1927. Ptilosis of the House Wren (*Troglodytes aedon aedon*). Auk 44(3): 387–414.

ten Cate, C.T., and D.R. Voss. 1999. Imprinting and evolutionary processes in birds: A reassessment. *In* Advances in the Study of Behavior, ed. P.J.B. Slater, J.S. Rosenblatt, C.T. Snowdon, and T.J. Roper, 28:1–31. Cambridge, MA: Academic Press.

Chapman, F.M. 1917. The Warblers of North America. 3rd ed. New York: D. Appleton.

Chesser, R.T., S.M. Billerman, K.J. Burns, C. Cicero, J.L. Dunn, A.W. Kratter, I.J. Lovette, N.A. Mason, P.C. Rasmussen, J.V. Remsen Jr., D.F. Stotz, and K. Winker. 2020. Checklist of North and Middle American Birds (online). American Ornithological Society. http://checklist.aou.org/taxa.

Clark, G.A., Jr. 1969. Oral flanges of juvenile birds. Wilson Bull. 81(3): 270–279.

Clotfelter, E.D., K.A. Schubert, V. Nolan Jr., and E.D. Ketterson. 2003. Mouth color signals thermal state of nestling Dark-eyed Juncos (*Junco hyemalis*). Ethology 109:171–182.

Crawford, R.D. 1978. Tarsal color of American Coots in relation to age. Wilson Bull. 90(4): 536–543.

Curson, J., D. Quinn, and D. Beadle. 1994. Warblers of the Americas. New York: Houghton Mifflin.

Davis, C.M. 1978. A nesting study of the Brown Creeper. Living Bird 17:237–263.

Dawson, W.L. 1923. The birds of California: A complete, scientific and popular account of the 580 species and subspecies of birds found in the state. 4 vols. Los Angeles: South Moulton Co.

de Ayala, R.M., N. Saino, A.P. Moller, and C. Anselmi. 2007. Mouth coloration of nestlings covaries with offspring quality and influences parental feeding behavior. Behav. Ecol. 18(3): 526–534.

Dickey, D.R. 1915. The hummers in a foothill valley. Country Life in America 28(2): 35–39.

Duerr, R.S., and L.J. Gage, eds. 2020. Hand-rearing birds. 2nd ed. Hoboken, NJ: John Wiley & Sons.

Dugas, M.B. 2015. Detectability matters: Conspicuous nestling mouth colours make prey transfer easier for parents in a cavity nesting bird. Biol. Lett. 11:20150771.

Dugas, M.B., and L.L. Dillow. 2013. Rictal flanges of nestling birds are most colorful near the gape. Wilson J. Ornithol. 125(2): 430–433.

Dunning, J.B. 2018. Body masses of North American birds. Ghadrdan M., ed. Eugene, OR: IWRC; 2018.

Emlen, J.T., Jr. 1942. Notes on a nesting colony of Western Crows. Bird-Banding 13(4): 143–154.

Emlen, S.T., Jr. 1967. Migratory Orientation in the Indigo Bunting (*Passerina cyanea*). Auk 84:309–342.

Emlen, S.T., J.D. Rising, and W.L. Thompson. 1975. A behavioral and morphological study of sympatry in the Indigo and Lazuli Buntings of the Great Plains. Wilson Bull. 87(2): 145–302.

Farner, D.S., J.R. King, and K.C. Parkes, eds. 1983. Avian Biology, vol. 7. New York: Academic Press.

Ficken, M.S. 1965. Mouth color of nestling passerines and its use in taxonomy. Wilson Bull. 77(1): 71–75.

Foster, M.S. 1967. Pterylography and age determination in the Orange-crowned Warbler. Condor 69(1): 1–12.

Franzreb, K.E., and S.A. Laymon. 1993. A reassessment of the taxonomic status of the Yellow-billed Cuckoo. Western Birds 24:17–28.

Grishaver, M.A., P.J. Mock, and K.L. Preston. 1998. Breeding behavior of the California Gnatcatcher in southwestern San Diego Co., CA. Western Birds 29:299–322.

Hadow, H.H. 1976. Growth and development of nestling Downy Woodpeckers. North American Bird Bander 1(4): 155–164.

Hamilton, W.J., Jr. 1935. Notes on nestling robins. Wilson Bull. 47:109–111.

Hauber, M.E., S.A. Russo, and P.W. Sherman. 2001. A password for species recognition in a brood parasitic bird. Proc. R. Soc. Lond. B 268:1041–1048.

Heinrich, B. 1999. Mind of the raven. New York: HarperCollins.

Hess, E.H. 1973. Imprinting: Early experience and the developmental psychobiology of attachment. New York: Van Nostrand Reinhold.

Hill, J.R., III. 1994. The growth of nestling Purple Martins. Purple Martin Update 3:1–3.

Holcomb, L.C., and G. Twiest. 1968. Red-winged Blackbird nestling growth compared to adult size and differential development of structures. Ohio J. Sci. 68(6): 277–284.

Horwich, R.H. 1966. Feather development as a means of aging young mockingbirds (*Mimus polyglottos*). Bird-Banding 37:257–267.

Howell, J.C. 1942. Notes on the nesting habits of the American Robin (*Turdus migratorius* L.). Am. Midl. Nat. 28:529-603.

Howell, S.N.G., and P. Pyle. 2015. Use of "definitive" and other terms in molt nomenclature: A response to Wolfe et al. (2014). Auk 132(2): 365–369.

Hudon, J., and A.H. Brush. 1989. Probable dietary basis of a color variant of the Cedar Waxwing. J. Field Ornithol. 60(3): 361–368.

Immelmann, K. 1972. Sexual and other long-term aspects of imprinting in birds and other species. *In* Advances in the Study of Behavior, ed. D.S. Lehrman, R.A. Hinde, and E. Shaw, 4:147–174. Cambridge, MA: Academic Press.

Immelmann, K., and S.J. Suomi. 1981. Sensitive phases in development. *In* Behavior development: The Bielefeld Interdisciplinary Project, ed. K. Immelmann, G.W. Barlow, L. Petrinovich, and M. Main, 395–431. Cambridge: Cambridge University Press.

Ingram, C. 1907. On tongue-marks in young birds. Ibis 1(4): 574-577.

Ingram, C. 1920. A contribution to the study of nestling birds. Ibis 2(4): 856-880.

Irwin, D.E., and T. Price. 1999. Sexual imprinting, learning and speciation. Heredity 82:347–354.

Jonsomjit, D., S.L. Jones, T. Gardali, G.R. Geupel, and P.J. Gouse. 2007. A guide to nestling development and aging in altricial passerines. Biol. Tech. Pub., FWS/BTP-R6008-2007, Washington, DC: U.S. Fish and Wildlife Service.

Keeton, W.T. 1969. Orientation by pigeons: Is the sun necessary? Science 165(3896): 922–928.

Kessel, B. 1957. A study of the breeding biology of the European Starling (*Sturnus vulgaris*) in North America. Am. Midl. Nat. 58(2): 257–331.

Kilner, R., and N.B. Davies. 1998. Nestling mouth colour: Ecological correlates of a begging signal. Anim. Behav. 56(3): 705–712.

Kroodsma, D. 2005. The singing life of birds: The art and science of listening to birdsong. Boston: Houghton Mifflin.

Lai, P., P. Pyle, K.R. Foster, and C.M. Goodwin. 2017. Identifying sparrows in juvenile plumage. Birding 49(6): 62–67.

Lepczyk, C.A., and W.H. Karasov. 2000. Effect of ephemeral food restriction on growth of House Sparrows. Auk 117:164–174.

Ligon, J.D., and M.D.J. Martin. 1974. Piñon seed assessment by the Piñon Jay, *Gymnorhinus cyanocephalus*. Anim. Behav. 22:421–429.

Linsdale, J.M. 1936. Coloration of downy young birds and of nest linings. Condor 38:111–117.

Loiseau, C., S. Fellous, C. Haussy, and O. Chastel. 2008. Condition-dependent effects of corticosterone on a carotenoid-based begging signal in House Sparrows. Hormones and Behavior 53(1): 266–273.

Lopes, R.J., J.D. Johnson, M.B. Toomey, M.S. Ferreira, P.M. Araujo, J. Melo-Ferreira, L. Andersson, G.E. Hill, J.C. Corbo, and M. Carneiro. 2016. Genetic basis for red coloration in birds. Curr. Biol. 26(11): 1427–1434.

Lorenz, K. 1935. Companions as factors in the bird's environment. Ornithol. 83:137-213, 289-413.

Lovett, I.J., and J.W. Fitzpatrick. 2016. The Cornell Lab of Ornithology handbook of bird biology, 3rd ed. Ithaca, NY: Cornell Lab of Ornithology.

Marin, M. 1997. Some aspects of the breeding biology of the Black Swift. Wilson Bull. 109(2): 290–306.

Marler, P. 1970. A comparative approach to vocal learning: Song development in White-crowned Sparrows. J. Comp. Physiol. Psychol. (monogr.) 71:1–25.

Marler, P. 1990. Song learning: The interface between behaviour and neuro-ethology. Philosophical Trans: Biol. Sci. 329:109–114.

McRae, S.B., P.J. Weatherhead, and R. Montgomerie. 1993. American robin nestlings compete by jockeying for position. Behav. Ecol. Sociobiol. 33:101–106.

Mueller, H.C. 1974. The development of prey recognition and predatory behaviour in the American Kestrel, *Falco sparverius*. Behaviour 49:313–324.

Mulligan, J.A. 1966. Singing behavior and its development in the Song Sparrow, *Melospiza melodia*. Univ. Calif. Publ. Zool. 81:1–76.

Murphy, M.T. 1981. Growth and aging of nestling Eastern Kingbirds and Eastern Phoebes. J. Field Ornithol. 52(4): 309–316.

Nelson, C.H. 1993. The downy waterfowl of North America. Deerfield, IL: Delta Station Press.

Nice, M.M. 1962. Development of behavior in precocial birds. Trans. Linn. Soc. N.Y. 8:1–211.

Nolan, V., Jr. 1978. The ecology and behavior of the Prairie Warbler, *Dendroica discolor*. Ornithological Monographs 26:1–596.

Oberholser, H.C. 1974. The bird life of Texas. Austin: University of Texas Press.

Orr, R.T. 1939. Observations on the nesting of the Allen Hummingbird. Condor 41(1): 17–24.

Parkes, K.C. 1984. An apparent hybrid Black-billed x Yellow-billed Cuckoo. Wilson Bull. General Notes 96(2): 294–296.

Parmalee, P.W. 1952. Growth and development of the nestling crow. Am. Midl. Nat. 47(1): 183–201.

Perry, A.E. 1965. The nesting of the Pine Siskin in Nebraska. Wilson Bull. 77(3): 243–250.

Pettingill, O.S. 1985. Ornithology in laboratory and field. 5th ed. Orlando, FL: Academic Press.

Power, H.W., III. 1966. Biology of the Mountain Bluebird in Montana. Condor 68(4): 351–371.

Preble, N.A. 1957. Nesting habits of the Yellow-billed Cuckoo. Am. Midl. Nat. 57(2): 474–483.

Pyle, P. 1997a. A further examination of wing and tail formulae in *Empidonax* and *Contopus* Flycatchers. *In* The era of Allan R. Phillips: A Festschrift, ed. R.W. Dickerman, Albuquerque, NM: Horizon Communications, 147–154.

Pyle, P. 1997b. Identification guide to North American birds. Pt. 1, Columbidae to Ploceidae. Point Reyes Station, CA: Slate Creek Press.

Pyle, P. 2008. Identification Guide to North American Birds, Pt. II, Anatidae to Alcidae. Point Reyes Station, CA: Slate Creek Press.

Pyle, P., C.M. Goodwin, and K.R. Foster. 2015. Identifying juvenile warblers: The fun really begins here. Birding 47(6): 58–69.

Ralph, C.J., and L.R. Mewaldt. 1975. Timing of site fixation upon the wintering grounds in sparrows. Auk 92:698–705.

Ricklefs, R.E. 1975. Patterns of growth in birds. Pt. 3, Growth and development of the Cactus Wren. Condor 77:34–45.

Ricklefs, R.E. 1983. Avian postnatal development. Avian Biology 7:1–83.

Ricklefs, R.E., D.F. Bruning, and G.W. Archibald. 1986. Growth rates of cranes reared in captivity. Auk 103:125–134.

Ridgway, R. 1886. A nomenclature of colors for naturalists and a compendium of useful knowledge for ornithologists. Boston, MA: Little, Brown.

Ridgway, R. 1907. The birds of North and Middle America: A descriptive catalogue of the higher groups, genera, species and subspecies of birds known to occur in North America. Pt. 4. U.S. National Mus. Bull., No. 50.

Ridgway, R. 1912. Color standards and color nomenclature. Self-published in Baltimore, MD.

Ritter, L.V. 1984. Growth of nestling scrub jays in California. J. Field Ornithol. 55(1): 48–53.

Saino, N., R. Ferrari, M. Romano, R. Martinelli, and A.P. Moller. 2003. Experimental manipulation of egg carotenoids effects immunity of Barn Swallow nestlings. Proc. R. Soc. Lond. B 270:2485–2489.

Saino, N., P. Ninni, S. Calza, R. Martinelli, F. De Bernardi, and A.P. Moller. 2000. Better red than dead: Carotenoid-based mouth coloration reveals infection in Barn Swallow nestlings. Proc. R. Soc. Lond. B 267:57–61.

Saunders, A.A. 1920. The color of natal down in passerine birds. Auk General Notes 37(2): 312.

Saunders, A.A. 1956. Descriptions of newly-hatched passerine birds. Bird-Banding 27:121–128.

Schimmel, L., and F. Wasserman. 1991. An interspecific comparison of individual and species recognition in the passerines *Turdus migratorius* and *Cyanocitta cristata*. Behaviour 118(1/2).

Sibley, D.A. 2014. The Sibley guide to birds. 2nd ed. New York: Alfred A. Knopf.

Skutch, A.F. 1945. Incubation and nestling periods of Central American birds. Auk 62:8–37.

Slagsvold, T., B.T. Hansen, L.E. Johannessen, and J.T. Lifjeld. 2002. Mate choice and imprinting in birds studied by cross-fostering in the wild. Proc. R. Soc. Lond. B 269:1449–1455.

Slagsvold, T., and K.L. Wiebe. 2011. Social learning in birds and its role in shaping a foraging niche. Phil. Trans. R. Soc. B 366:969–977.

Sluckin, W. 1972. Imprinting and early learning. London: Methuen.

Sluckin, W., and E.A. Salzen. 1961. Imprinting and perceptual learning. Q. J. Exp. Psychol. 8:65–77.

Smith, S.M. 1972. The ontogeny of impaling behavior in the Loggerhead Shrike, *Lanius ludovicianus*. L. Behavior 42:232–247.

Spencer, R.O. 1943. Nesting habits of the Black-billed Cuckoo. Wilson Bull. 55(1): 11–22.

Suzuki, Y., and E.A. Miller. 2004. Avian Ophthalmology. Wildl. rehabil. bull. 22(2).

Timmermans, S.T.A., G.E. Craigie, and K.E. Jones. 2004. Common Loon pairs rear four-chick broods. Wilson Bull. 116(1): 97–101.

Tyrrell, W.B. 1945. A study of the northern raven. Auk 62:1–7.

Van Tyne, J., and A.J. Berger. 1976. Fundamentals of Ornithology. 2nd ed. New York: John Wiley & Sons.

Waters, H. 2019. The future for birds. Audubon. Fall 2019:14-19.

Welty, J.C., and L. Baptista. 1988. The Life of Birds. 4th ed. Orlando, FL: Harcourt Brace Jovanovich.

Weston, H.G., Jr. 1947. Breeding behavior of the Black-headed Grosbeak. Condor 49:54–73.

Wetherbee, D.K. 1957. Natal plumages and downy pterylosis of passerine birds of North America. Bull. Am. Mus. Nat. Hist. 113(5): 339–436.

Wetherbee, D.K. 1958. New descriptions of natal pterylosis of various bird species. Bird Banding 24:232–236.

Wetherbee, D.K. 1960. Egg weight, juvenile weight, pterylosis of neonates. *In* The Kirtland's Warbler, by H. Mayfield. Bulletin 40. Bloomfield Hills, MI: Cranbrook Institute of Science.

Wetherbee, D.K. 1961. Observations on the developmental conditions of neonate birds. Amer. Midl. Nat. 65(2): 413–435.

Wetherbee, D.K., and N.S. Wetherbee. 1961. Artificial incubation of eggs of various bird species and some attributes of neonates. Bird Banding 32:141–159.

Whitmore, K.D., and J.M. Marzluff. 1998. Hand-rearing corvids for reintroduction: Importance of feeding regime, nestling growth, and dominance. J. Wildlife Management, 62(4): 1460–1479.

Yamaguchi, A. 1998. A sexually dimorphic learned birdsong in the Northern Cardinal. Condor 100:504–511.

Yom-Tov, Y., and A. Ar. 1993. Incubation and fledging durations of woodpeckers. Condor 95:282–287.

INDEX

Page numbers in bold indicate species description. Numbers in italics indicate illustrations.